Microfluidics for Medical Applications

RSC Nanoscience & Nanotechnology

Editor-in-Chief:
Paul O'Brien FRS, *University of Manchester, UK*

Series Editors:
Ralph Nuzzo, *University of Illinois at Urbana-Champaign, USA*
Joao Rocha, *University of Aveiro, Portugal*
Xiaogang Liu, *National University of Singapore, Singapore*

Honorary Series Editor:
Sir Harry Kroto FRS, *University of Sussex, UK*

How to obtain future titles on publication:
A standing order plan is available for this series. A standing order will bring
delivery of each new volume immediately on publication.

For further information please contact:
Book Sales Department, Royal Society of Chemistry, Thomas Graham House,
Science Park, Milton Road, Cambridge, CB4 0WF, UK
Telephone: +44 (0)1223 420066, Fax: +44 (0)1223 420247
Email: booksales@rsc.org
Visit our website at www.rsc.org/books

Microfluidics for Medical Applications

Edited by

Albert van den Berg
University of Twente, Enschede, The Netherlands
Email: A.vandenBerg@ewi.utwente.nl

Loes Segerink
University of Twente, Enschede, The Netherlands
Email: l.i.segerink@utwente.nl

ROYAL SOCIETY
OF CHEMISTRY

THE QUEEN'S AWARDS
FOR ENTERPRISE:
INTERNATIONAL TRADE
2013

RSC Nanoscience & Nanotechnology No. 36

Print ISBN: 978-1-84973-637-4
PDF eISBN: 978-1-84973-759-3
ISSN: 1757-7136

A catalogue record for this book is available from the British Library

Published by The Royal Society of Chemistry,
Thomas Graham House, Science Park, Milton Road,
Cambridge CB4 0WF, UK

Registered Charity Number 207890

For further information see our website at www.rsc.org

Printed and bound by CPI Group (UK) Ltd, Croydon, CR0 4YY

Preface

Ever since the beginning of the research in the lab on chip (LOC) field, people have been searching for the "killer app".[1] Unfortunately, unlike a field such as microelectronics, where Moore's law for memories and processors has driven the development or, rather, set the agenda, over several decades, in the LOC area no such big application has been identified so far. Rather a very diversified field of applications has emerged over time, with areas such as LOC for analytical chemistry,[2] drug development,[3] cell biology,[4] DNA sequencing and analysis,[5] chemical microreactors,[6] and medical applications.[7] Of all these, probably still the highest expectations are found in the latter area. This area is still emerging as indicated by the numerous scientific journals that publish articles relevant to this field, such as *Lab on a Chip, Analytical Chemistry, Microfluidics and Nanofluidics, Biomedical Microdevices, Biomicrofluidics* and *Integrative Biology*. Every week a lot of exiting new research is published, aiming to be applied for innovative medical devices, treatments, or diagnostics. While it is impossible to show all work performed in this field, with this volume we try to give an overview and perspective of this research field. Therefore we give an overview of the state-of-the-art given by a collection of world-wide top-level researchers, with contribution in three different, recently emerged subareas of this field: tissue and organs on chip, microfluidic tools for medicine, and point of care diagnostics.

For engineering microtissues on chip, 3D constructs are required, which serve as scaffolds for the different components of the tissue. Fibers are promising structures to make these constructs, which can be adapted in ways suited for the application. In Chapter 1 different types of fiber fabrication using microtechnology are discussed. Besides tissue engineering on chip, organs on chip is another emerging field in microfluidics. An example of this is the kidney on chip, which is discussed in Chapter 2. This chapter

RSC Nanoscience & Nanotechnology No. 36
Microfluidics for Medical Applications
Edited by Albert van den Berg and Loes Segerink
© The Royal Society of Chemistry 2015
Published by the Royal Society of Chemistry, www.rsc.org

was written by Kahp-Yang Suh, who totally unexpectedly and to our deep sadness passed away after finishing the chapter. As a tribute we decided to retain his chapter, to give the reader a nice insight into the brilliant work he did in the organs on a chip field. An overview of the current state-of-the-art is given with respect to this organ. Not only can organs be modelled with microfluidic systems, also the functioning of parts of the body can be modeled using such a system. An example is the functioning of the blood-brain barrier. Although these microfluidic systems are not yet as good as the conventional models, first steps are being made in the development of a reliable model. In Chapter 3 the current status of these microfluidics models is described as well as which parts need to be improved to end up with a model that will be better than the conventional one. An example of how these models can be used to understand a brain-related disease such as Alzheimer's disease is given in Chapter 4. In this chapter some fundamental questions regarding this disease are raised, which can be possibly answered with the help of microfluidics.

Besides the use of microfluidics to model the functioning of organs and as a tool to create constructs for tissue engineering, it can also be used in more general terms for medicine, which is the subject of Chapters 5 to 7. An example of this is the generation of bubbles for both contrast enhancement in ultrasound as well as drugs delivery at a specific spot. For the production of these monodisperse bubbles, also microfluidics can be used, which will be discussed in Chapter 5. The use of other spherical particles, the magnetic particles, is described in Chapter 6. These particles can be actuated with magnets and used for several assay steps in diagnostic devices. Chapter 7 shows the use of lab on a chip systems for assisted reproductive technologies, such as *in-vitro* fertilization and intracytoplasmic sperm injection. These treatments can benefit from miniaturization since it can improve the gamete selection, but also the procedures involved in cryopreservation and embryo development.

The latter section of the book covers some examples of point-of-care diagnostics using lab-on-a-chip systems. This is not solely restricted to traditional microfluidic systems, since the use of paper-based microfluidic tests is especially useful for diagnostics in the developing world. Chapter 8 includes examples of these paper-based devices, but also gives the requirements for testing in low-resource settings. A different example of point-of-care diagnostics is the detection of circulating tumor cells (CTCs). Although it is major challenge to increase the throughput of microfluidics systems to detect these rare cells (a few cells $(10 \text{ mL})^{-1}$), it has potential to improve the detection limit (Chapter 9). A way to perform this detection makes use of electrical impedance measurements, which will be discussed in more detail in Chapter 10. Besides the detection of CTCs, the use of microfluidic impedance cytometry is shown for a full blood count. In addition to the detection of cells in a fluid, microfluidics can also be used to measure analytes in blood. A widespread application of this is the measurement of glucose for diabetics. Chapter 11 covers this example, but also shows other routine clinical laboratory tests that are nowadays used. The last chapter of this book shows the development of a lab on a chip for ion measurements in biological fluids using capillary electrophoresis on chip. Here the steps are described that need to be taken to get a new microfluidic device ready for point-of-care measurements and practical application and market introduction.

We have tried to give you an overview of the diverse applications microfluidic technology can be advantageous in for medical applications. Some topics are still in the research phase, while others are currently incorporated in the hospital or patient's daily life. Furthermore in some cases it will serve as a tool to test drugs; others are used as a tool to detect certain disease markers. Finally we hope that with these examples, you get (more) inspired and enthusiastic to work in this wonderful multidisciplinary field and help to find the killer app!

Loes Segerink and Albert van den Berg

References

1. H. Becker, *Lab on a Chip*, 2009, **9**, 2119–2122.
2. A. Manz, N. Graber and H. M. Widmer, *Sensor. Actuator. B Chem.*, 1990, **1**, 244–248.
3. P. Neuzil, S. Giselbrecht, K. Lange, T. J. Huang and A. Manz, *Nat. Rev. Drug Discov.*, 2012, **11**, 620–632.
4. E. K. Sackmann, A. L. Fulton and D. J. Beebe, *Nature*, 2014, **507**, 181–189.
5. J. Liu, M. Enzelberger and S. Quake, *Electrophoresis*, 2002, **23**, 1531–1536.
6. J. P. McMullen and K. F. Jensen, in *Annual Review of Analytical Chemistry*, Vol. 3, ed. E. S. Yeung and R. N. Zare, 2010, vol. 3, pp. 19–42.
7. A. Floris, S. Staal, S. Lenk, E. Staijen, D. Kohlheyer, J. Eijkel and A. van den Berg, *Lab on a Chip*, 2010, **10**, 1799–1806.

Contents

RSC Nanoscience & Nanotechnology No. 36
Microfluidics for Medical Applications
Edited by Albert van den Berg and Loes Segerink
© The Royal Society of Chemistry 2015
Published by the Royal Society of Chemistry, www.rsc.org

References   126

Chapter 7   Microfluidics for Assisted Reproductive Technologies   131
              *David Lai, Joyce Han-Ching Chiu, Gary D. Smith and
              Shuichi Takayama*

              7.1   Introduction   131
              7.2   Gamete Manipulations   132
                    7.2.1   Male Gamete Sorting   133
                    7.2.2   Female Gamete Quality Assessment   137
              7.3   *In Vitro* Fertilization   139
              7.4   Cryopreservation   141
              7.5   Embryo Culture   144
              7.6   Embryo Analysis   146
              7.7   Conclusion   148
              References   148

**Chapter 8   Microfluidic Diagnostics for Low-resource Settings:
              Improving Global Health without a Power Cord   151**
              Joshua R. Buser, Carly A. Holstein and Paul Yager

              8.1   Introduction: Need for Diagnostics in Low-resource
                    Settings   151
                    8.1.1   Importance of Diagnostic Testing   151
                    8.1.2   Limitations in Low-resource Settings   152
                    8.1.3   Scope of Chapter   152
              8.2   Types of Diagnostic Testing Needed in Low-resource
                    Settings   153
                    8.2.1   Diagnosing Disease   153
                    8.2.2   Monitoring Disease   158
                    8.2.3   Counterfeit Drug Testing   161
                    8.2.4   Environmental Testing   162
              8.3   Overview of Microfluidic Diagnostics for Use at the
                    Point of Care   162
                    8.3.1   Channel-based Microfluidics   163
                    8.3.2   Paper-based Microfluidics   164
              8.4   Enabling All Aspects of Diagnostic Testing in
                    Low-resource Settings: Examples of and Opportunities
                    for Microfluidics (Channel-based and Paper-based)   171
                    8.4.1   Transportation and Storage of Devices in
                            Low-resource Settings   172
                    8.4.2   Specimen Collection   173
                    8.4.3   Sample Preparation   174
                    8.4.4   Running the Assay   176

CHAPTER 1

Microtechnologies in the Fabrication of Fibers for Tissue Engineering

MOHSEN AKBARI,[a,b,c,d,†] ALI TAMAYOL,[a,b,c,d,†] NASIM ANNABI,[a,b,c]
DAVID JUNCKER[d,e] AND ALI KHADEMHOSSEINI*[a,b,c]

[a] Biomaterials Innovation Research Center, Department of Medicine,
Brigham and Women's Hospital, Harvard Medical School, Cambridge
MA 02139, USA; [b] Harvard-MIT Division of Health Sciences and
Technology, Massachusetts Institute of Technology, Cambridge MA 02139,
USA; [c] Wyss Institute for Biologically Inspired Engineering, Harvard
University, Cambridge MA 02139, USA; [d] Biomedical Engineering
Department, McGill University, Montreal H3A 0G1, Canada; [e] Department
of Neurology & Neurosurgery, McGill University, Montreal H3A 2B4,
Canada
*Email: alik@rics.bwh.harvard.edu

1.1 Introduction

Tissue engineering is a multidisciplinary field that brings together researchers with backgrounds in engineering, biology, medicine, and chemistry to build tissue-like constructs for patient treatment or research. The ultimate goal of many research efforts in tissue engineering is to create biological replacements for diseased and damaged organs in the human

---

[†] Both authors have contributed equally to the work.

RSC Nanoscience & Nanotechnology No. 36
Microfluidics for Medical Applications
Edited by Albert van den Berg and Loes Segerink
© The Royal Society of Chemistry 2015
Published by the Royal Society of Chemistry, www.rsc.org

body. Such constructs should mimic the physiological environment including the structural and physicochemical features of native tissues.[1] Therefore, fabrication tools that allow for the creation of biocompatible complex 3D structures with controlled internal architecture and cell distribution and an effective vascular network are required.

Fiber-based techniques, which include textile technologies (*i.e.* weaving, braiding, knitting, embroidering), electrospinning, and direct writing, hold great promise for engineering 3D biomimetic tissue-like constructs. These techniques enable tuning the mechanical and structural properties of the fabricated constructs with interconnected pores and controlling the distribution of different cell lines in the constructs.[3] Creating biopolymeric fibers with topographical properties that vary spatiotemporally on the micro- or nanoscale is the initial step for any fiber-based tissue engineering approach. In addition, fibers can serve as carriers for biomolecules and microorganisms. The biological and mechanical properties of the fabricated fibers are essential for the functionality of the resultant tissue constructs.[2] Surface topology of the fibers also plays an important role in directing cell growth within the tissue construct.

Recent developments in microtechnologies along with the fast pace of growth of biopolymer science have allowed for the fabrication of fibers with amenable biomechanical properties for tissue engineering. In this chapter, we describe fiber fabrication techniques used in tissue engineering while emphasizing the role of microfluidics and microtechnologies. We categorize current fiber formation techniques into four methods: i) co-axial flow systems, ii) wetspinning, iii) meltspinning (extrusion), and iv) electrospinning. These methods are popular and have been enhanced by microtechnologies. We discuss the operational principles of these techniques and explore their advantages and limitations in tissue engineering.

1.2   Fiber Formation Techniques

1.2.1   Co-axial Flow Systems

Co-axial flow in microsystems is achieved by creating two or more flow streams in parallel. Due to the laminar nature of the flow in micro-channels, the interface of fluids remains stable and mixing only occurs due to molecular diffusion across the interface between the fluids. As a result, fibers with uniform cross-section can be fabricated. Co-axial flow-based microfluidic systems have been recently used for creating micron-size fibers featuring different shapes and sizes and containing different cell types and chemicals. This section describes the principle and theory of co-axial fiber fabrication and explores the current state-of-the-art in creating hydrogel fibers using microfluidic systems.

The fabrication of single layer hydrogel fibers in a co-axial flow format is shown in Figure 1.1a. The microfluidic system contains a central channel that delivers a pre-polymer solution (core) into a main channel. The

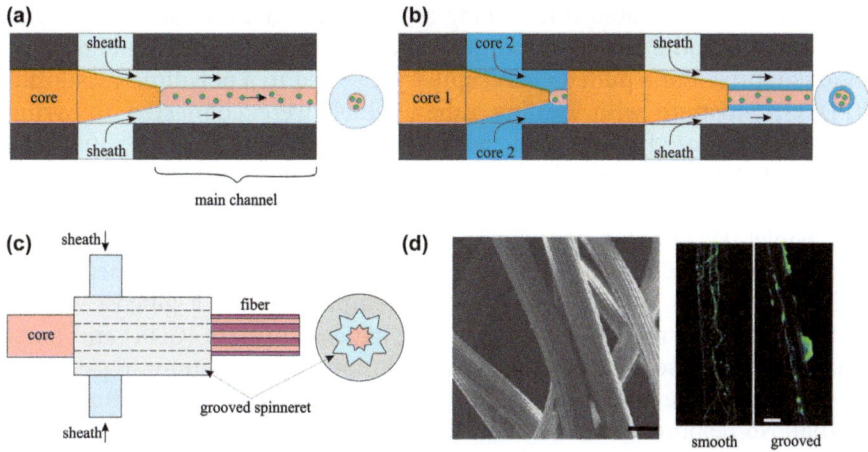

Figure 1.1 Principle of co-axial flow fiber formation. (a) A hydrogel fiber is created by polymerizing a pre-polymer (core) along a main channel. A sheath solution flows around the core stream and facilitates the fiber formation. The core stream can contain cells or chemicals. (b) Fabrication of composite fibers by forming three streams of two pre-polymers (core 1 and core 2) and a sheath flow. (c) Schematic of creating grooved fibers by using a microfluidic grooved spinneret. (d) Grooved fibers fabricated by the microfluidic spinneret (left) and fluorescent micrograph of neurons grown on the grooved fibers (right).[4] Scale bars show 20 μm in (d) left and 50 μm in (d) right.
Images in (d) are adapted with permission from Macmillan Publishers Ltd [Nature Materials], copyright (2011).

delivered solution from two side-channels forms a sheath flow around the core stream. Polymerization of the core solution (hydrogel formation) occurs downstream of the flow either by cross-linkers directly from the neighboring fluids or by light irradiation. The core solution can be loaded with cells or chemicals for different biomedical applications.[4] The sheath flow acts as a lubricant and facilitates fiber formation by preventing channel clogging during the hydrogel formation. Moreover, due to the short length of micro-channels containing the co-axial flow, cells are only exposed to a high shear stress and cross-linking reagents for a short time; this property helps the formation of hydrogel fibers containing viable and functional cells.

Fiber dimensions can be tuned by changing the ratio between the core and sheath flow rates and their relative viscosities. The fiber diameter, when the sheath and core viscosities are identical, is obtained from the following relationship:[5]

$$D_{fiber} = D_{channel} \left[1 - \left(\frac{Q_{sh}}{Q_{core} + Q_{sh}} \right)^{1/2} \right]^{1/2}, \qquad (1)$$

where D_{fiber} is the fiber diameter, $D_{channel}$ is the main channel diameter, and Q_{sh} and Q_{core} are the sheath and core flow rates, respectively. Jeong *et al.* for the first time fabricated fibers using co-axial flow in a microfluidic system.[5] Their microfluidic device was similar to the schematic shown in Figure 1.1a and comprised of a pulled glass capillary inserted in a polydimethylsiloxane (PDMS) substrate with feeding tubes, which were connected to syringe pumps. They used a photopolymerizable pre-polymer (4-hydroxybutyl acrylate (4-HBA)) as the core fluid and a mixture of 50% (v/v) polyvinyl alcohol (PVA) and 50% (v/v) deionized water (DI). They exposed the outlet channel to ultraviolet (UV) light in order to photopolymerize the core solution "on-the-fly". They showed that eqn (1) can be used for predicting the diameter of the fabricated fibers within ± 8%. In an attempt to create a glucose sensing microfiber, they mixed two enzymes, *i.e.* glucose oxidase (GOX) and horse-radish peroxidase (HRP) in the core solution. Fibers containing enzymes responded to the glucose existing in the solution by emitting a fluorescent signal. No response was detected in the fibers without enzymes.[5]

In another study, Hwang *et al.* used a similar microfluidic device and created microfibers from poly(lactic-*co*-glycolic acid) (PLGA) to investigate the effects of the microfibers' diameter on the orientation of mouse fibroblasts of L929 cells.[6] The core solution was 10% (w/v) PLGA dissolved in dimethyl sulfoxide (DMSO) and the sheath solution was a mixture of 50% (v/v) glycerine in water. The exchange of DMSO and water at the interface between the core and sheath solution solidified the PLGA and fibers in the range of 10–242 μm were collected at the device outlet on a motorized rotating glass slide. Hwang *et al.* showed that cells tend to orient themselves along the long axis of these fibers more as the fiber diameter decreases.[6]

Shin *et al.* created alginate fibers using a microfluidic device, schematically shown in Figure 1.1a.[7] They used sodium alginate as the core and calcium chloride (CaCl$_2$) as the sheath solutions. Consequently, Ca^{2+} ions diffused from CaCl$_2$ into the alginate central stream along the flow direction, forming calcium alginate cell-laden fibers before exiting the microfluidic device. They showed that for a constant sheath flow rate, the fiber diameters increased as the core flow rate was increased. However, as the core flow rate increased, instability in the flow occurred and spiral curls were formed. To assess the potential use of the alginate fibers fabricated by co-axial flow configuration, Shin *et al.* loaded the fibers with protein and mammalian cells by mixing bovine serum albumin (BSA) and human fibroblast cells (L292) in sodium alginate, respectively.[7] The *in vitro* cell viability assay confirmed that the fabrication process was not harmful to the cells.

Inspired by the work of Shin *et al.*,[7] Ghorbanian *et al.*[8] developed a microfluidic direct writer (MFDW) to construct 3D cell-laden alginate structures containing interconnected pores. The MFDW was mounted on a motorized stage and was automatically controlled and moved at a speed synchronized with the speed of fiber fabrication. To avoid channel blockage, they designed a declogging mechanism that injected a degelling agent (*e.g.* ethylenediaminetetraacetic acid) to dissolve the clogged gel. They formed a simple 3D construct by layer-by-layer deposition of these cell-laden alginate fibers on a glass slide. Using a standard live/dead assay, they showed that the writing process was not harmful to mammalian cells.

To improve cell adhesion properties of alginate for tissue engineering applications, Lee *et al.* used a co-axial flow microfluidic system to create chitosan-alginate composite fibers.[9] They used water-soluble chitosan and sodium alginate mixture as the core and $CaCl_2$ as the sheath streams. It was found that the bi-component fibers offer a superior cell viability over pure alginate fibers, evidenced by their live/dead assay results for human hepatocellular carcinoma (HepG2) cells over 7 days of incubation.[9] Although water-soluble chitosan is mechanically weak, the mechanical strength of the composite fibers did not change significantly.

Multiple-layer fibers can be fabricated by adding more streams to the microfluidic device. For example, Figure 1.1b shows the process of creating a two-layer composite fiber. First pre-polymer (core 1) surrounded by the second pre-polymer (core 2) enters the main channel and a sheath solution is formed around them. Cross-linking of the pre-polymers occurs downstream of the main channel either chemically or optically. The diameter of the created two-layer composite fiber can be estimated using the following relationship:[10]

$$D_{fiber} = D_{channel} \left[1 - \left(\frac{Q_{sh}}{Q_{core1} + Q_{core2} + Q_{sh}} \right)^{1/2} \right]^{1/2}, \qquad (2)$$

where Q_{core1} and Q_{core2} are the volumetric flow rates of the first and second pre-polymer solutions, respectively.

Lee *et al.* fabricated a microfluidic device, similar to the schematic shown in Figure 1.1b, to create hollow alginate fibers.[11] They used $CaCl_2$ as core 1, sodium alginate as core 2, and another stream of $CaCl_2$ as the sheath stream. They encapsulated HIVE-25 cells in the hollow fibers and implanted them in a composite hydrogel (mixture of agar, gelatin, and fibronectin).[11] The composite hydrogel was loaded with human smooth muscle cells (HIVS-125) to closely mimic a tissue with a stable microvessel network.

In another study, Hu *et al.* devised a triple-orifice spinneret to create co-axial flow of three different types of hydrogels.[10] They used a wide range of hydrogel materials including enzymatically cross-linking gelatin-hydroxyphenylpropionic acid (Gtn-HPA), alginate, poly-(*N*-isopropyl acrylamide) (poly(NIPAAM)), and polysulfone. With their triple-orifice spinneret, Hu *et al.* fabricated hollow and multilayer composite cell-laden fibers. The ability to change the flow rate of each stream during the fiber fabrication process enabled them to change the total fiber diameter and thickness of each layer "on the fly".[10]

The morphology of the fabricated fibers can be determined by changing the cross-sectional shape of the main channel. For example, Kang *et al.* used a grooved round channel to fabricate artificial tubuliform fibers with grooves on their surfaces (Figure 1.1c).[4] They showed that the number of grooves and their sizes could be tuned by changing the shape of the channel and adjusting the flow rates. The grooved fibers were then used to investigate the effect of mechanical cues on rat embryonic neurons (Figure 1.1d). It was shown that the neural cells aligned themselves along the ridges of the nanogrooved fibers, indicating that the surface morphology of fibers played an important role on the cellular behaviour of the neurons.

Hydrogel fibers that are heterogeneous in chemical and physical microstructure are of interest to better mimic the microenvironment of natural tissues. Yamada *et al.* fabricated a PDMS-based microfluidic system to continuously synthesize chemically and physically anisotropic calcium–alginate fibers.[12] Their device comprised a main core stream (propylene glycol + alginate + cells) co-axed with three streams of propylene glycol + alginate, buffer solution, and $CaCl_2$ solution. Cell-laden fibers were collected by a rotating roller that was partially dipped in a bath of $CaCl_2$. They added polyglycolic acid (PGA) to the alginate solution to adjust the stiffness of the local region of the alginate fibers and form sandwich-type solid-soft-solid structures. They showed that these structures provided better control of the cellular proliferation and networking. Inspired by the silk-spinning process in spiders, Kang *et al.* developed a microfluidic chip consisting of six hydrogel streams and one sheath flow (Figure 1.2a).[4] They used

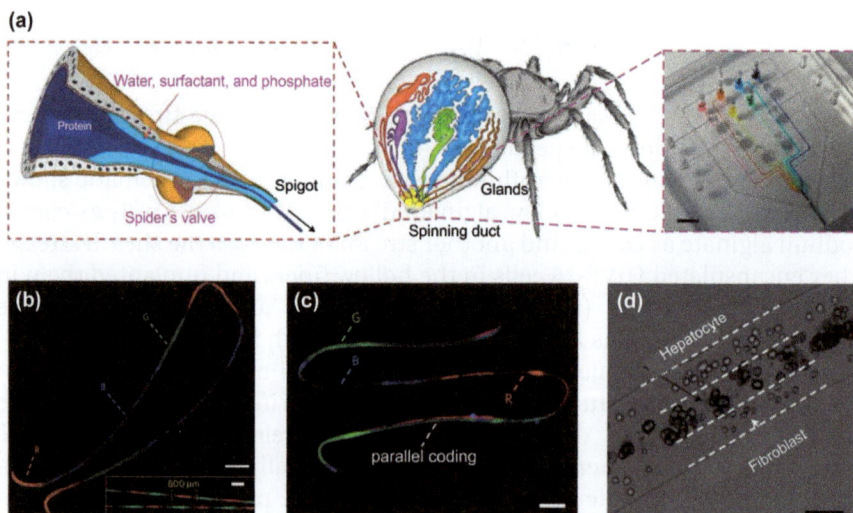

Figure 1.2 Digitally tunable physicochemical coding of microfibers. (a) A microfluidic device inspired by silk-spinning system of spiders. The chip consisted of six hydrogel streams and one sheath ($CaCl_2$) flow. Pneumatic valves controlled the flow rate of each stream to adjust the composition of the fabricated fibers. A motorized spool collected the fabricated fibers at the microchip outlet. (b) A serially coded fiber (inset shows coding with micrometer resolution). (c) A parallel coded fiber. (d) Co-culture of hepatocytes and fibroblasts embedded in alginate fibers. Scale bars: 5 mm (a), 1 mm (b) and (c), 400 μm (b), inset, and 20 μm (d).
Images are adapted with permission from Macmillan Publishers Ltd [Nature Materials], copyright (2011).

sodium alginate solutions loaded with different chemicals and cell types and CaCl$_2$ as the curing agent. The composition of the fabricated fibers was controlled using pneumatic valves. With this configuration, they controlled the spatial topography and chemical composition along the fibers. For example, they coded their fibers in serial (Figure 1.2b) and parallel (Figure 1.2c) configurations. They used the parallel coding feature to co-culture hepatocytes and fibroblasts on a single fiber. This shows the robustness of using microfluidic platforms for fabrication of cell-laden fibers for tissue engineering applications.

Fiber fabrication using co-axial flow in microfluidic systems holds great promise as it enables continuous fabrication of fibers with tunable morphological, structural, and chemical features. Various hydrogels including chemically and optically cross-linkable materials can be used in the co-axial format. Moreover, the incorporation of cells and chemicals in single- and multilayer fibers during the manufacturing process is possible. However, the biopolymers that are currently in use cannot form mechanically strong fibers to be easily manipulated.[2] Furthermore, the process of fiber fabrication is relatively slow, which makes creating 3D structures a time-consuming process.

1.3 Wetspinning

In wetspinning, polymer fibers are formed by continuously injecting a pre-polymer solution into one or multiple coagulation baths to polymerize and form long fibers. The system is usually simple and includes a reservoir for the pre-polymer, a spinneret, and a coagulation bath (Figure 1.3a). Pre-polymers can be injected manually,[13,14] gravitationally,[15] using a syringe pump,[16,17] or using pressurized air.[18–20] To achieve chemically and mechanically stable fibers, the pre-polymer and the final polymer should be insoluble in the coagulation solution.[16] Fibers fabricated by this method can be deposited randomly or in a predefined pattern in the coagulation bath to form a porous scaffold or can be collected on a motorized spool and assembled in a consequent process. Wetspinning has been widely used for fabrication of fibers from various biocompatible materials including alginate, collagen-alginate composite,[9] collagen,[13] chitosan,[21] poly ε-caprolactone (PCL),[16] starch-PCL composite,[22] chitosan-tripolyphosphate composite,[23] and calcium phosphate cement-alginate composite.[9] Fiber diameter can be adjusted by controlling the polymer composition and viscosity, the injection flow rate, and the diameter of the spinneret.[9,24] In general, wetspun fibers have been fabricated with a wide a range of diameters from ∼30 to 600 μm.[25] In addition, the velocity of the pre-polymer jet entering the coagulation bath and the relative viscosity of both solutions should be optimized to prevent instability in the fiber (polymer) stream, which can affect the quality of the fabricated fibers.

For example, Tuzlakoglu *et al.* created chitosan fibers using the wetspinning process for bone tissue engineering.[21] They dissolved chitosan in acetic

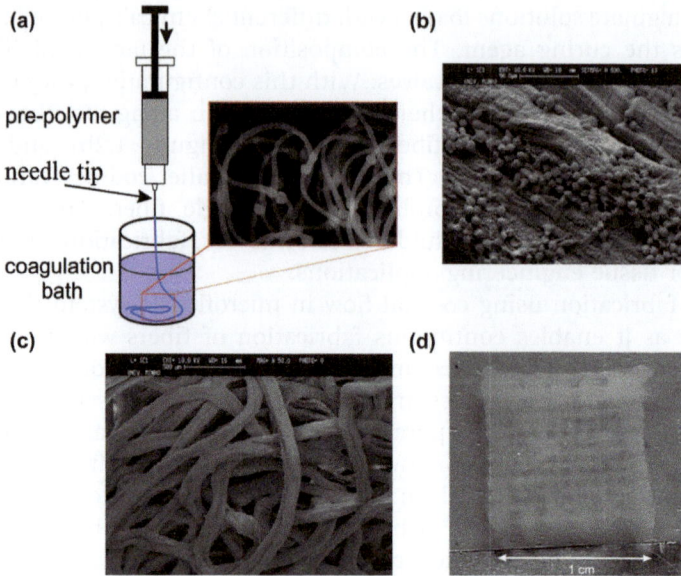

Figure 1.3   Principle of wetspinning. (a) A hydrogel fiber is created by polymerizing a
pre-polymer in a coagulation bath containing the cross-linking reagents.
The pre-polymer could contain cells, drugs, or growth factors. (b) Attachment of osteoblast-like cells to wetspun collagen fibers. (c) SEM images
showing a typical chitosan scaffold fabricated by random deposition of
fibers in a coagulation bath.[21] (d) A fibrin + alginate scaffold fabricated
by depositing fibers in the coagulation bath in a predefined pattern.[26]

acid and injected the solution in a bath of 30% 0.5 M Na_2SO_4, 10%
1 MNaOH, and 60% water and created randomly deposited fibers.[21] They
kept fibers in the coagulation bath for 1 day to fully cross-link. Tuzlakoglu
et al. seeded the scaffolds with osteoblast-like cells, which proliferated
(Figure 1.3b,c). In another work, Neves *et al.* created chitosan/PCL composite
fibers by injecting the pre-polymer solution into a methanol bath, which
formed fibers.[27] They formed randomly deposited 3D scaffolds with porosity
in the range of 64% to 83%. The scaffolds were seeded with bovine chon-
drocytes for cartilage tissue engineering. They showed that the presence of
PCL resulted in rough fiber surface, which led to better cell attachment. Pati
et al. formed fibers by adding chitosan to two different coagulation baths
containing sodium triphosphate and NaOH to form chitosan-TPP and
regular chitosan fibers, respectively.[23] Fibers were deposited over each other
to form a scaffold with porosities up to 89%. Chitosan-TPP scaffolds had a
higher degradation rate than chitosan scaffolds. Chitosan-TPP scaffolds also
offered a better cell viability and a faster degradation rate.

   DeRosa *et al.* cross-linked type I collagen solution in acetic acid by a bath
of acetone and ammonium hydroxide at a pH of 9.[28] They coated the fibers
with poly-L-lysine (PLL) and then seeded with Schwann cells for neural tissue
engineering. Cell attachment and proliferation was much higher on the fiber

with surface treatment. Puppi *et al.* used wetspinning and cross-linked star PCL using an ethanol bath.[16] The fabricated fibers were porous and had diameters in the range of 100–300 µm. The fibers were randomly deposited to form a 3D structure, which was then seeded with murine pre-osteoblasts for bone tissue engineering. They also encapsulated levofloxacin and enrofloxacin within the fibers as antimicrobial drugs.

Landers *et al.* injected a mixture of sodium alginate 5% (w/v) and gelatin 1% (w/v) into a $CaCl_2$ bath to form solid fibers.[26] They also fabricated alginate-fibrin fibers by a adding mixture of alginic acid with fibrinogen to a solution of $CaCl_2$, thrombin, and sodium chloride. These fibers were deposited using a bioplotter to form a 3D fibrous mesh with a predefined fiber arrangement (Figure 1.3d). They seeded the scaffolds with mouse connective tissue fibroblasts. Lee *et al.* cross-linked α-tricalcium phosphate/alginate fibers with $CaCl_2$.[25] By changing the needle tip size they fabricated fibers with diameters in the range of 200–600 µm. The fibers were seeded with mesenchymal stem cells (MSCs) for bone tissue engineering. The scaffolds supported the differentiation of MSCs along the osteogenic lineage.

To improve the mechanical properties of wetspun hydrogel fibers carbon nanotubes (CNTs) or graphene oxide have been added to the pre-polymer solution. He *et al.* injected sodium alginate/graphene oxide (GO) into a bath of $CaCl_2$.[29] The maximum tensile strength and Young's modulus were increased by several fold at 4 wt% GO loading (from 0.32 and 1.9 to 0.62 and 4.3 GPa, respectively). Fibers were seeded with rabbit cartilage cells and showed a good attachment and proliferation. Spinks *et al.* added CNT to wetspun chitosan fibers to improve their mechanical properties.[17] They showed that the Young's moduli of fibers were improved from 4.25 GPa for pure chitosan fibers to 10.25 GPa for those containing 5% CNT. CNTs can also enhance the electrical conductivity of hydrogel fibers. Electrically conductive fibers can help stimulation of cells such as cardiomyocytes. MacDonald *et al.* showed that the electrical conductivity of fibroblast-laden collagen increases from 3 Ms cm^{-1} to 7 mS cm^{-1} upon addition of 4% CNTs.

Scaffolds can be made by wetspinning fibers and randomly depositing them on a substrate,[9,16,22,23,25] by rolling them up,[26] or by patterning the fibers in an ordered arrangement.[26] Since the fibers fabricated with wetspinning are relatively thick, the pore size of the formed scaffolds is large ($\sim$250–500 µm).[27] The porosity of scaffolds fabricated by wetspinning ranges from 15% to 92%.[23] Lee *et al.* randomly deposited wetspun α-tricalcium phosphate/alginate fibers and created scaffolds with low (13.6%), medium (34.0%), and high (53.7%) porosities.[25] They showed that the mechanical properties of the fabricated scaffolds increased by decreasing their porosities. For example, the elastic moduli of the scaffolds with high, medium, and low porosities were 96 ± 63 MPa, 398 ± 63 MPa, and 573 ± 87 MPa, respectively. These values fall in the biological range for trabecular bone (50–500 MPa). Reducing the porosity, however, increases the diffusional mass transfer resistance and reduces cell infiltration to the inner parts of the scaffold. Pati *et al.* deposited chitosan-TPP fibers to form 3D scaffolds with a high porosity (up to 89%) and seeded them with 3T3 fibroblasts. Lander *et al.* patterned wetspun alginate/fibrin fibers to form 3D structures (Figure 1.3d). They also showed that the porosity of the fabricated scaffolds can be determined by the following relationship:[26]

$$\text{Porosity} = 1 - \frac{\pi d_1^2}{4 d_2 d_3}, \tag{3}$$

where d_1 is the fiber diameter, d_2 and d_3 are the spacing between adjacent parallel fibers in horizontal and vertical directions.

Incorporation of cells within the fibers is possible using wetspinning. However, long exposure of the materials to cross-linking reagents, which are usually not cell-friendly, can limit the fabrication of cell-laden fibers from some materials. The most common material for fabrication of cell-laden fibers is sodium alginate because the pre-polymer solution and the gelation agent (calcium chloride) are both compatible with cells. For example, Arumuganathar *et al.* formed cell-laden fibers using a three-needle pressure-assisted system to fabricate multi-compositional structures that carried living cells in an inner layer.[30] The encapsulating medium flowed in the outer needle and provided a sheath for the suspension layer.[30–32]

In general, wetspinning is a simple method for fabrication of many natural and synthetic polymer fibers. As a multiple spinneret can be used this technique is scalable and can be used for industrial fabrication of fibers. However, the long time required for cross-linking of wetspun fibers has prevented fabrication of cell-laden fibers from a wide range of polymers.

1.4   Meltspinning (Extrusion)

Meltspinning or extrusion is the process of creating fibers by injecting a melted polymer through a micron size spinneret to form continuous fiber strands (Figure 1.4). The fiber strands are cooled after spinning on cooled drums (cold-drawn). Another heating step can be added to the process (hot-drawn) to improve the fiber properties. Meltspun fibers exhibit relatively high mechanical properties. Yuan *et al.* showed that the tensile strength of meltspun poly(L-lactic acid) (PLLA) fibers was in the range of 300–600 MPa.[33] They also showed that hot drawing of fibers significantly affects the mechanical properties of meltspun fibers. For example, Young's modulus of hot-drawn PLLA fibers (3.6–5.4 GPa) was significantly higher than their cold-drawn counterparts. In another study, Gomes *et al.* used meltspun fibers from starch with ε-polycaprolactone (SPCL) and starch with poly(lactic acid) (SPLA).[34] They created scaffolds by a fiber bonding process that includes bonding of meltspun fibers in an additional annealing process. They showed that fabricated scaffolds exhibited compression modulus in the range of 1.8 MPa–9.61 MPa.

The cross-section of meltspun fibers and their texture can be tailored by using spinnerets with predefined cross-sections. This feature is particularly beneficial for applications where the fiber texture is important for

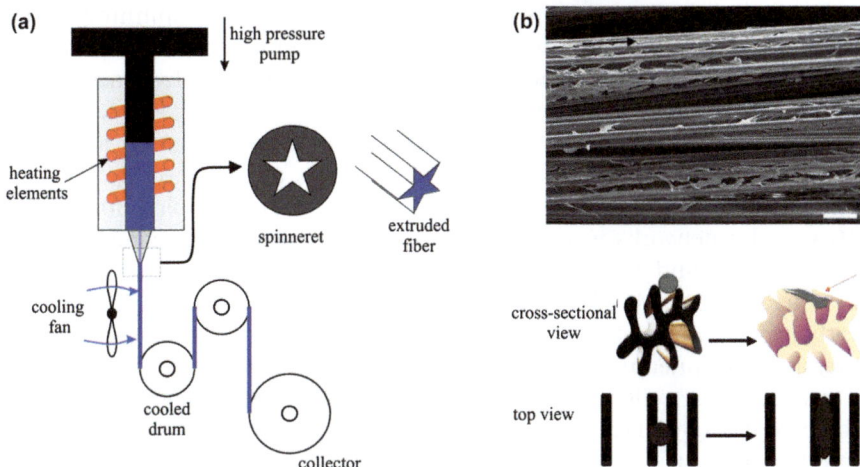

Figure 1.4   Process of fiber fabrication with extrusion. (a) Schematic of the meltspinning process. A polymer is heated to its melting point and injected with
high pressure through a spinneret with a predefined geometry. The fiber
strands are collected on cooled drums after the extrusion process.
(b) Fibroblast cells (NHDF) grown on grooved meltspun PET fibers. All
cells aligned themselves along the grooves direction. However, cells were
unable to fit into the grooves when the width of the grooves was smaller
than the cells size and flattened on top of the groove.
Images in (b) are adapted with permission from John Wiley & Sons Inc.
[J. Biomedical Materials Research: Part A], copyright (2010).

orientating the cells. Lu *et al.* used this capability to create grooved fibers for
mechanical guidance of rat skin fibroblasts (RSFs) and rat aortic smooth
muscle cells (RASMCs).[35] Poly(lactic acid) (PLA) and polyethylene ter-
ephtalate (PET) grooved fibers were fabricated using the process shown in
Figure 1.4. They used a spinneret with grooves 5–15 µm deep and 10 µm wide
to fabricate fibers with the nominal diameter of 50 µm. They showed that
cultured cells attached and extended their cytoplasmic lamellapodia within
the grooves after 4 weeks of incubation. Moreover, the cells aligned themselves parallel to the direction of the grooves.

Sinclair *et al.* showed that the dimension of the grooves played an important role in cell alignment.[36] They produced meltspun PET grooved fibers
with different groove widths (6–53 µm) and depths (6–35 µm). Normal
human dermal fibroblast (NHDF) cells were cultured on the fibers for a
period of 2 weeks. Their results indicated that all grooved fibers were capable
of supporting cellular attachment and proliferation, nuclear elongation
parallel to the fiber's longitudinal axis, and aligned matrix synthesis. However, when the width of the grooves was smaller than the cells size, cells were
unable to fit into the grooves. As a result, cells flattened on the top of the
groove and extended parallel to the fiber's longitudinal axis.

Fabrication of hollow fibers with meltspinning was also demonstrated.
Hinüber *et al.* used a special spinneret to create hollow fibers from

poly(3-hydroxybutyrate) (PHB).[37] They showed that meltspinning at low process temperatures without additives led to the formation of well-defined hollow PHB fibers. These hollow fibers can be used to deliver oxygen and nutrients to cells during the culturing period.

In an interesting study, Miller *et al.* combined meltspinning and direct writing processes to create 3D filament networks of carbohydrate glasses.[38] They used this cytocompatible network as a sacrificial template for creating 3D vascular networks in tissue constructs. Their device was comprised of a glass syringe and steel nozzles. The syringe was mounted on a custom-modified RepRap Mendel 3D printer with associated electronics. They heated carbohydrate glass up to 100 °C and printed the molten glass under nitrogen pressure with pneumatic control to create a sacrificial network. After pouring a suspension of cells in ECM pre-polymer on the network and cross-linking, the carbohydrate glass filaments were dissolved to form vessels while their interfilament fusions become intervessel junctions. They showed that the perfused vascular network created by this method sustained the metabolic function of primary rat hepatocytes in engineered tissue constructs.

Meltspinning is a relatively fast process that allows the high-throughput fabrication of micron size fibers with irregular cross-sections. For example, Yuan *et al.* fabricated 10 meters of PLLA fibers with diameter of 110–160 μm in 10 minutes (1 m min^{-1}) and Gomes *et al.* created 100 meters of SPCL and SPLA with diameter 120–500 μm in 5 minutes (20 m min^{-1}).[33,34] However, meltspinning is a sophisticated method that requires high temperatures in the range of 150 °C to 295 °C.[34,36] These high temperatures prevent the encapsulation of cells, proteins, and temperature-sensitive chemicals in the fibers. Moreover, the melted polymers are usually highly viscous (*e.g.* viscosity of PLA at melting point is 10^4 times higher than water[39]), therefore high pressures are required to push the polymer through the spinneret.[40] Mass loss, rapid decrease of viscosity during the process, and thermal degradation of some materials are the other main challenges of the meltspinning process that often have considerable influence on the mechanical properties of the resultant fibers.[41] In general, fibers larger than a few micrometers can be fabricated with meltspinning.

1.5   Electrospinning

Electrospinning is a fiber fabrication technique that uses an electrical field to draw a viscoelastic polymer from a spinneret and deposit the fibers on a collector plate.[42,43] The essential components of an electrospinning device include a viscous polymer solution, a pumping system (injector) with a metallic tip, and a collector plate at a distance from the tip (Figure 1.5a). As a result of the applied electrical field, the polymer stream breaks into smaller branches, which in turn form fibers that deposit on the collector surface (Figure 1.5). The properties of the polymer, the applied voltage, the distance from the tip of the injector to the collector plate, and the properties of the collector plate could affect the microstructure of the final fibrous construct.

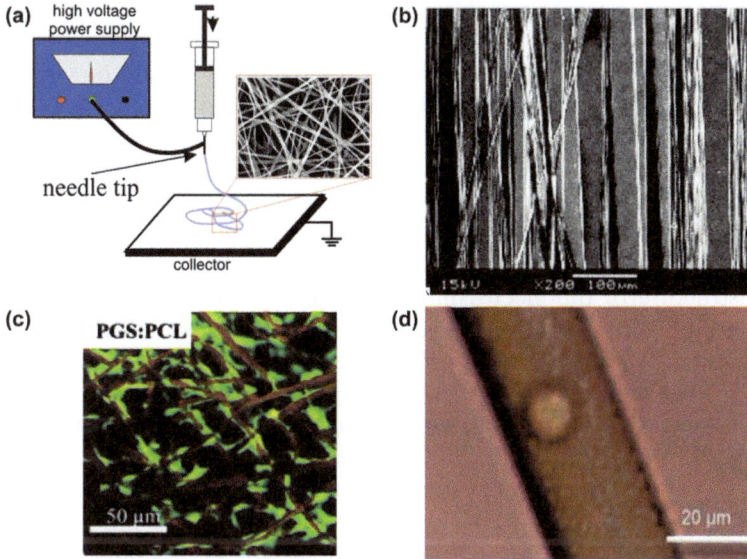

Figure 1.5 Electrospinning for tissue engineering applications. (a) A typical electro-spinning system, which is comprised of a pumping system (injector), high voltage source, and a grounded collector plate. The inset shows a typical random fibrous mat which is the outcome of regular electrospinning.[45] (b) Using a collector plate on a rotating drum, aligned electrospun fibers can be formed.[45] (c) Confocal laser scanning images of HUVECs seeded on a PCL-PGS scaffold after 7 days of culture.[46] (d) An optical micrograph of a cell-laden electrospun fiber.[47]
Images (a) and (b) reprinted with permission from ref. 45 with permission from Elsevier. Image (c) reprinted from ref. 46 with permission from John Wiley and Sons. Image (d) reprinted with permission from ref. 47 with permission from Sage publications.

Depending on the polymer physical properties (*i.e.* viscosity, surface tension, and electrical conductivity), fibers with diameters in the range of a few nanometers to several micrometers can be created.[44] Fibrous scaffolds can be created by random deposition of fibers on a collector plate. Fibers can also be collected on a rotating collector to form mats with aligned fibers. The relative simplicity of the method as well as the ease of controlling the key process parameters such as flow rate and voltage are the main reasons for the popularity of this technique.[43]

Electrospinning has been widely used for tissue engineering applications as scaffolds fabricated by this method mimic the microstructure of the extracellular matrix of native tissues. Electrospun scaffolds can be fabricated from natural and synthetic polymers including silk fibroin, chitosan, collagen, gelatin, hyaluronic acid, fibrinogen, poly(ester urethane) urea (PEUU), poly(glycerol sebacate) (PGS), PCL, polyurethane (PU), PLGA, and PGA[44] for applications such as cardiac graft, wound dressing, and cartilage and bone tissue engineering.[3]

Selection of the proper solvent to dissolve the polymer prior to electro-spinning is essential. The solvent should be volatile to evaporate before the fibers hit the collector plate. Moreover, the solution should be viscoelastic. For example, it has been observed that if the solution is not viscous enough, fibers do not form and particles would form instead. Some of the solvents that have been used include 1,1,1,3,3,3-hexafluoro-2-propanol for collagen PEUU, and silk fibroin, tetrahydrofolate for chitosan, anhydrous chloroform-ethanol mixture for PCL and PGS, and water for gelatin and hyaluronic acid.

The applied voltage affects the diameter of the formed fiber and the microstructure of the electrospun scaffolds.[48] Low applied voltage cannot initiate the electrospinning process. As the applied electric field (larger than the threshold) increases, the polymer stream breaks into more branches and the fiber diameter is reduced initially. Sant *et al.* fabricated electrospun PGS-PCL sheets for cardiac tissue engineering (Figure 1.5c). They observed that at 2 : 1 PGS:PCL ratio, a minimum fiber diameter was achieved by applying a 17.5 kV voltage. Their results showed that the fiber diameter decreased from 4 μm at 12.5 kV to 3 μm at 17.5 kV and then increased to 8 μm at 20 kV.[46]

Injection flow rate is another important parameter that affects the quality and geometry of the fabricated fibers. High flow rates (injection velocities) lead to wet and non-polymerized fibers as the solvent does not have enough time to evaporate. The unevaporated solvent could be harmful to cells. In general, fiber diameter has a direct relationship with the volumetric flow rate. The distance from the needle tip to the collector plate is also important as it should be large enough that the solvents evaporate prior to fiber attachment to the collector plate.

In regular electrospinning on aluminum foil collectors the fibers randomly distribute and orient in the scaffold (Figure 1.5a). Other types of collectors include rotating disk or mandrel, pin, mesh, and parallel wires. Also, it is possible to collect the fibers in a coagulation bath to cross-link the fibers. Rotating disks have been used to control the alignment of the fibers.[49] Yang *et al.* electrospun PLLA fibers on a disk with a rotation of 1000 rpm and a flow rate of 1 mL hr^{-1} and an applied voltage of 12 kV.[45] By changing the concentration of the polymer they fabricated aligned fibers with diameter in the rage of 150–3000 nm (Figure 1.5b). In another notable study, Amoroso *et al.* electrospun PEUU fibers on a mandrel with rotation speed of 266 rpm and rastering speed of 0.3–30 cm s^{-1}. Simultaneously, they electrosprayed cell culture medium and cells on the scaffold during the fabrication process to cut down the time for penetration of cells between various layers of fibers. The mechanical properties of the fabricated electrospun sheets were comparable to those of native porcine pulmonary valve.

In spite of all the progresses made in electrospinning, manufacturing of thick 3D complex scaffolds is difficult.[50] In addition, the high fiber packing density and the small pore sizes ($\sim$10–15 μm) limits cell infiltration into the scaffolds.[51] The latter issue has triggered an array of techniques including salt/polymer leaching,[52] wet electrospinning using a bath collector[53] or an ice crystal collector,[54] and laser/UV irradiation[55] to produce electrospun scaffolds with both large pores and high porosity.

Encapsulation of living cells in electrospun fibers is a difficult process as: i) the employed solvents are usually cytotoxic; and ii) the high electrical field could damage the cells. Despite these challenges fabrication of cell-laden fibers though electrospinning has been demonstrated in several studies[47,56,57] (Figure 1.5d). In these studies, a co-axial needle configuration was used, in which each needle was connected to a separate syringe pump. The central stream contained a cell and the sheath stream contained a polymer. The encapsulation of living cells in electrospun fibers faced many challenges as the diameter of the fibers is typically smaller than a single cell.[56,57] Another shortcoming of the electrospinning of cell encapsulated constructs is the lack of control over cell distribution in the volume.

1.6 Conclusions

Fiber-based tissue engineering is an emerging area that holds a great promise for creating artificial organs for transplantation. Fibers are the building blocks of these engineered constructs. Owing to recent advancements of microtechnologies, fabrication of sophisticated fibers with engineered mechanical properties, topography, and composition are now possible. These fibers are usually formed using wetspinning, co-axial flow systems, meltspinning, and electrospinning. Here, we discussed various fiber fabrication techniques that utilize microtechnologies. Among fiber fabrication techniques, electrospinning has become popular for fabricating tissue scaffolds owing to its relative simplicity, similarity of the fabricated structures to native tissues, and ease of control over the key process parameters. However, the fabrication of cell-laden fibers with precise control over cell distribution using electrospinning is still a major challenge.

Wetspinning is probably the simplest method for fiber fabrication and wetspun fibrous structures offer a tunable porosity, but larger pore sizes in comparison to electrospun structures that make them more amenable for cell penetration. However, long exposure of the fibers to harmful cross-linking reagents has limited fabrication of cell-laden fibers to alginate fibers. In addition, due to the large size of the fabricated fibers the wetspun fibers cannot mimic the extracellular matrix.

Co-axial systems for fiber spinning that encompass those using a microfluidic chip can fabricate fibers with tunable morphological, structural, and chemical features, continuously. Moreover the size of fibers fabricated by co-axial systems is usually smaller than wetspinning. The main drawback of these systems is their reliability as they are vulnerable to clogging during the manufacturing process.

Meltspinning facilitates the fabrication of fibers with irregular texture and cross-section; however, the high temperature required during the fabrication process prevents encapsulation of cells within the fibers. Meltspun fibers have the potential to be used sacrificially for forming a vasculature network within hydrogel constructs.

Acknowledgements

Financial support from NSERC, CIHR, CHRP, CFI, Genome Canada, and Genome Quebec is gratefully acknowledged. M.A and A.T. acknowledge NSERC postdoctoral fellowships. D.J. acknowledges support from a Canada Research Chair. The authors declare no conflict of interests in this work. A.K. acknowledges funding from the National Science Foundation CAREER Award (DMR 0847287), the office of Naval Research Young National Investigator Award, and the National Institutes of Health (HL092836, DE019024, EB012597, AR057837, DE021468, HL099073, EB008392).

References

1. N. Annabi, A. Tamayol, J. A. Uquillas, M. Akbari, L. E. Bertassoni, C. Cha, *et al.*, *Adv. Mat.*, 2014, **26**, 85–124.
2. M. Akbari, A. Tamayol, V. Laforte, N. Annabi, A. Khademhosseini, A. H. Najafabadi, *et al.*, *Adv. Funct. Mat.*, 2014, **24**, 4060–4067.
3. A. Tamayol, M. Akbari, N. Annabi, A. Paul, A. Khademhosseini and D. Juncker, *Biotechnology Advances*, 2013, **31**(5), 669–687.
4. E. Kang, G. S. Jeong, Y. Y. Choi, K. H. Lee, A. Khademhosseini and S.-H. Lee, *Nat. Mater.*, 2011, **10**, 877–883.
5. W. Jeong, J. Kim, S. Kim, S. Lee, G. Mensing and D. J. Beebe, *Lab Chip*, 2004, **4**, 576–580.
6. C. Hwang, Y. Park, J. Park, K. Lee, K. Sun, A. Khademhosseini and S. Lee, *Biomed. Microdevices*, 2009, **11**, 739–746.
7. S.-J. Shin, J.-Y. Park, J.-Y. Lee, H. Park, Y.-D. Park, K.-B. Lee, C.-M. Whang and S.-H. Lee, *Langmuir*, 2007, **23**, 9104–9108.
8. S. Ghorbanian, M. A. Qasaimeh, M. Akbari, A. Tamayol and D. Juncker, *Biomed. Microdevices*, 2004, **16**(3), 387–395.
9. B. R. Lee, K. H. Lee, E. Kang, D.-S. Kim and S.-H. Lee, *Biomicrofluidics*, 2011, **5**, 022208.
10. M. Hu, R. Deng, K. M. Schumacher, M. Kurisawa, H. Ye, K. Purnamawati and J. Y. Ying, *Biomaterials*, 2010, **31**, 863–869.
11. K. H. Lee, S. J. Shin, Y. Park and S. H. Lee, *Small*, 2009, **5**, 1264–1268.
12. M. Yamada, S. Sugaya, Y. Naganuma and M. Seki, *Soft Matter*, 2012, **8**, 3122–3130.
13. D. Enea, F. Henson, S. Kew, J. Wardale, A. Getgood, R. Brooks and N. Rushton, *J. Mater. Sci. Mater. Med.*, 2011, **22**(6), 1569–1578.
14. Y. P. Kato, D. L. Christiansen, R. A. Hahn, S. J. Shieh, J. D. Goldstein and F. H. Silver, *Biomaterials*, 1989, **10**, 38–42.
15. M. R. Williamson and A. G. A. Coombes, *Biomaterials*, 2004, **25**, 459–465.
16. D. Puppi, D. Dinucci, C. Bartoli, C. Mota, C. Migone, F. Dini, G. Barsotti, F. Carlucci and F. Chiellini, *J. Bioact. Compat. Polym.*, 2011, **26**, 478–492.

17. G. M. Spinks, S. R. Shin, G. G. Wallace, P. G. Whitten, S. I. Kim and S. J. Kim, *Sensors Actuator B Chem.*, 2006, **115**, 678–684.
18. S. Arumuganathar and S. N. Jayasinghe, *Biomacromolecules*, 2008, **9**, 759–766.
19. S. Arumuganathar, S. N. Jayasinghe and N. Suter, *Soft Matter*, 2007, **3**, 605–612.
20. S. N. Jayasinghe and N. Suter, *Biomicrofluidics*, 2010, **4**, 014106.
21. K. Tuzlakoglu, C. M. Alves, J. F. Mano and R. L. Reis, *Macromol. Biosci.*, 2004, **4**, 811–819.
22. I. B. Leonor, M. T. Rodrigues, M. E. Gomes and R. L. Reis, *J. Tissue Eng. Regener. Med.*, 2011, **5**, 104–111.
23. F. Pati, B. Adhikari and S. Dhara, *J. Mater. Sci. Mater. Med.*, 2012, **23**(4), 1085–1096.
24. N. E. Fedorovich, L. Moroni, J. Malda, J. Alblas, C. A. Blitterswijk and W. J. A. Dhert, *Cell and Organ Printing*, eds. B. R. Ringeisen, B. J. Spargo and P. K. Wu, Springer, Netherlands, 2010, pp. 225–239.
25. G.-S. Lee, J.-H. Park, U. S. Shin and H.-W. Kim, *Acta Biomaterialia*, 2011, **7**, 3178–3186.
26. R. Landers, A. Pfister, U. Hübner, H. John, R. Schmelzeisen and R. Mülhaupt, *J. Mater. Sci.*, 2002, **37**, 3107–3116.
27. S. C. Neves, L. S. Moreira Teixeira, L. Moroni, R. L. Reis, C. A. Van Blitterswijk, N. M. Alves, M. Karperien and J. F. Mano, *Biomaterials*, 2011, **32**, 1068–1079.
28. K. DeRosa, M. Siriwardane and B. Pfister, *Proceeding of IEEE Bioengineering Conference (NEBEC)*, April 1–3, Troy, NY, 2011, pp. 1–2.
29. Y. He, N. Zhang, Q. Gong, H. Qiu, W. Wang, Y. Liu and J. Gao, *Carbohydr. Polymer.*, 2012, **88**(3), 1100–1108.
30. S. Arumuganathar and S. N. Jayasinghe, *Macromol. Rapid Comm.*, 2007, **28**, 1491–1496.
31. S. Arumuganathar, S. Irvine, J. R. McEwan and S. N. Jayasinghe, *J. Appl. Polymer Sci.*, 2008, **107**, 1215–1225.
32. S. Arumuganathar and S. N. Jayasinghe, *NANO*, 2007, **2**, 213–219.
33. X. Yuan, A. F. T. Mak, K. Kwok, B. K. O. Yung and K. Yao, *J. Appl. Polymer Sci.*, 2001, **81**, 251–260.
34. M. Gomes, H. Azevedo, A. Moreira, V. Ellä, M. Kellomäki and R. Reis, *J. Tissue Eng. Regener. Med.*, 2008, **2**, 243–252.
35. Q. Lu, A. Simionescu and N. Vyavahare, *Acta Biomaterialia*, 2005, **1**, 607–614.
36. K. D. Sinclair, K. Webb and P. J. Brown, *J. Biomed. Mater. Res. A*, 2010, **95A**, 1194–1202.
37. C. Hinüber, L. Häussler, R. Vogel, H. Brünig and C. Werner, *Macromol. Mater. Eng.*, 2010, **295**, 585–594.
38. J. S. Miller, K. R. Stevens, M. T. Yang, B. M. Baker, D.-H. T. Nguyen, D. M. Cohen, E. Toro, A. A. Chen, P. A. Galie, X. Yu, R. Chaturvedi, S. N. Bhatia and C. S. Chen, *Nat. Mater.*, 2012, **11**, 768–774.
39. M. Ajioka, K. Enomoto, K. Suzuki and A. Yamaguchi, *J. Polymer. Environ.*, 1995, **3**, 225–234.

40. M. Akbari, M. Bahrami and D. Sinton, *Int. J. Heat Mass Tran.*, 2011, **54**, 3970–3978.
41. S. Fakirov and D. Bhattacharyya, *Handbook of Engineering Biopolymers: Homopolymers, Blends and Composites*, Hanser Gardner Pubns, 2007.
42. R. L. Mauck, B. M. Baker, N. L. Nerurkar, J. A. Burdick, W.-J. Li, R. S. Tuan and D. M. Elliott, *Tissue Eng. B Rev.*, 2009, **15**, 171–193.
43. M. Deng, R. James, C. Laurencin and S. Kumbar, *IEEE Transactions on NanoBioscience*, 2012, **11**(1), 3–14.
44. N. Bhardwaj and S. C. Kundu, *Biotechnol. Adv.*, 2010, **28**, 325–347.
45. F. Yang, R. Murugan, S. Wang and S. Ramakrishna, *Biomaterials*, 2005, **26**, 2603–2610.
46. S. Sant, C. M. Hwang, S.-H. Lee and A. Khademhosseini, *J. Tissue Eng. Regener. Med.*, 2011, **5**, 283–291.
47. Y. H. Shih, J. C. Yang, S. H. Li, W. C. V. Yang and C. C. Chen, *Textil. Res. J.*, 2012, **82**, 602–612.
48. E. Zussman, *Polymer. Adv. Tech.*, 2011, **22**, 366–371.
49. L. M. Bellan and H. G. Craighead, *Polymer. Adv. Tech.*, 2011, **22**, 304–309.
50. C. M. Hwang, A. Khademhosseini, Y. Park, K. Sun and S.-H. Lee, *Langmuir*, 2008, **24**, 6845–6851.
51. M. F. Leong, W. Y. Chan, K. S. Chian, M. Z. Rasheed and J. M. Anderson, *J. Biomed. Mater. Res. A*, 2010, **94**, 1141–1149.
52. Y. H. Lee, J. H. Lee, I. G. An, C. Kim, D. S. Lee, Y. K. Lee and J. D. Nam, *Biomaterials*, 2005, **26**, 3165–3172.
53. Y. Yokoyama, S. Hattori, C. Yoshikawa, Y. Yasuda, H. Koyama, T. Takato and H. Kobayashi, *Mater. Lett.*, 2009, **63**, 754–756.
54. M. Simonet, O. D. Schneider, P. Neuenschwander and W. J. Stark, *Polymer Eng. Sci.*, 2007, **47**, 2020–2026.
55. H. woon Choi, J. K. Johnson, J. Nam, D. F. Farson and J. Lannutti, *J. Laser Appl.*, 2007, **19**, 225.
56. A. Lopez-Rubio, E. Sanchez, Y. Sanz and J. M. Lagaron, *Biomacromolecules*, 2009, **10**, 2823–2829.
57. A. Townsend-Nicholson and S. N. Jayasinghe, *Biomacromolecules*, 2006, **7**, 3364–3369.

CHAPTER 2

Kidney on a Chip

LAURA HA,[a] KYUNG-JIN JANG*[b] AND KAHP-YANG SUH[a,c]

[a] Interdisciplinary Program of Bioengineering, Seoul National University, Seoul, 151-742, Korea; [b] Wyss Institute for Biologically Inspired Engineering at Harvard University, Boston, MA 02115, USA; [c] School of Mechanical and Aerospace Engineering, Seoul National University, Seoul, 151-742, Korea
*Email: kyung-jin.jang@wyss.harvard.edu

2.1 Introduction

There is a great need for *in vitro* kidney models for investigating renal functions and medical applications. To date, most studies on kidney have utilized conventional static two-dimensional culture systems. In recent years, however, it is generally accepted that the two-dimensional cell culturing scaffolds are not representative of the cellular environment found in organisms. In fact, under such biased and simplified conditions, tissue-specific properties such as architecture, biochemical cues, and cell-cell communication disappear. Therefore, to prove the significance of the scientific findings, it is inevitable to perform additional *ex vitro* experiments such as animal testing afterwards.

To overcome such a drawback of two-dimensional scaffolds, *in vitro* microenvironments mimicking engineered scaffolds have been the subject of massive research because their features include tissue–tissue interfaces, spatiotemporal gradients of chemicals, and oxygen. The mechanically active microenvironments that are central to the functions of virtually all living organs are crucial for their functions.[1] Due to considerable advances in the area of microsystems engineering, microfabrication techniques have

RSC Nanoscience & Nanotechnology No. 36
Microfluidics for Medical Applications
Edited by Albert van den Berg and Loes Segerink
© The Royal Society of Chemistry 2015
Published by the Royal Society of Chemistry, www.rsc.org

adapted to enable precise control of cell shape position function and tissue organization in a highly structured scaffold. In particular, microfluidics has enabled fine control of dynamic fluid flows and pressure on the micrometer scale; therefore, it is now possible to create a microenvironment that presents cells with organ-relevant chemical gradients and mechanical cues, which promotes cells to express a more differentiated ordinary phenotype. Such a merger of microfluidics and cell biology has led to the advent of "organ on chips" that can restructure the tissue organization and functional complexity of living organs using cells cultured in microfluidics devices.[2]

Early attempts to micro-engineer cell culturing systems have been focused on fabricating adhesive substrates to control cell behaviors including cell shape, cell growth, cell position, and cell differentiation.[3,4] Progress of the soft lithography-based microfabrication and microfluidics has made it possible to create a complicated cell culture platform that reconstructs three-dimensional micro-architecture of tissues and organs. Employing a microfluidic cell culture system has modified the way in which either a two-dimensional or a three-dimensional environment can be precisely formed with living cells.[4] Moreover, development of the micro-engineering approaches has opened new possibilities to build *in vitro* models that reorganize more complicated three-dimensional organ level structures and to incorporate more vital dynamic mechanical and chemical signals to the cell culture models.[5] Through this approach, many microfabrication integrated models of bones, blood, brain, vessels, liver, lung, airways, gut, muscles, liver, and kidney have been developed.[6–20] In conjunction with these efforts, tissue-tissue interfaced organ chip models that can recapitulate renal functions *via* various cues such as hormone, metabolites, cytokines, nutrient, chemical, and physical signal transfer across the interface of neighboring tissues have been introduced.[13,21,22]

In this chapter, a general overview of progress made on the kidney on a chip will be introduced. We will focus on the understanding of kidney structure and function, and application of micro-engineered kidney-mimetic microsystems for the study of kidney function and disease *in vitro*. We will also discuss the potential use of these "kidney chips" for biomedical pharmaceutical and environmental safety testing applications as well as challenges for the field that must be overcome to translate these technologies into useful products in the future.

2.2   Kidney Structure and Function

The kidney is a paired bean shaped organ, encapsulated by a smooth and easily removable thick tough fibrous capsule. It is positioned on each side of the vertebral column and usually positioned more caudal. In the human body, the weight of each kidney ranges from 125 g to 170 g in an adult male and from 115 g to 155 g in a female, measuring roughly 11 cm long, 5 cm to

7.5 cm wide, and 2.5 cm to 3 cm thick. The mammalian kidney is divided into three sections: renal cortex, outer renal medulla, and an inner renal medulla. It contains up to 18 lobes with each lobule being composed of nephrons, the functional units of the kidney. Each kidney is known to compose more than one million of nephrons and each nephron consists of glomerulus and systemic tubules.

The kidney is supplied with blood by a single renal artery that is divided into five segmental arteries at the renal hilum. Within the kidney, the segment further branches into several lobular arteries and interlobular arteries; therefore, blood flows through the afferent arteriole into the glomerulus and blood flows out of the glomerulus through the efferent arteriole. The afferent and efferent arterioles control the glomerulus capillary pressure to make it selectively contracting and dilating.

A unique high pressure mass of capillary, the glomerulus, is encapsulated by Bowman's capsule, which is a double-walled capsule; the space between the glomerulus and Bowman's capsule is called Bowman's space. Kidney filtration starts as a plasma-like blood filtered through capillary to Bowman's space as blood pressure forces fluid from the blood in the glomerulus into the lumen of Bowman's capsule. Bowman's capsule is permeable to water and small solutes, but not to blood cells or large molecules such as plasma protein; thus the filtrate contains salts, glucose, amino acids, vitamins, nitrogenous wastes, and other small molecules and forms primary urine.

The filtrates then proceed into the nephron. In this segment, some substances are added to the filtrate to form urine and some substances are reabsorbed back to the nephron. Nephron tubules are separated into four segments: proximal convoluted tubule, loop of Henle, distal convoluted tubule, and collecting tubule. The proximal convoluted tubule has functions of draining filtrates from Bowman's capsule; it also reabsorbs important nutrient substances almost completely. Throughout the segment of the loop of Henle, water and ions are reabsorbed from urine; the concentration of urine is controlled in this section. The distal convoluted tubule plays a role in regulating potassium, sodium, and pH, such that further dilution of urine occurs. The last section, called the collecting tubule, joins to several tubules to collect filtrate and sodium regulation function takes place.

In vertebrates and some other chordates, the functions such as osmoregulation, an active regulation of the bodily fluidic osmotic pressure to maintain the homeostasis, and excretion are regulated by the kidney. The kidney's main functions include eliminating metabolic wastes by filtering the blood, and selectively re-absorbing filtered ions and water to maintain the normal composition of the blood. Through the adjustment of blood composition, the kidney successfully maintains blood volume and pressure, and ensures balance of various ions such as sodium, chloride, potassium, calcium, hydrogen, phosphate, and pH. Moreover, it takes a role in eliminating products of metabolism such as urea, uric acid, and creatinine.

In the human body, blood is supplied to each kidney by a renal artery and drained by a renal vein. Interestingly, this small organ accounts for less than

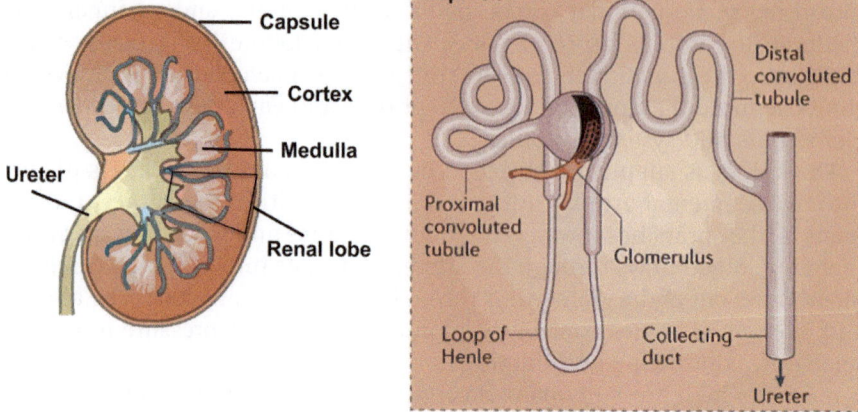

Figure 2.1 Kidney structure. The kidney is surrounded by the renal capsule and divided into renal cortex and renal medulla. These structures contain lobes and each lobule is composed of a nephron, the functional unit of the kidney. The nephron consists of the glomerulus, proximal tubule, loop of Henle, distal tubule, and collecting duct.[101,102]

1% of human body mass but receives roughly 25% of the blood exiting the heart (Figure 2.1).[23,51,65,101]

2.3   Mimicking Kidney Environment

2.3.1   Extracellular Matrix

The importance of the Extracellular Matrix (ECM) in cell biology has been well documented. Many kidney-related biological studies have revealed that the ECM has significant roles in cell adhesion, cell migration, molecular filtration, and cell differentiation, and that not all basement membranes hold the same ECM components. In fact, renal epithelial basement membranes illustrate a molecular heterogeneity that is suspected to cause functional specificity of distinct nephron segments.[24]

There are several primary components that form a renal basement membrane ECM: collagen IV, laminin, entactin/nidogen, and sulfated proteoglycans.[24] Collagen IV is approximately 180 kDa in length with six genetically distinct α chains (α1 to α6).[25,26] For the nephron, the pattern of collagen IV chain deposition varies from species to species.[27–31]

Laminin is a trimer containing chains (α,β,γ) and there are currently 12 heterotrimers reported (from laminin-1 to laminin-12). Laminin-1 is a rarely found component in the basement membrane of other organs. Interestingly, it is known to be the major component in adult kidney proximal tubular basement membranes and the loop of Henle basement membrane.[32,33] Laminin-1 seems to have a critical role in transitioning mesenchyme to epithelium at the onset of nephrogenesis.[34] Low levels of laminin-2 are also seen in tubular basement membranes.[33,35] The most abundant trimer in the mature kidney is laminin-10, which is found in all tubular and collecting duct basement membranes.[33,35]

Entactin/nidogen, which is known as a linker of laminin and collagen IV networks in the basement, is a 150 kDa long molecule with three globular domains. It is ubiquitous in the kidney and it remains to be determined whether these molecules have particular functions or general roles in arrangement and preservation of all basement membranes.[25,36,37]

Proteoglycans consist of protein cores with attached heparan sulfate, chondroitin sulfate, and dermatan sulfate side-chains.[38,39] Due to the long carbohydrate chains the basement membrane consists of negative charge and this unique property of the membrane is suspected to contribute charge selective filtration properties of the glomerular basement membrane. In addition to the filtration function, proteoglycans are believed to link laminin, collagen IV and entactin/nidogen to stabilize the basement membrane.[40]

The ECM, which covers basement membranes, not only contributes to their structure and function, but is also involved in glomerular and tubulointerstitial disease.[24] Therefore, we can expect that better understanding of the basement membrane components would provide improved tools to mimic the renal tubular environment and study complex renal physiological processes (Figure 2.2).[24,101,102]

2.3.2 Mechanical Stimulation

The most prominent mechanosensory cue for flow subjecting cells would be fluid shear stress. Over decades, extensive research has been conducted on the biochemical and ultrastructural effects of fluid shear on cells. For example, Dewy *et al.* first published a method of culturing vascular endothelial cells under carefully controlled condition of fluid shear.[41]

To mimic the dynamic renal fluid environment, the flow system is supposed to simulate *in vitro* fluid shear stress. A cell monolayer is subjected to fluid shear stress by generating a pressure gradient over the cell surface. To calculate the fluid shear stress on the cells, the mathematical model assumes a Newtonian fluid in which the shear tensor is proportional to the deformation tensor. For steady flow between infinitely wide parallel plates, wall shear stress τ is calculated. The equation $\tau = 6\mu Q/bh^2$ is derived from the Navier–Stokes momentum equation of two-dimensional flow, where Q is the flow rate, μ is the medium viscosity, and b and h are channel width and height, respectively.[42,43]

Based on many studies, it has been well known for many years that the fluid shear stress allows renal tubular cells to modulate proximal Na^+, HCO_3^-, Cl^-, and water reabsorption as well as distal Na^+ absorption and K^+ secretion.[44–49] However, the basic mechanism of how tubule epithelial cells sense axial flow had remained unexplained until Guo *et al.* first proposed a hydrodynamic mechanosensory hypothesis for brush border microvilli in 2000. In the paper, Guo claimed that both Na^+ reabsorption and HCO_3^-

A

PT
Lam-1, –10
[Lam–2]
Col α1, α2(IV)
[α3–α5(IV)]
E/N, P, B, a

GBM
Lam-11
Col α3–α5(IV)
[α1, α2(IV)]
E/N, A, p

BC
Lam-1, –10
[lam–2]
Col α1, α2(IV)
α5, α6(IV)
[α3, α4(IV)]
E/N, P, B, a

MM
Lam-1, –2, –10
[Lam–4, –11]
Col α1, α2(IV)
E/N, P, B

DT
Lam-10
[Lam–2]
Col α1–α6(IV
E/N, P, B, a

CD
Lam-10
[Lam–2]
Col α1, α2(IV)
α5, α6(IV)
E/N, P, B

Lam:	Laminin
Col:	Collagen
E/N:	Entactin/Nidogen
A:	Agrin
P:	Perlecan
B:	Bamacan

LH
Lam-1, –10
[Lam–2]
Col α1, α2(IV)
E/N, P, B, a

B Glomerular cells

Podocyte
Endothelial cell
Mesangial cell

Proximal tubule cells   Distal tubule cells

Loop of Henle cells   Collecting duct cells

C   Shear Stress   Tubular fluid

Extracellular matrix
Osmotic gradient
Hormonal stimulation
Interstitial fluid

Blood

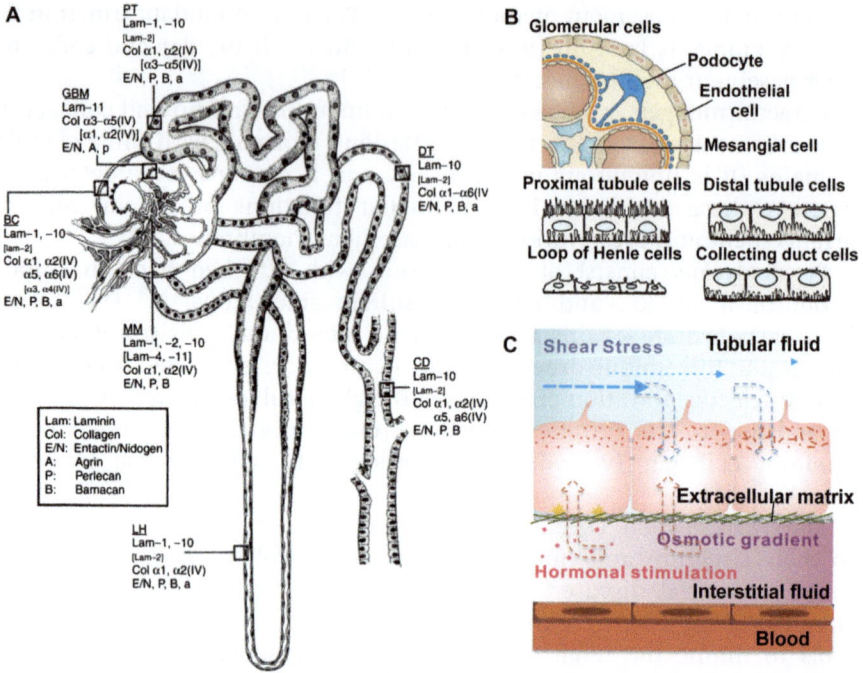

Figure 2.2   Mimicking the kidney environment. (A) Schematic of a nephron with the components of various basement membranes. Abbreviations are: Lam, laminin; Col, collagen; A, agrin; P, perlecan; B, bamacan; GBM, glomerular basement membrane; MM, mesangial matrix; BC, Bowman's capsule; PT, proximal tubule; LH, loop of Henle; DT, distal tubule; CD, collecting duct.[24] (B) Various kidney cell types.[101,102] (C) The *in vivo* kidney tubular environment is composed of tubular fluid, kidney tubular cell, extracellular matrix, interstitial fluid, and blood. The kidney cells are subjected to fluid shear stress and control reabsorption and secretion of water and ions by complex endocrine mechanisms.[43]

transport depend on fluid flow induced bending moment (torque) of the brush boarder microvilli.[50]

2.3.3   Various Kidney Cells

There are various distinct cell types in the kidney, such as glomerular cells, proximal tubule cells, loop of Henle cells, thick ascending limb cells, distal tubule cells, collecting duct cells, interstitial kidney cells, and renal endothelial cells. Each cell type performs unique functions and these cells offer an essential tool to investigate kidney function.

2.3.3.1   *Glomerular Cells*

Under a microscope, it is possible to examine a thin fenestrate endothelium lined along glomerular capillaries. The nucleus of an endothelial cell lies adjacent to the mesangium and the rest of the cell portion is positioned irregularly around capillary lumen.[51] Endothelial cells perform as the primary barrier of the constitutions of blood from capillary lumen to Bowman's space.[52] Therefore the cells provide the charge-selective properties of the glomerular capillary wall. Due to the presence of glycocalyx rich glycoproteins, the surface of glomerular endothelial cells exhibits negative charge, and this negative charge offers the charge-selective properties of the glomerular capillary wall; it offers an important role as the filtration barrier.[53,54]

The largest cell located in the glomerulus is the podocyte. The long cytoplasmic process of podocytes extends from the main cell body and then divides into individual pedicels that make direct contact with the glomerular basement membranes. A large number of microtubules, microfilaments, and intermediate filaments are present in the cytoplasm; especially abundant actin filaments are shown in the foot process region of the cell.[55,56] Intact slit diaphragm and podocyte anionic sites are known to have an important function in establishing the selective properties of the filtration barrier.

Mesangium, which constitutes mesangial cells and their surrounding matrix, is separated from the capillary lumen by the endothelium. Mesangial cells hold many functional properties of smooth muscle cells; therefore it often represents a specialized pericyte.[57] In addition to structural support, mesangial cells provide contractile properties that contribute to regulation of glomerular filtration.[57]

The visceral epithelium is continuous with the parietal epithelium, which is located in the outer wall of Bowman's capsule. The parietal epithelial cell functions as the last barrier for the urinary filtrate and macromolecules can escape into the space between the parietal cell and the basement membrane of Bowman's capsule.[58] Interestingly, a recent study reported that parietal epithelial cells have the ability to transdifferentiate into podocytes and repopulate the glomerular tuft.[59,60]

The cells that are located in the walls of the afferent and efferent arterioles are called juxtaglomerular granular cells.[61,62] Due to the smooth muscle cell-like features, such as existence of myofilaments inside juxtaglomerular granular cells, they also have been called myoepithelial cells; juxtaglomerular granular cells are believed to represent modified smooth muscle cells. Extraglomerular mesangium, which is located between the afferent and efferent arterioles in close contact with the macula densa, is uninterrupted with intraglomerular mesangium.[61] Other glomerular cells called periporal cells are believed to be components of juxtaglomerular apparatus (JGA).

2.3.3.2 Proximal Tubule Cells

The most distinctive feature of the proximal tubule is the presence of a brush border formed by densely packed microvilli.[63] The presence of microvilli in the proximal tubule greatly increases the luminal surface area of the cells, therefore facilitating their reabsorptive function as well as a putative flow sensing within the lumen.[64] The cytoplasms of the proximal tubule cells are densely packed with mitochondria. The presence of mitochondria not only gives the cells an acidophilic appearance, it also supplies the energy for the active transport of sodium ion out of the proximal tubule.[64] The cells contain a large number of lysosomes that involve the degradation of material absorbed.[65] Now it is generally believed that the reabsorption of numerous proteins and polypeptides by the proximal tubule is mediated by a multi-ligand endocytic receptor called megalin, which is placed in the brush border.[66,67] Extensive lateral interdigitations of neighboring cells are examined in lining the proximal tubule epithelial cells.

2.3.3.3   Distal Tubule Cells

Distal convoluted tubule cells contain numerous elongated mitochondria, appearing to be tall and cuboid. Distal tubule cell nuclei occupy a central to apical position and, unlike proximal tubule cells, they lack the well-developed brush border and the extensive endocytic apparatus. The distal convoluted tubule luminal surface is covered with abundant small microprojections (microplicae); on the other hand, the lateral distal convoluted tubule cell margins are straight, without the interdigitations between neighboring cells. The individual cells own one primary cilium, which is centrally placed on the apical surface. The distal tubule epithelium is distinguished not only by extensive invaginations of the basolateral plasma membrane, but also by interdigitations between adjacent cells.

2.3.3.4   Collecting Duct Cells

Traditionally two types of cells, principal and intercalated cells, have been identified in the mammalian collecting duct. Principal cells are, as the name implies, the major cell type in the collecting duct; they were originally thought to exist throughout the collecting duct. Principal cells have a relatively smooth membrane covered with short stubby microvilli and a single primary cilium that is known to mediate on collecting the duct's influence on sodium and potassium balance *via* sodium channels and potassium channels located on the cell's apical membrane.[68] Under a microscope, a few cell organelles are examined in principal cells; mitochondria are small and scattered randomly in the cytoplasm and a few lysosomes are present.[69]

Another type of collecting duct cells, intercalated cells, disappear at different lengths along the inner medullary collecting duct (IMCD) depending on the species. There is both structural and functional evidence that the cells in the terminal part of the IMCD comprise a different population of IMCD cells.[70] Intercalated cells are described by the existence of tubulovesicular membrane in the cytoplasm and well-known microprojections on the luminal surface.[69]

In addition, numerous mitochondria, a well-developed Golgi apparatus, and abundant polyribosomes are also the key features of the intercalated cells. Currently, at least two configurations of intercalated cells (α-intercalated cells and β-intercalated cells) have been described in the cortical collecting duct. Those two types of intercalated cells engage in acid-base homeostasis, thus the intercalated cells are believed to play an important role in the kidney's acidosis and alkalosis responses.

2.3.4 Extracellular Environment

Both nervous and hormonal controls regulate the osmosis function of the kidney. One of the key hormones that manage the control of the circuit is the antidiuretic hormone (ADH), which is also called vasopressin. The hypothalamus secretes the ADH and the posterior pituitary gland stores it until osmoreceptor cells (located in the hypothalamus) monitor the osmolarity and stimulate the release of ADH from the posterior pituitary gland.

Once the osmolarity increases above the set point of 300 mOsm L^{-1}, more ADH is released into the bloodstream. The main targets of released ADH in the kidney are the distal tubule and collecting duct; ADH makes the epithelium become more permeable to water and results in increased water reabsorption, which concentrates urine, reduces urine volume, and lowers blood osmolarity back toward the set point. As the osmolarity of the blood collapses, a negative feedback mechanism diminishes the activity of osmoreceptor cells in the hypothalamus and ADH secretion is lessened.

Water-selective channels formed by aquapoins (AQP) are influenced by ADH to uptake water. ADH binding to receptor molecules causes a transitory increase of AQP molecules in the membranes of collecting duct cells, which produces a urine volume reducing phenomenon.

Another regulatory mechanism that assists to maintain kidney homeostasis is the renin-angiotensin-aldosterone system (RAAS). The juxtaglomerular apparatus, which is located near the afferent arteriole, has a critical role in the RAAS. When blood pressure in the afferent arteriole drops, the JGA immediately secretes an enzyme called renin, which initiates a chemical reaction that yields a peptide called angiotensin II. This functional hormone, angiotensin II, decreases blood flow and raises the blood pressure including the kidney by constricting the arterioles. In addition to that it stimulates the adrenal glands to release aldosterone, a hormone that acts on nephron tubules to reabsorb more sodium (Na^+) and water and increase blood volume and pressure.

To maintain homeostasis, the RAAS checks and balances itself as a part of a complex feedback circuit. Decrease in blood volume promotes the JGA to secrete renin, while an increase in blood volume stimulates angiotensin II and aldosterone to release, thus blocking the release of renin. Aterial natriuretic peptide (ANP) is a hormone that counters the RAAS system. Increased blood pressure stimulates ANP released from the heart atria walls to inhibit the release of renin from the JGA. Renin inhibition causes NaCl reabsorption by collecting ducts and diminishes aldosterone release from the adrenal glands, eventually leading towards a series of steps to lower blood volume and pressure. The precise regulatory function of ANP is still under active research.[23]

2.4 Kidney on a Chip

2.4.1 Microfluidic Approach for Kidney on a Chip

The mammalian kidney is a complicated organ where reabsorption of water molecules and small solutes is managed by endocrine mechanisms. To maintain homeostasis, renal tubular epithelial cells have confined the renal transporters and water channels at the apical and basolateral membranes subjected to a urinary flow along the nephron.[71-73] Like vascular endothelial cells, renal tubular epithelial cells experience mechanical forces called fluid flow induced shear stress in response to variations in urinary flow rate. The estimated flow rate within the renal tubular kidney is in the rage of 0.2–20 dyn cm^{-1}.[2,72]

Due to the various fluidic related studies that have been carried out over many years, some hidden functions and phenomena of renal tubular cells have recently been revealed. For example, Duan *et al.* and Essig *et al.* illustrated that shear stress can cause cytoskeleton reorganization and junction reformation in renal proximal tubular epithelial.[74,75] Another study done by Cai *et al.* reported that in the IMCD cells, nitric oxide production is modulated by shear stress.[76]

Based on the announced studies, it has become certain that when culturing the renal tubular cells *in vitro* supplying controlled fluidic environments is crucial to form polarized membrane traffic and cell polarity. Therefore, a special microfluidics-integrated device was invented to provide a fluidic environment to the kidney tubular cells and to culture the cells on a porous membrane, which can enhance cell junction formation and height improvement by providing nutrients from the basolateral region; under this condition, cells were expected to become highly differentiated and polymerized.[7,43]

In this section we will introduce methods of fabricating various kidney chips using microfabrication techniques.

2.4.2 Fabrication of Kidney on a Chip

Microfluidic devices, a well-established microscale technology in biological applications, have been exercised for controlling small amounts of sample in a quick, high-resolution, and cheap manner. A simple definition of micro-channel, the most commonly used configuration for microfluidic systems, is a microfabricated channel on the micrometer scale. Micro-channel is known to provide many unique fluidic properties, offering new opportunities to explore fundamental phenomena such as fluidic transport

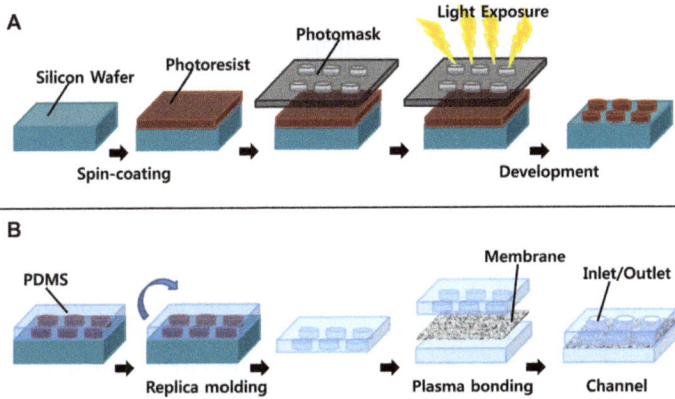

Figure 2.3 Fabrication of kidney on a chip basement. (A) Photolithograph is a microfabrication technique that selectively transfers a microscale pattern to a photosensitive material through radiation exposure. A thin layer of a photoresist (light-sensitive material) is spin coated uniformly and then covered with a photomask that contains a pattern defined by an opaque chrome layer. After that, a high intensity of UV light is exposed over the masked photoresist. Due to the photomask, the region where UV light is not exposed becomes soluble with a developer solution and dissolves away during the development process leaving behind a desired microscale pattern etched into the photoresist. (B) Soft lithography is a fabrication technique that involves the use of an elastomeric material and replica molding technique. Firstly, uncured PDMS is poured on the pattern of the photoresist (silicon master mold) that was produced from photolithography; then the PDMS is baked in an oven for 1 hour. Once the PDMS is completely cured, it is removed from the silicon master mold; a precise microscale pattern of the original material is generated on the PDMS. Through careful bonding of two PDMS replica molds, a PDMS multilayer microfluidic device (MMD) is formed.[43,77,78,81,90]

and molecular behavior and to deliver tools for biomedical applications (Figure 2.3).[77,78]

Conventional microfabrication techniques are used to fabricate microchannels in wafers or film substrates by standard photolithography patterning, a process used in microfabrication to selectively remove parts of a thin film or the bulk of a substrate, followed by wet/dry substrate etching.[79,80] This technique transfers a geometric pattern on a photomask to a light-sensitive chemical photoresist coated on a substrate using high intensity light.[81,82]

A single process of photolithography combines several sequential steps: cleaning, preparation, photoresist deposition, light exposure and development, etching, and photoresist removal. In addition to the photolithography technique, there are many methods used to generate micro-channels. The patterning of narrower channels, which requires higher resolution, can be attained by using extreme ultraviolet light.[81,84] E-beam lithography (EBL) and focused-ion beam (FIB) lithography are attractive alternatives to create highly accurate and dependable channel patterns without using the photomask. Other maskless patterning methods for micro-channel fabrication include interferometric lithography (IL), laser patterning, and surface machining.[85–89]

Once the silicon channel master is generated, soft lithography is performed to fabricate a multilayer microfluidic device (MMD). Soft lithography is a collective name for a set of lithographic techniques with soft polydimethylsiloxane (PDMS): replica molding, microcontact printing, micromolding in capillaries, microtransfer molding, solvent-assisted micromolding, and near-field conformal photolithography. These techniques use a type of elastomer such as PDMS as the mold, stamp, or mask to generate or transfer the desired patterns.[90,91]

For applications where patterning of non-planar substrates, unusual materials, or large-area patterning are the major concerns, soft lithography provides instant advantages over photolithography and other microfabrication techniques. It is known to be promising for microfabrication of relatively simple, single-layer structures likely applied in microanalytical systems, cell culture, sensors, and applied optics. The success of soft lithography implies that it has a potential to become an important technique in the micro- and nanofabrication field.[78,90,91]

2.4.3   Various Kidney Chips

Most of the developed kidney chips have been focused on one specific goal: mimicking an *in vitro*-like cellular condition inside the chip in order to enable more accurate physiological responses of the cell *in vitro*. Renal cells and nephrons have already been stimulated with microfluidic devices in many trials. In fact, the previous cell culture models led to new insights into cell and organ function. In this section, we will overview some of the kidney chips developed to investigate various aspects of kidney structure and functions *in vitro*.

2.4.3.1   Collecting Duct on Chip

A sandwiched assembly of a PDMS microfluidic channel, a polyester membrane (pore size of 0.4 μm and thickness of 10 μm), and PDMS were bonded to generate an MMD. After the bonded MMD was coated with fibronectin, the primary cultured rat IMCD cells were seeded and cultured in the channel. Then a range of shear stress (the range of 0.2–20 dyn cm^{-2}) was applied over various time ranges. Using this chip, the cell's shear stress induced physiological changes including cytoplasm F-actin reorganization and protein (AQP2 and Na-K pump) translocation were examined.[7,43]

The experimental data indicated that both the magnitude and duration of the luminal fluid shear stress (FSS) had a significant effect on actin depolymerization. The study also suggested that the FSS alone can induce AQP2 translocation to apical plasma membrane in collecting duct cells.[43] In addition, a combination of 1 dyn cm^{-2} FSS for 5 h and 30 minutes of AVP

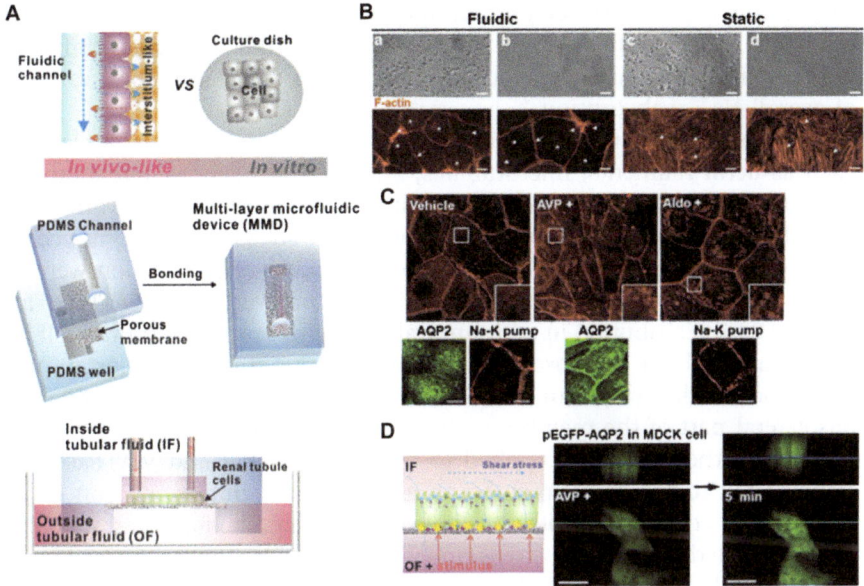

Figure 2.4 Collecting duct on a chip.[7,43] (A) Schematic resembling renal tubular environments *in vitro*. To mimic renal tubular environments, an interstitium-like porous membrane substrate was inserted in a sandwich format between a PDMS microfluidic channel and a PDMS well (mass reservoir) to enable the transport of molecules and stimulus as well as the application of a fluidic shear stress to the cultured cells. (B) Reorganization of cytoskeleton of IMCD cells under four different conditions: MMD (fluidic, a), PDMS channel + glass substrate (fluidic, b), transwell (static, c), and glass substrate (static, d). In fluidic condition, cells are subjected to 1 dyn cm^{-2} of fluid shear stress for 5 h. Optical (top panel) and confocal microscopy images of IMCD cells stained with antibodies for F-actin (red). (C) Immunofluorescence images of F-actin, AQP2, and Na-K-pump in response to hormonal stimulations (AVP+ and Aldo+). The images for the vehicle are also shown as a control. Each inset shows a magnified view of the boxed region (10×10 μm). (D) Schematic of drug screening and test of the real-time AQP2 trafficking using AQP2-GFP transfected MDCK cells. After stimulation of AVP+ for 5 min., AQP2-GFP translocated to the plasma membrane. Upper images show the x–z optical sectioned confocal microscope images into the green line. The blue line indicates the middle plane of the cell. Scale bar, 10 μm.

stimulation demonstrated even more enhanced AQP translocation to apical plasma membrane (Figure 2.4).[7,43]

2.4.3.2 Proximal Tubule on Chip

The studies done by Du *et al.* in 2004 and 2006 investigated the role of fluid shear stress in modulating Na$^+$ and HCO$_3^-$ reabsorption in mouse proximal tubule. Their studies reported that both NHE3 and V-ATPase activities were improved by increases in luminal flow rate.[46,92] In 2007, Weinstein *et al.*

suggested a mathematical equation that includes a torque-dependent solute transport in a compliant tubule; the model has predicted that coordinated luminal and peritubular transporter regulation is mandatory for a variation of overall Na^+ reabsorption.[93]

Based on the studies suggesting a close relation of cytoskeletal elements such as actin filaments, microtubules, and associated molecular motors to mechanical signaling, dynamic remodeling of the cytoskeleton in various cell types has reported underexposure of fluid shear stress for 3–5 h; noticeable changes in actin cytoskeletal and junctional protein reorganization were observed over shear stress exposure time.[50,92–95] Moreover, NHE, V-ATPASE and Na/K-ATPase distribution in an immortalized mouse proximal tubule cell line were examined. In the absence of the stress, the apical transporters were spread across the cytoplasm, and sodium pumps were mostly observed at the basolateral part of the membrane, interacting with actin stress fibers. However, when the stress was applied to cells, both NHE3 and V-ATPase were translocated to the apical membrane and sodium pump expression greatly improved in comparison with the control. Therefore, the following statement was concluded: fluid shear stress provides a significant task for the cytoskeletal network trafficking of the transporters to their functional membrane locations. An intact actin cytoskeleton is important for regulation of flow-dependent Na^+ and HCO_3^- absorption. A trafficking of flow induced NHE3 and Na/K-ATPase, whereas the intact microtubule network is important for the flow-induced V-ATPase trafficking.[6]

2.4.3.3 Surface Topography Controlled Kidney Chip

A study was carried out to precisely control the topographic microenvironment of *in vitro* tissues.[96] By aligning and assembling PDMS microfluidic channels with dimensions of 18 mm length, 3 mm width, and 150 μm depth arranged in a 3×1 array, and topographical substrates with dimensions of 0.75 μm width, 0.75 μm depth, and 1.5 μm pitch, a microscale tissue modeling device (MTMD) was fabricated. Then, after the MTMD was coated with fibronectin, the human renal proximal tubule epithelial cell line (HK-2) was cultured in the MTMD. Once the HK-2 cells formed a confluent monolayer, various intensities of FSS (0 dyn cm^{-2}, 0.02 dyn cm^{-2}, 1 dyn cm^{-2}) were applied to the channel, and the cell's shear stress induced alignment and tight junction formation were examined.

This study suggested that FSS alone does not induce alignment of cells, which should be assisted by surface topography. Moreover, it was concluded that topography alone appears to increase ZO-1 continuity in cell monolayers, but this change becomes more pronounced when FSS is coupled with the topography.[96]

2.4.3.4 Wearable Peritoneal Dialysis Chip

The wearable peritoneal dialysis system was developed to assist kidney failure patients.[97] The system composes the artificial kidney filter pad and conduits for the circulation of peritoneal dialysis fluid. The absorption capacity of the filter pad for urea and the middle molecules is typically in the range of 30–70% of its own weight, but can reach 100% for specific molecules. A wearable peritoneal dialysis system has an advantage in that it forms a disposable and replacement part of the device. It can be replaced by a fresh filter pad when it has been saturated with toxic substances.

2.5 Future Opportunities and Challenges

Animal testing has been known as one of the obligated steps to determining therapeutic significances and safety of newly formulated drugs and pharmaceutical products. It is currently the major bottleneck which biotechnology and pharmaceutical industries are facing because it takes up too much money and time, while being confronted with ethical issues. Moreover, even if the industry takes economical and ethical risks, the traditional animal testing approach often fails to predict the invented drug's toxicity to the human model. Therefore, there is an intense pressure to find ways to identify suitable alternatives that can circumvent the necessities for animal studies while improving the success of the drug development process.[98]

The kidney-on-chip microsystems are expected to have a considerable impact on the future of the pharmaceutical industry by screening the newly invented drug toxicity and replacing animal studies. By creating human relevant disease models inside the small chip, this microsystem will successfully offer an alternative approach for efficacy testing. Therefore, such a model is speculated to have a significant impact on various stages of the drug development process, including recognition and prioritization of the lead compound as well as target validity. In addition, collections of integrated organ chips *via* microfluidic linkages are predicted to provide a way to model physiological interplay between different organs.

Kidney-on-chip technology also has a significant impact on environmental toxicology and industrial safety applications. More predictive and reproducible *in vitro* analysis using human cells containing kidneys on chips is expected to be able to evaluate the number of environmental toxicants with reduced time, cost, and animals involved in safety testing. If successful, this new screening technology will allow for shortening testing time, decreasing testing costs, and faster decision making with more predictive human relevant data.

As mentioned before, the development of the microfluidic system is expected to provide many advantages in our current industries. However, despite the many promising advantages that the microfluidic system would provide, there are some current challenges that are restricting the widespread use and development of the kidney chip. One of the challenges is the base material (PDMS) used in the kidney chip. The use of PDMS in the fabrication of the kidney chip can raise problems for drug discovery applications. Due to the natural property of PDMS, small hydrophobic molecules including certain drugs, fluorescent dyes, or cell signaling molecules can be absorbed. Therefore, this may result in the reduction of effective drug concentrations, cross contamination, lower detection sensitivities and high background fluorescence. In addition, PDMS is not

A

Collecting-duct
-on-a-chip

Multi-layer microfluidic device (MMD)

Kidney tubular epithelial cells

Inside
tubular fluid (IF)

Renal tubule
cells

Outside
tubular fluid (OF)

B

B

PDMS channel

Cell adhesion area

Topographical substrate

C

wall

PDMS channel

Renal tubule cells

C

cells

membrane

substrate

Glomerulus

Tubule

blood into
device

1.5cm

2cm

blood out

urine out of
device

extracellular fluid

Loop of Henle

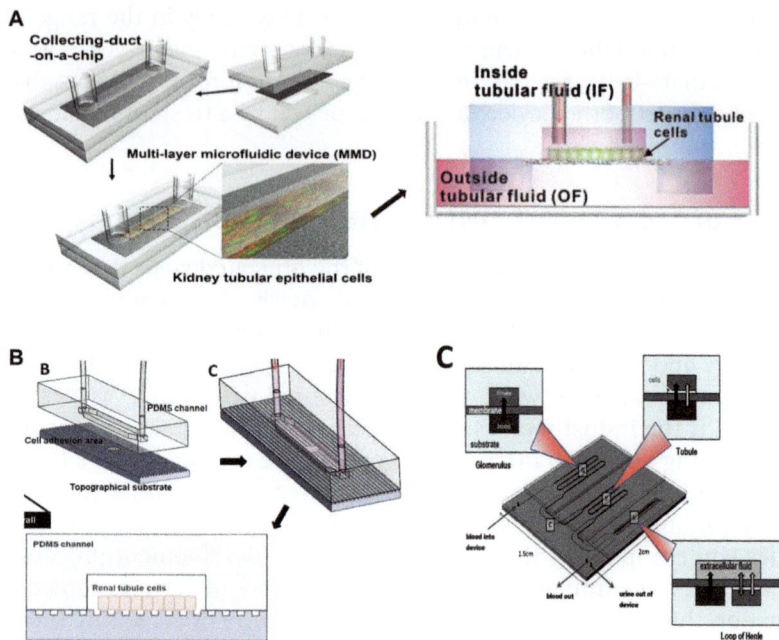

Figure 2.5   Various kidney on chips. (A) Collecting duct on chip/proximal tubule on chip formation.[7,43] (B) Surface topography controlled kidney chip.[96] (C) Nephron on a chip device.[83]

the most favorable material for scale-up manufacturing of microfluidic devices. Consequently, it is possible that residual uncross-linked oligomers may leach from PDMS and interact with cells and culture medium.[99,100] Therefore, it is desired to identify alternative materials, such as thermoplastic materials with similar properties, particularly for drug discovery applications in the future.

Lastly, it is important that the ultimate, fully developed kidney chip should provide a user interface that is easy to use, enables automation, and ensures reproducibility of the data. These types of complicated technologies are likely to be inappropriate for high-throughput chemical screening. Instead, it presents the capacity to produce high quality and high content data at key stages of the drug discovery process to enable important decision making and to replace unpredictable animal models. Another benefit of kidney chip technologies is that they offer the potential to construct sophisticated models of human disease to identify disease targets and establish drug effectiveness in human relevant systems.[16]

Being aware that virtually all organs are connected in the human body, we should admit that there is still a lot to be accomplished in the reconstruction of complex three-dimensional models that reconstitute whole kidney physiology and their relevant functions. Therefore, despite the considerable progresses in the creation of micro-engineered tissue and kidney model, more studies would be needed to better understand kidney functions and reconstruct *in vitro*-like microenvironments inside a small chip (Figure 2.5).[7,43,83,96]

References

1. F. Pampaloni, E. G. Reynaud and E. H. K. Stelzer, *Nat. Rev. Mol. Cell Bio.*, 2007, **8**, 839.
2. I. Meyvantsson and D. J. Beebe, *Annu. Rev. Anal. Chem.*, 2008, **1**, 423.
3. J. Voldman, M. L. Gray and M. A. Schmidt, *Annu. Rev. Biomed. Eng.*, 1999, **1**, 401.
4. J. El-Ali, P. K. Sorger and K. F. Jensen, *Nature*, 2006, **442**, 403.
5. D. Huh, G. A. Hamilton and D. E. Ingber, *Trends Cell Biol.*, 2011, **21**, 745.
6. K. J. Jang, G. A. Hamilton, L. McPartlin, A. Bahinski, H. N. Kim, K. Y. Suh and D. E. Ingber, Human kidney proximal tubule-on-a-chip for drug transporter studies and nephrotoxicity assessment, presented at *Miniaturized Systems for Chemistry and Life Science*, Seattle, Washington, USA, 2011.
7. K. J. Jang, H. S. Cho, D. H. Kang, W. G. Bae, T. H. Kwon and K. Y. Suh, *Integr. Biol. UK*, 2011, **3**, 134.
8. R. Baudoin, L. Griscom, M. Monge, C. Legallais and E. Leclerc, *Biotechnol. Progr.*, 2007, **23**, 1245.
9. G. J. Mahler, M. B. Esch, R. P. Glahn and M. L. Shuler, *Biotechnol. Bioeng.*, 2009, **104**, 193.
10. S. G. Harris and M. L. Shuler, *Biotechnol. Bioproc. E*, 2003, **8**, 246.
11. J. W. Park, B. Vahidi, A. M. Taylor, S. W. Rhee and N. L. Jeon, *Nat. Protoc.*, 2006, **1**, 2128.
12. S. R. Khetani and S. N. Bhatia, *Nat. Biotechnol.*, 2008, **26**, 120.
13. M. J. Powers, K. Domansky, M. R. Kaazempur-Mofrad, A. Kalezi, A. Capitano, A. Upadhyaya, P. Kurzawski, K. E. Wack, D. B. Stolz, R. Kamm and L. G. Griffith, *Biotechnol. Bioeng.*, 2002, **78**, 257.
14. P. J. Lee, P. J. Hung and L. P. Lee, *Biotechnol. Bioeng.*, 2007, **97**, 1340.
15. A. Carraro, W. M. Hsu, K. M. Kulig, W. S. Cheung, M. L. Miller, E. J. Weinberg, E. F. Swart, M. Kaazempur-Mofrad, J. T. Borenstein, J. P. Vacanti and C. Neville, *Biomed. Microdevices*, 2008, **10**, 795.
16. D. Huh, B. D. Matthews, A. Mammoto, M. Montoya-Zavala, H. Y. Hsin and D. E. Ingber, *Science*, 2010, **328**, 1662.
17. M. Shin, K. Matsuda, O. Ishii, H. Terai, M. Kaazempur-Mofrad, J. Borenstein, M. Detmar and J. P. Vacanti, *Biomed. Microdevices*, 2004, **6**, 269.
18. J. W. Song, W. Gu, N. Futai, K. A. Warner, J. E. Nor and S. Takayama, *Anal. Chem.*, 2005, 77, 3993.
19. M. T. Lam, Y. C. Huang, R. K. Birla and S. Takayama, *Biomaterials*, 2009, **30**, 1150.
20. K. Jang, K. Sato, K. Igawa, U. I. Chung and T. Kitamori, *Anal. Bioanal. Chem.*, 2008, **390**, 825.

21. M. J. Powers, D. M. Janigian, K. E. Wack, C. S. Baker, D. B. Stolz and L. G. Griffith, *Tissue Eng.*, 2002, **8**, 499.

22. N. J. Douville, P. Zamankhan, Y. C. Tung, R. Li, B. L. Vaughan, C. F. Tai, J. White, P. J. Christensen, J. B. Grotberg and S. Takayama, *Lab Chip*, 2011, **11**, 609.

23. N. A. Campbell, J. B. Reece, L. A. Urry, M. L. Cain, M. L. Wasserman, P. V. Minorsky and R. B. Jackson, *Biology*, ed. B. Wilbur, Pearson Education Inc., San Francisco, CA, p. 963.

24. J. H. Miner, *Kidney Int.*, 1999, **56**, 2016.

25. R. Timpl, *Eur. J. Biochem.*, 1989, **180**, 487.

26. M. Paulsson, *Crit. Rev. Biochem. Mol.*, 1992, **27**, 93.

27. J. H. Miner and J. R. Sanes, *J. Cell Biol.*, 1994, **127**, 879.

28. M. Desjardins, F. Gros, J. Wieslander, M. C. Gubler and M. Bendayan, *Lab Invest.*, 1990, **63**, 637.

29. Y. Ninomiya, M. Kagawa, K. Iyama, I. Naito, Y. Kishiro, J. M. Seyer, M. Sugimoto, T. Oohashi and Y. Sado, *J. Cell Biol.*, 1995, **130**, 1219.

30. R. J. Butkowski, J. Wieslander, M. Kleppel, A. F. Michael and A. J. Fish, *Kidney Int.*, 1989, **35**, 1195.

31. B. Peissel, L. Geng, R. Kalluri, C. Kashtan, H. G. Rennke, G. R. Gallo, K. Yoshioka, M. J. Sun, B. G. Hudson, E. G. Neilson and J. Zhou, *J. Clin. Invest.*, 1995, **96**, 1948.

32. L. M. Sorokin, S. Conzelmann, P. Ekblom, C. Battaglia, M. Aumailley and R. Timpl, *Exp. Cell Res.*, 1992, **201**, 137.

33. L. M. Sorokin, F. Pausch, M. Durbeej and P. Ekblom, *Dev. Dynam.*, 1997, **210**, 446.

34. G. Klein, M. Langegger, R. Timpl and P. Ekblom, *Cell*, 1988, **55**, 331.

35. J. H. Miner, B. L. Patton, S. I. Lentz, D. J. Gilbert, N. A. Jenkins, N. G. Copeland and J. R. Sanes, *J. Cell Biol.*, 1997, **137**, 685.

36. R. Timpl, *Curr. Opin. Cell Biol.*, 1996, **8**, 618.

37. A. E. Chung and M. E. Durkin, *Am. J. Resp. Cell Mol. Biol.*, 1990, **3**, 275.

38. R. Timpl, *Experientia*, 1993, **49**, 417.

39. J. R. Couchman, R. Kapoor, M. Sthanam and R. R. Wu, *J. Biol. Chem.*, 1996, **271**, 9595.

40. J. Vandenborn, A. A. Vankraats, M. A. H. Bakker, K. J. M. Assmann, L. P. W. J. Vandenheuvel, J. H. Veerkamp and J. H. M. Berden, *Diabetologia*, 1995, **38**, 161.

41. C. F. Dewey, S. R. Bussolari, M. A. Gimbrone and P. F. Davies, *J. Biomech. Eng.*, 1981, **103**, 177.

42. R. G. Bacabac, T. H. Smit, S. C. Cowin, J. J. W. A. Van Loon, F. T. M. Nieuwstadt, R. Heethaar and J. Klein-Nulend, *J. Biomech.*, 2005, **38**, 159.

43. K. J. Jang and K. Y. Suh, *Lab Chip*, 2010, **10**, 36.

44. J. Schnerma, M. Wahl, G. Liebau and H. Fischbach, *Pflug Arch. Eur. J. Phy.*, 1968, **304**, 90.

45. G. Giebisch and E. E. Windhager, *Am. J. Med.*, 1963, **34**, 1.

46. Z. P. Du, Q. S. Yan, Y. Duan, S. Weinbaum, A. M. Weinstein and T. Wang, *Am. J. Physiol. Renal*, 2006, **290**, F289.
47. G. Malnic, R. W. Berliner and G. Giebisch, *Am. J. Physiol.*, 1989, **256**, F932.
48. L. M. Satlin, S. H. Sheng, C. B. Woda and T. R. Kleyman, *Am. J. Physiol. Renal*, 2001, **280**, F1010.
49. Y. Duan, A. M. Weinstein, S. Weinbaum and T. Wang, *Proc. Natl Acad. Sci. USA*, 2010, **107**, 21860.
50. P. Guo, A. M. Weinstein and S. Weinbaum, *Am. J. Physiol. Renal*, 2000, **279**, F698.
51. F. Jørgensen, Copenhagen, Ejnar Munksgaard, 1966, p. 221.
52. K. Ichimura, R. V. Stan, H. Kurihara and T. Sakai, *J. Am. Soc. Nephrol.*, 2008, **19**, 1463.
53. J. Sorensson, A. Bjornson, M. Ohlson, B. J. Ballermann and B. Haraldsson, *Am. J. Physiol. Renal*, 2003, **284**, F373.
54. M. Jeansson and B. Haraldsson, *Am. J. Physiol. Renal*, 2006, **290**, F111.
55. D. Vasmant, M. Maurice and G. Feldmann, *Anat. Rec.*, 1984, **210**, 17.
56. P. M. Andrews and S. B. Bates, *Anat. Rec.*, 1984, **210**, 1.
57. D. Schlondorff, *FASEB J.*, 1987, **1**, 272.
58. T. Ohse, A. M. Chang, J. W. Pippin, G. Jarad, K. L. Hudkins, C. E. Alpers, J. H. Miner and S. J. Shankland, *Am. J. Physiol. Renal*, 2009, **297**, F1566.
59. D. Appel, D. B. Kershaw, B. Smeets, G. Yuan, A. Fuss, B. Frye, M. Elger, W. Kriz, J. Floege and M. J. Moeller, *J. Am. Soc. Nephrol.*, 2009, **20**, 333.
60. T. Ohse, M. R. Vaughan, J. B. Kopp, R. D. Krofft, C. B. Marshall, A. M. Chang, K. L. Hudkins, C. E. Alpers, J. W. Pippin and S. J. Shankland, *Am. J. Physiol. Renal*, 2010, **298**, F702.
61. L. Barajas, *Am. J. Physiol.*, 1979, **237**, F333.
62. L. Barajas, *J. Ultra Mol. Struct. Res.*, 1970, **33**, 116.
63. A. B. Maunsbach, *J. Ultra. Res.*, 1966, **16**, 239.
64. T. Wang, *Curr. Opin. Nephrol. Hypertens.*, 2006, **15**, 530.
65. E. I. Christensen and S. Nielsen, *Semin. Nephrol.*, 1991, **11**, 414.
66. D. Kerjaschki, L. Noronhablob, B. Sacktor and M. G. Farquhar, *J. Cell Biol.*, 1984, **98**, 1505.
67. D. Kerjaschki and M. G. Farquhar, *Proc. Natl Acad. Sci. Biol.*, 1982, **79**, 5557.
68. A. C. Guyton, *Textbook of Medical Physiology*, Elsevier/Saunders, Philadelphia, 2006.
69. K. M. Madsen, J. W. Verlander and C. C. Tisher, *J. Electron Microsc. Tech.*, 1988, **9**, 187.
70. K. M. Madsen, W. L. Clapp and J. W. Verlander, *Kidney Int.*, 1988, **34**, 441.
71. C. Palfrey and A. Cossins, *Nature*, 1994, **371**, 377.
72. S. Nielsen, J. Frokiaer, D. Marples, T. H. Kwon, P. Agre and M. A. Knepper, *Physiol. Rev.*, 2002, **82**, 205.

73. Y. J. Lee, I. K. Song, K. J. Jang, J. Nielsen, J. Frokiaer, S. Nielsen and T. H. Kwon, *Am. J. Physiol. Renal*, 2007, **292**, F340.
74. Y. Duan, N. Gotoh, Q. S. Yan, Z. P. Du, A. M. Weinstein, T. Wang and S. Weinbaum, *Proc. Natl Acad. Sci. USA*, 2008, **105**, 11418.
75. M. Essig, F. Terzi, F. M. Burtin and G. Friedlander, *Am. J. Physiol. Renal*, 2001, **281**, F751.
76. Z. Q. Cai, J. D. Xin, D. M. Pollock and J. S. Pollock, *Am. J. Physiol. Renal*, 2000, **279**, F270.
77. B. H. Jo, L. M. Van Lerberghe, K. M. Motsegood and D. J. Beebe, *J. Microelectromech. Syst.*, 2000, **9**, 76.
78. P. Kim, K. W. Kwon, M. C. Park, S. H. Lee, S. M. Kim and K. Y. Suh, *Biochip J.*, 2008, **2**, 1.
79. D. Mijatovic, J. C. T. Eijkel and A. van den Berg, *Lab Chip*, 2005, **5**, 492.
80. P. Mao and J. Y. Han, *Lab Chip*, 2005, **5**, 837.
81. H. N. Chapman, A. K. Ray-Chaudhuri, D. A. Tichenor, W. C. Replogle, R. H. Stulen, G. D. Kubiak, P. D. Rockett, L. E. Klebanoff, D. O'Connell, A. H. Leung, K. L. Jefferson, J. B. Wronosky, J. S. Taylor, L. C. Hale, K. Blaedel, E. A. Spiller, G. E. Sommargren, J. A. Folta, D. W. Sweeney, E. M. Gullikson, P. Naulleau, K. A. Goldberg, J. Bokor, D. T. Attwood, U. Mickan, R. Hanzen, E. Panning, P. Y. Yan, C. W. Gwyn and S. H. Lee, *J. Vac. Sci. Technol. B*, 2001, **19**, 2389.
82. P. Naulleau, K. A. Goldberg, E. H. Anderson, D. Attwood, P. Batson, J. Bokor, P. Denham, E. Gullikson, B. Harteneck, B. Hoef, K. Jackson, D. Olynick, S. Rekawa, F. Salmassi, K. Blaedel, H. Chapman, L. Hale, P. Mirkarimi, R. Soufli, E. Spiller, D. Sweeney, J. Taylor, C. Walton, D. O'Connell, D. Tichenor, C. W. Gwyn, P. Y. Yan and G. J. Zhang, *J. Vac. Sci. Technol. B*, 2002, **20**, 2829.
83. E. Weinberg, M. Kaazempur-Mofrad and J. Borenstein, *Int. J. Artif. Organs*, 2008, **31**, 508.
84. P. Naulleau, *J. Vac. Sci. Technol. B*, 2002, **20**, 2829.
85. J. L. Pearson and D. R. S. Cumming, *Microelectron. Eng.*, 2005, **78–79**, 343.
86. C. Danelon, C. Santschi, J. Brugger and H. Vogel, *Langmuir*, 2006, **22**, 10711.
87. K. Wang, S. Yue, L. Wang, A. Jin, C. Gu, P. Wang, H. Wang, X. Xu, Y. Wang and H. Niu, *IEE P. Nanobiotechnol.*, 2006, **153**, 11.
88. M. J. O'Brien, P. Bisong, L. K. Ista, E. M. Rabinovich, A. L. Garcia, S. S. Sibbett, G. P. Lopez and S. R. J. Brueck, *J. Vac. Sci. Technol. B*, 2003, **21**, 2941.
89. A. P. Han, N. F. de Rooij and U. Staufer, *Nanotechnology*, 2006, **17**, 2498.
90. G. M. Whitesides, E. Ostuni, S. Takayama, X. Y. Jiang and D. E. Ingber, *Annu. Rev. Biomed. Eng.*, 2001, **3**, 335.
91. X. M. Zhao, Y. N. Xia and G. M. Whitesides, *J. Mater. Chem.*, 1997, **7**, 1069.

92. Z. P. Du, Y. Duan, Q. S. Yan, A. M. Weinstein, S. Weinbaum and T. Wang, *Proc. Natl Acad. Sci. USA*, 2004, **101**, 13068.

93. A. M. Weinstein, S. Weinbaum, Y. Duan, Z. P. Du, Q. S. Yan and T. Wang, *Am. J. Physiol. Renal*, 2007, **292**, F1164.

94. N. L. Cowger, E. Benes, P. L. Allen and T. G. Hammond, *J. Appl. Physiol.*, 2002, **92**, 691.

95. M. J. Morton, K. Hutchinson, P. W. Mathieson, I. R. Witherden, M. A. Saleem and M. Hunter, *J. Am. Soc. Nephrol.*, 2004, **15**, 2981.

96. E. M. Frohlich, X. Zhang and J. L. Charest, *Integr. Biol. UK*, 2012, **4**, 75.

97. S. F. Schilthuizen, L. F. Batenburg, F. Simonis, F. F. Vercauteren, US Patent 20100100027A1, 2010, April 22.

98. D. C. Swinney and J. Anthony, *Nat. Rev. Drug Discov.*, 2011, **10**, 507.

99. M. W. Toepke and D. J. Beebe, *Lab Chip*, 2006, **6**, 1484.

100. R. Mukhopadhyay, *Anal. Chem.*, 2007, **79**, 3248.

101. S. Nielsen, T. H. Kwon, R. A. Fenton, and J. Prætorius, *Brenner and Rector's The Kidney*, Elsevier, Philadelphia, 9th ed. 2011, vol. 1, Anatomy of Kidney, 31.

102. C. Kurts, U. Panzer, H.-J. Anders and A. J. Rees, *Nat. Rev. Immunol.*, 2013, **13**, 738.

CHAPTER 3

Blood-brain Barrier (BBB): An Overview of the Research of the Blood-brain Barrier Using Microfluidic Devices

ANDRIES D. van der MEER,*[a] FLOOR WOLBERS,*[b]
ISTVÃN VERMES[c] AND ALBERT van den BERG[c]

[a] Wyss Institute for Biologically Inspired Engineering, Harvard University, Boston, USA; [b] Eindhoven University of Technology, Microsystems Group, Department of Mechanical Engineering and ICMS Institute for Complex Molecular Systems, The Netherlands; [c] University of Twente, BIOS Lab on a Chip group, MESA$^+$ Institute for Nanotechnology, Enschede, The Netherlands
*Email: andries.vandermeer@wyss.harvard.edu; f.wolbers@tue.nl

3.1 Introduction

The brain is the central organ of the nervous system. It has key roles in sensory processing, motor function control, memory, and behavior. The brain has a tremendous energy demand: up to 20% of the basal metabolism in the human body localizes to the brain.[1] Because of its extraordinary metabolic activity, the brain has a high density of blood vessels. Around 15% of the cardiac output flows through the brain. The cerebral blood vessels not only serve as a blood supply, they are also essential for controlling transport of molecules and cells into and out of the brain. The endothelial tissue that

RSC Nanoscience & Nanotechnology No. 36
Microfluidics for Medical Applications
Edited by Albert van den Berg and Loes Segerink
Published by the Royal Society of Chemistry, www.rsc.org

forms the inner lining of the cerebral blood vessels differs significantly from that of the peripheral vasculature. Cerebral endothelium forms a very tight layer, ensuring that most molecules will not enter the brain compartment by passive transport. Virtually all transport into and out of the brain compartment is receptor-mediated. This well-controlled, endothelial barrier is what is referred to as the "blood-brain barrier". The blood-brain barrier (BBB) is important in maintaining brain homeostasis, preventing brain toxicity and regulating brain inflammation. In addition, dysfunction of the BBB plays a prominent role in many neurological disorders, such as Alzheimer's disease, multiple sclerosis, and brain epilepsy.

3.2   Blood-brain Barrier

3.2.1   Neurovascular Unit

In principle, the barrier function of the BBB is maintained by the brain endothelium. The endothelium forms the tight seal between the blood compartment and the brain compartment. The endothelium regulates all transport into and out of the brain. Still, the endothelium acquires and maintains its specialized phenotype by interacting with other cell types in its vicinity: pericytes, astrocytes, and neurons. This unit of interacting cell types is referred to as the "neurovascular unit" (NVU). The endothelial cells are in direct contact with pericytes and astrocyte endfeet – long micrometer-sized protrusions of the star-shaped cells. The pericytes and astrocytes induce the specialized phenotype in the endothelial cells by various signaling pathways. For example, astrocytes stimulate endothelial cells with molecules like Transforming Growth Factor-beta and basic Fibroblast Growth Factor and pericytes secrete Angiopoietin-1 and Transforming Growth Factor-beta, all of which have clear effects on endothelial cells and vessel maturation.[2,3] The neurons do not interact directly with the endothelial cells of the capillaries, but are always found in close proximity to a vessel. There is some evidence that neurons have inductive effects on formation of a tight BBB, but the exact nature of this interaction between neurons and endothelial cells is not quite clear.[4] In summary, the functionality of the BBB is a result of the interaction between the various cell types that form the NVU.

3.2.2   Transport

Transport over the BBB can be characterized as either passive or active. Because of the tight barrier, most transport over the BBB is driven by active processes; mainly receptor-mediated uptake and excretion. The most important molecules that are actively transported into the brain are glucose and amino acids, but more specialized proteins like insulin and transferrin are also shuttled into the brain by active transport.[3] Passive transport of hydrophilic molecules is almost completely abrogated due to the tightness of the endothelial layer. On the other hand, small lipophilic molecules do manage to penetrate the BBB by passively diffusing into and out of the lipid membrane of the endothelial cells.[3] Interestingly, large classes of molecules are aspecifically, but actively, transported away from the brain and into the blood by multidrug resistance proteins.

3.2.3   Multidrug Resistance

Multidrug resistance (MDR) is defined as resistance to a broad range of structurally and functionally unrelated cytotoxic substances.[5] MDR is evoked by the (over)expression of efflux transporters in cellular membranes, which (actively) pump out drugs from the cell's interior, thereby reducing their cytotoxicity.[6] Specific for the brain, these efflux transporters prevent the brain from exposure to potential toxicants and maintain brain homeostasis.[7] Mainly due to the presence of efflux transporters, treatments for brain diseases fail, as drugs cannot pass the BBB and hence cannot be efficient in the brain. The efflux of drugs is mediated by multidrug resistance proteins, most of which belong to the family of ATP-binding cassette (ABC) transporters, such as P-glycoprotein (Pgp), multidrug resistance proteins (MRPs), and breast cancer resistance protein (BCRP). ABC transporters utilize the energy of the hydrolysis of ATP to transport substances over the cellular membrane.[5] Each ABC transporter has its own preferential substrates, though there is overlap.[5] Apart from ABC transporters, drug efflux in the brain is mediated by the solute-linked carrier (SLC) family,[8] consisting of organic cation transporters (OCT), organic anion transporters (OAT), and organic anion transporting polypeptides (OATP). These transporters do not use ATP, and therefore cannot transport a drug against a concentration gradient. OATP, OAT, and OCT serve as exchangers, hence they exchange a drug for another molecule or ion.[5,7,9]

The effect and function of efflux transporters were revealed in knockout mice, which lack a specific drug efflux transporter. Additionally, specific inhibitors of drug efflux transporters enhance drug penetration or inhibit drug efflux, which nowadays is of increasing interest for the pharmaceutical industry.[5] However, caution has to be used when administering these drugs as these transporters serve a protective role in many tissues. The use of immunoliposomes and drug-containing nanoparticles could circumvent the need for efflux inhibitors.[5]

Pgp was the first efflux transporter discovered in endothelial cells of the BBB and serves to protect the brain from lipophilic and amphiphilic compounds, which, due to their composition, can otherwise easily penetrate the BBB *via* simple diffusion.[5-7,10] Pgp is primarily expressed at the luminal side of the endothelial cell, hence drugs which serve as a Pgp substrate are directly pumped back to the blood.[5,6,9,10] Pgp is also localized in specialized micro-domains of the plasma membrane, called caveolae. In capillary endothelial cells, caveolae are involved in the transport of macromolecules across the cell *via* transcytosis.[6,8] Parallel expression of Pgp and caveolin-1 at both astrocyte endfeet and endothelium supports the hypothesis that these transport mechanisms work cooperatively.[8,11,12]

The efflux transporter breast cancer resistance protein (BCRP) is also expressed at the luminal side of the endothelial cell and is involved in drug resistance to chemotherapeutics.[7,11] BCRP has substrate overlap with Pgp, though the substrates that overlap bind with lower affinity.[8]

MRP preferentially transport anions, but also cations and neutral compounds. Their transport is facilitated by binding of glutathione to the transporter molecule.[5,9,10] MRP are mainly expressed at the luminal side, but also present at the abluminal side of the endothelial cell.[5,9] MRP shares substrate similarity with Pgp.[10] The human MRP consist of at least nine members (MRP1–9), of which MRP1 and 2 are best characterized.[9]

For some of the OAT and OATP transporters, their function and localization on the membrane is known.[10] OATP 1 and 2 are not efflux transporters, but uptake receptors, hence take substrates from the blood and transport them to the brain.[7] However, uptake is limited by Pgp. OAT3 is an efflux transporter, expressed on the luminal and abluminal side.[7]

3.2.4 Neurodegenerative Diseases – Loss of BBB Function

Neurodegenerative disease are characterized by neuronal cell damage, which, because neurons cannot be replaced, causes loss of function.[13] The global burden of neurodegenerative diseases is ever increasing, as the population is aging. In 2040, the WHO predicts that neurodegenerative diseases will be the world's second leading cause of death, after cardio-vascular diseases.[13] Alterations in the BBB can initiate, propagate, and hence increase the neuronal damage in the brain, as pathogens and immune cells can enter the brain in an uncontrolled way.[14,15] Interestingly, pericytes, which contribute to the formation of the BBB tight junctions, decrease with age, leading to BBB breakdown.[16] Moreover, in traumatic brain injury, the damage in an older brain is larger and the recovery slower, as compared to a younger brain.[17] Additionally, in some neurodegenerative diseases, the expression of efflux transporters is up-regulated, which hinders drug transport and therapy. In this section, we focus on Alzheimer's disease, multiple sclerosis, and epilepsy.

3.2.4.1 Alzheimer's Disease

Alzheimer's disease (AD) is the most common type of dementia, progressively impairing the cognitive functions and causing memory loss.[13] A hallmark of AD is the accumulation of amyloid, caused by impaired clearance of amyloid and brain hypoperfusion.[14,15,18,19] Pgp is identified as one of the β-amyloid efflux pumps in Alzheimer's disease. Hence a decline in Pgp activity and expression diminishes β-amyloid clearance.[5,18,20,21] Accumulation of amyloid in cerebral vessels results in vascular defects, which aggravate the neurodegenerative process and inflammatory response.[22] Dysfunction of the BBB plays an important role in the pathogenesis of AD; however, it is still under debate whether disruption of the BBB is the cause (*i.e.* neurovascular dysfunction leads to amyloid accumulation and neurovascular inflammation)[17,18] or consequence of the disease (*i.e.* the formation of amyloid plaques causes the redistribution of tight junctions, resulting in BBB disruption).[15]

3.2.4.2   Multiple Sclerosis

In multiple sclerosis (MS), an active (auto-) immune system damages the myelin sheaths, surrounding the neurites.[13] Myelin is essential for neuron–neuron communication. MS starts with inflammation in the brain venules, causing an increased transport of immune cells (activated T-lymphocytes) over the BBB, resulting in demyelination.[14,15,17,18,23,24] Moreover, the penetration of the BBB by immune cells initiates the production cytokines and chemokines, sustaining the inflammatory response[25] and enhancing the BBB permeability.[23] At the site of vascular inflammation, there is a change in expression and localization of the tight and adherens junction proteins, modifying BBB integrity.[14,24,25]

3.2.4.3   Epilepsy

Epilepsy is a chronic neurological disorder, affecting approximately 1% of the world's population, and characterized by recurring seizures.[26] Although anti-epileptic therapy effectively controls seizures in most patients, it is estimated that therapy fails in one-quarter of the cases.[11] The overexpression of the multidrug resistance proteins Pgp and MRP directly relates to this refractoriness, as anti-epileptic drugs are substrates for these efflux transporters, and hence lower brain uptake.[5,8,11] Additionally, it was recently suggested that there is a positive feedback cycle between inflammation and epilepsy.[26,27] Prolonged seizures up-regulate adhesion molecules for leukocyte extravasation and increase the synthesis and release of pro-inflammatory proteins. Together with BBB breakdown, this contributes to neuronal hyper-excitability by decreasing the seizure threshold, and hence increasing seizure frequency.[26,27]

3.3   Modeling the BBB *in Vitro*

In recent years, much effort has been made to unravel the mechanisms behind neurodegenerative diseases and link these to BBB dysfunction, as described in the previous paragraph. However, a complete overview is still lacking. To continue research and find the missing pieces of the puzzle and place all the occurring processes in the right order, it is of utmost importance to have a reliable and realistic model of the BBB. Most of the models today rely on a transwell culture system, in which monolayers of endothelial cells are cultured alone or in combination with other cells from the neurovascular unit. Some systems have been developed to culture cells under physiologically relevant fluid shear stress. General overviews of these models are given in numerous review papers in literature.[28–34]

3.3.1 Microfluidic *in Vitro* Models of the BBB: the "BBB-on-Chip"

A specific class of *in vitro* models of biological tissues relies on cultures in microfluidic chips – micrometer-sized devices that contain fluid-filled compartments with defined geometries. These models have been called "organs-on-chips" to emphasize their microfluidic nature, as well as their organ-level functionality.[35,36] Development of organs-on-chips is based on the concept of "microenvironment engineering", in which the cells and tissues are grown in a microenvironment that mimics the environment in the modeled human organ in terms of geometry, chemical composition, mechanical stimuli, and soluble gradients. By employing this concept of microenvironment engineering, *in vitro* models can be engineered that are more realistic and relevant for studying human physiology and pathophysiology.[37,38]

The first attempts to mimic the microenvironment of the human BBB by designing a "BBB-on-Chip" have been undertaken by several groups (Table 3.1 and Figure 3.1). The micro-engineered models always rely on a strategy in which a device is fabricated that has two compartments, separated by a semi-permeable barrier. The compartments typically have cross-sections with dimensions that are smaller than a millimeter. One of the compartments is supposed to mimic the lumen of a blood vessel, while the other compartment is meant to represent the brain. By culturing endothelial cells on the semi-permeable barrier, the two compartments are separated. First attempts are made to induce a differentiated BBB by engineering the microenvironment: co-cultures of astrocytes are introduced in the brain compartments and/or shear stress is applied to the endothelial surface.

Table 3.1 Overview of microfluidic models of the blood-brain barrier.

Article	Endothelial cells	Co-cultured cells	Shear stress (dyne cm^{-2})	TEER (Ω cm^2)	Permeability dextran (cm s^{-1})
Booth and Kim, 2012[39]	b.End3 (mouse)	C8D1A Astrocytes (mouse)	0.02	250	2×10^{-6}
Yeon *et al.*, 2012[42]	HUVEC (human)	None	0.3–7	Not determined	Not quantified
Griep *et al.*, 2013[40]	hCMEC/D3 (human)	None	5.8	40	Not determined
Achyuta *et al.*, 2013[41]	RBE4 (rat)	Embryonic cortical mix (rat)	None	Not determined	Not quantified
Prabhakarpandian *et al.*, 2013[43]	RBE4 (rat)	None	0.04	Not determined	Not quantified

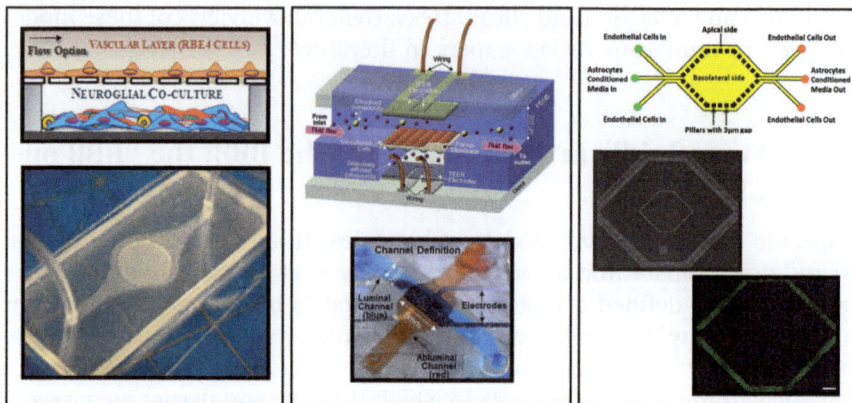

Figure 3.1 Three examples of BBB-on-Chips, showing the two specific configurations. In the classic configuration, the blood compartment sits on top of the brain compartment, separated by a porous plastic membrane (left and middle pictures). In the other model, the blood and brain compartments are positioned side-by-side, separated by a microgap barrier (right picture). Left shows the neurovascular unit on chip from Achyuta *et al.*, adapted from ref. 41. Middle presents the µBBB of Booth and Kim, adapted from ref. 39. Right is the synthetic microvasculature model of the BBB (SyM-BBB) of Prabhakarpandian *et al.*, adapted from ref. 43.

So far, two specific configurations of BBB-on-Chip models have been reported. Some of the models rely on a "classic" configuration in which the blood compartment sits on top of the brain compartment and the two are separated by a porous plastic membrane that is "sandwiched" between them.[39–41] In other models, the blood compartment and brain compartment are positioned side-by-side and the semi-permeable barrier is formed by microgaps in the wall between the two compartments.[42,43] Both configurations have their advantages. In a side-by-side configuration, all the cells and compartments are easily accessible by microscopy. On the other hand, a "sandwich" configuration has a far greater effective surface area of endothelial cells and co-cultured tissue.

It must be said that these first attempts of designing a BBB-on-Chip model are relatively primitive in terms of cell culture conditions and functionality when compared to the current *in vitro* models that are based on transwell systems. Many groups working with transwell models report transendothelial electrical resistance (TEER) values that are over 500 Ω cm^2 and permeability coefficients for low-molecular weight (3–4 kDa) dextran of 1×10^{-6} cm s^{-1}. These values are better than anything that has been accomplished with BBB-on-Chip models so far. However, the majority of the current BBB-on-Chip models do not incorporate basic parameters, such as astrocyte co-culture, yet. Moreover, it is to be expected that culture conditions that have been optimized for transwell models cannot be transferred directly to the microfluidic systems; further optimization may be necessary. The most important message that must be taken from the studies with the first BBB-on-Chip systems is that the concept of a micro-engineered BBB model is viable. By including more microenvironment parameters in the models, it is to be expected that the BBB-on-Chip systems will eventually be more on par with current models in terms of functionality. Subsequently, full advantage can be taken from the unique features of the organ-on-chip-like technology.

In the following sections, an overview is given of which main micro-environment parameters can be engineered, included, or optimized in future generations of BBB-on-Chip systems.

3.3.2   Cellular Engineering

3.3.2.1   *Endothelial Cells*

The main and most important component of the BBB is the brain endo-thelium.[32,44] The brain endothelial cells control the permeability of the BBB and regulate the transport over the BBB. For the development of an *in vitro* model of the BBB, researchers have used/use cells of various origins, either non-brain or brain, animal- or human-derived.[31–34] Ideally, the BBB con-struct should mimic the *in vivo* situation (anatomically and/or functionally), and hence consist of primary brain-derived endothelial cells, preferably of human origin. However, these resources are scarce (or even restricted), expensive, require specific skills for isolation, and are not always well characterized. Additionally, under culture conditions, primary endothelial cells dedifferentiate, because of lack of exposure to natural physiological stimuli, which affect the intrinsic BBB characteristics and function.[31–34] Hence, in some cases, it is better to use a standardized cell line of brain-origin. Nevertheless, it is essential to validate the used *in vitro* model, *i.e.* to characterize the cells (*e.g.* for the expression of tight junctions and efflux transporters, quantify TEER and permeability), correlate it with the human *in vivo* situation (*i.e.* are important BBB characteristics retained), and address why this cell type is applicable for the research to be performed (*e.g.* permeability, transport, inflammatory studies). Table 3.2 gives a brief over-view of the cells commonly used to model the BBB *in vitro*. The large number of different *in vitro* models states that until now there has been no "ideal" model of the BBB, and it depends on the research question which model is advantageous, which also accounts for microfluidic models of the BBB.

3.3.2.2   *Astrocytes*

Although brain endothelial cells are the principal components of the BBB, they cooperate with other cells, such as astrocytes, pericytes, and neurons, to maintain BBB function.[3,59,60] Astrocytes secrete a wide range of chemical agents which induce and modulate the characteristic BBB properties. A co-culture of endothelial cells interacting with astrocytes show high TEER and coinciding low permeability values. *Via* the direct interaction with the

Table 3.2   Overview of the cells used to model the BBB *in vitro*.[a]

	Cell type	Origin	Specified	Ref
Non-brain	Epithelial	Canine	MDCK	31
	Endothelial	Human	HUVEC	42
	Stem cells	Human	hPSC	45, 46
Brain	Endothelial	Human	hCMEC/D3	40, 47–49
			HBMEC	50, 51
		Bovine	BBCEC	52
		Porcine	PBEC	53–55
		Murine	RBE4	32, 41, 43
			RBEC	56–58
			b.end 3–5 (mouse)	32, 39

[a]Abbreviations: MDCK – Madin-Darby Canine Kidney cell line; HUVEC – human umbilical vein endothelial cells; hPSC – human pluripotent stem cells; hCMEC/D3 – human cerebral microvascular endothelial cell line; HBMEC – human brain microvascular endothelial cells; BBCEC – bovine brain capillary endothelial cells; PBEC – porcine brain endothelial cells; RBE4 – rat brain endothelial cells; R(M)BEC – rat (microvascular) brain endothelial cells.

astrocyte endfeet, tight junctions are up-regulated, resulting in a tighter barrier and significant improvement of the efflux transporter Pgp.[47,54,56,58] To circumvent the use of astrocytes, researchers have used astrocyte conditioned medium (ACM) as an alternative.[42,43] Incubation of brain endothelial cells with ACM in BBB-on-Chips significantly decreased the permeability and increased the expression of tight junction proteins and the efflux transporter PgP compared to untreated endothelial cells.[42,43] Co-culture with astrocytes is shown in two-compartment based BBB-on-Chips;[39,41] however, only one expressed clear quantitative data,[39] which emphasizes the need for more sophisticated experimental models on chip.

3.3.2.3   Pericytes

Pericytes are the cells closest to the brain endothelial cells; however, in only a few (Transwell) studies pericytes are incorporated in the BBB model, to determine their significant role in BBB functionality.[48,56] A triple co-culture of endothelial cells, pericytes, and astrocytes significantly increased TEER, as compared to an endothelial-astrocyte co-culture.[48,56] There is still debate over whether direct contact between endothelial cells and pericytes is a prerequisite for high TEER values. Nevertheless, the expression of the tight junction proteins occludin, claudin-5, and ZO-1 increased in the triple co-culture model.[56] Until now, no attempts have been made to develop a neurovascular unit on chip, which requires a combination of cellular and biochemical engineering, as discussed in the next section.

3.3.3   Biochemical Engineering

3.3.3.1   Extracellular Matrix

The abluminal side of the endothelium is tethered to an underlying basement membrane. This basement membrane consists of a thin (~ 50 nm) sheet of connective extracellular matrix proteins, such as collagen IV, laminin, heparan sulfate proteoglycan, and fibronectin.[61] The basement membrane of capillaries in the brain differs from basement membranes in the rest of the body,[62] and there are clear indications that the exact composition of the basement membrane affects the permeability of the layers of brain endothelial cells that are grown on it.[63] The differentiating effect of the basement membrane is also exemplified by the fact that changes in the composition of the basement membrane *in vivo* are localized to sites of endothelial dysfunction in *e.g.* Alzheimer's disease.[64]

Apart from the basement membrane, the rest of the extracellular matrix of the brain has a unique composition compared to peripheral tissues. The matrix of the brain is very rich in proteoglycans, which are proteins with attached glycosaminoglycans (a type of polysaccharide chain). Specific proteoglycans, such as lecticans, have been implicated in organizing the brain matrix and in maintaining neural and glial physiology.[65]

The current BBB-on-Chip models mimic the basement membrane by coating with either fibronectin or collagen I (Table 3.1). Based on the indications that the exact composition of the basement membrane can have important effects on endothelial differentiation, other matrix coatings should be used in future versions of the BBB-on-Chip models. Specifically, collagen IV and laminin seem to have a clear effect on barrier tightness.[63] Additionally, if a model incorporates astrocytes for co-culture, it is recommended to incorporate a realistic matrix for this cell type as well. Based on literature, a matrix rich in proteoglycans and hyaluronic acid may provide a good start.[65] One of the most interesting things to explore when designing a matrix for the astrocytic compartment of the devices is whether a three-dimensional culture of astrocytes will have any effect on BBB physiology. To achieve this, one of the compartments of a micro-device may be filled with a cell suspension inside a hydrogel.[66] This concept may even be taken a step further and the entire model may be engineered using a patterned hydrogel with a capillary-like lumen that is lined by endothelial cells. Several examples of such micro-engineered models of human capillaries have already appeared in literature.[67,68]

3.3.3.2 Local and Systemic Bioactive Ligands

As is clear from current work in *in vitro* models of the BBB, soluble bioactive molecules have a major impact on barrier functionality. The most popular soluble molecules that are added to *in vitro* cultures of cerebral endothelial cells are anti-inflammatory hormones and signaling molecules, such as hydrocortisone and cyclic adenosine monophosphate (cAMP).[69] Of course, the addition of astrocyte-conditioned media is also a clear example of how to modulate BBB barrier function by soluble factors.

Interestingly, not all work on soluble factors in BBB models focuses on improving barrier function.[70] The addition of inflammatory factors, like histamine or tumor necrosis factor-alpha, is also popular in order to model dysfunction of the BBB and to mimic disease processes as found in *e.g.* cancer and Alzheimer's disease.

When considering soluble factors, the main advantage of using microfluidic systems is that they can be used to induce dynamic changes in media composition.[71] By employing microfluidic mixers and coupling them to a microfluidic BBB-on-Chip, complex temporal concentration profiles can be created. Such profiles can potentially be used to *e.g.* mimic acute and long-term phases of cerebral disease.

3.3.4   Biophysical Engineering

3.3.4.1   Shear Stress

In the human body, endothelial cells are constantly subjected to fluid flow over their surface. The result of this flow is a tangential force, or shear stress. It is well known in the field of vascular biology that endothelial cells can sense shear stress and that shear stress patterns have a profound impact on endothelial function.[72]

Work on cerebral endothelial cells in models that incorporate fluid flow shows that BBB permeability is dramatically decreased by introducing physiological levels of shear stress in the system.[47]

Microfluidic models are very well suited for stimulating endothelium with realistic levels of fluid shear stress, because of their small size. A small channel needs relatively modest levels of fluid flow ($\sim$1 milliliter per hour or less) to generate a physiologically relevant shear stress, which is easily an order of magnitude lower when compared to flow rates (tens of milliliters per hour) in conventional flow chambers. Many studies have confirmed the suitability of microfluidic set-ups for studying the endothelial response to shear stress.[73] Only one of the BBB-on-Chip models that have been published so far shows a strong enhancement of the barrier function of cerebral endothelial cells when they are cultured under physiological levels of shear stress.[40] This initial study demonstrates both the importance of shear stress as a culture parameter and the feasibility of including it routinely in microfluidic models of the BBB.

3.3.4.2   Electrophysiology

Because of the close interaction of the BBB with brain tissue, it would be highly advantageous to build models that combine BBB-like cultures with neuronal cultures. In such models, it would be possible to study the electrophysiology of neurons in the context of the neurovascular unit.

Conventional *in vitro* models in which endothelial cells and brain tissue slices were kept in culture together have demonstrated the power of this type of co-culture. In these studies, only neuroactive molecules that could penetrate the modeled BBB affected the electrophysiology of the brain tissue slice.[74]

Interestingly, micro-engineered models of neuronal cultures on patterned electrodes have already been reported in literature.[75,76] By combining such models with BBB-on-Chip models, it should be possible to perform more comprehensive studies of the electrophysiology of the neurovascular unit.

3.4 Measurement Techniques

3.4.1 Transendothelial Electrical Resistance

One of the most popular measurements for BBB barrier function *in vitro* is the quantification of transendothelial electrical resistance (TEER). By applying a known potential difference over a layer of endothelial cells, and measuring the electrical current as it is carried by ions through the paracellular space, the electrical resistance per area of endothelium can be determined. The TEER provides a reliable way to compare various BBB models, to keep track of the integrity of the cell layer in a particular model, and to register experimentally induced changes in the barrier properties. High TEERs are considered a hallmark of well-differentiated cerebral endothelial cells.

Measuring electrical resistance in microfluidic systems can be challenging due to the inherently small fluid volumes and due to limitations on the size of the electrodes that are compatible with such small systems. Two of the currently published BBB-on-Chip models report measurements of TEER values.[39,40] Both systems rely on integrated electrodes and alternating current measurements in order to reduce ion path lengths and fouling of electrode surfaces, respectively.

The integration of electrodes means that the BBB-on-Chip systems are amenable to continuous and real-time monitoring of the barrier function. Studies with conventional *in vitro* models of the BBB have shown that this can be a powerful tool in studying BBB physiology.[77,78]

3.4.2 Permeability

Apart from measuring the TEER, permeability assays are standard methods to validate barrier integrity.[79] With the use of molecule tracers that are labeled fluorescently or radioactively, the amount of molecule that passes the barrier can be quantified. The molecules that are generally used in these assays are either small molecules, such as sucrose, or larger polysaccharides like dextran. By using differently sized molecules, the quality of the tight junctions can be evaluated. For example, a dextran with high molecular weight might have low permeability in even immature layers of endothelium, while a small molecule like sucrose will only be efficiently excluded by highly differentiated cell layers. Typical values for permeability coefficients in *in vitro* assays are 1–30×10^{-6} cm s^{-1} for sucrose[79] and 5–10×10^{-6} cm s^{-1} for 4-kDa dextran.[80] By adding tracers to the vascular compartment of a BBB-on-Chip model and then collecting effluent from the brain compartment, permeability can be estimated. The feasibility of this concept was shown by several groups (Table 3.1); however, in only one model quantitative permeability coefficients were measured, which allows comparison with other *in vitro* models and the *in vivo* situation.[39]

Another important way to assess the functionality of a modeled BBB is to study the efflux of known substrates for multidrug resistance pumps. A popular way to assess this is to load the endothelial cells with a fluorescent substrate for Pgp, such as rhodamine 123, and to follow the fluorescence intensity of the cells over time. Additionally, drug permeability assays can be performed and the measured values correlated to known *in vitro/in vivo* permeability data.[56] The current BBB-on-Chip models hardly assess this functionality,[42,43] but the compatibility of these devices with microscopy or other equipment will make it easy to do so.

3.4.3 Fluorescence Microscopy

Microscopy analysis is extensively used to study the morphology and the structure of the BBB.[32] Especially fluorescence microscopy is practical to validate the BBB model used and to address why this model is applicable for the performed research. With fluorescence microscopy BBB-related proteins, such as adherens and tight junctions (ZO-1, occludin, claudin-5), efflux transporters, and the different cell types of the neurovascular unit (*e.g.* endothelial cells with Von Willebrand factor, astrocytes with GFAP, and pericytes with α-smooth muscle actin), can be visualized in detail. In three of the current BBB-on-Chips, the structure and morphology of the barrier is checked.[39–41] Overall, to validate the fabricated BBB and state to have a realistic model, a combination of cell imagining with TEER and permeability measurements must be performed.

3.5 Conclusion and Future Prospects

It is clear that the BBB is important in brain physiology, neurological pathology, and the challenges involved with drug delivery to the brain. Because of its importance, *in vitro* models of the BBB have attracted a lot of attention. This chapter has highlighted a new class of micro-engineered *in vitro* models of the BBB, the so-called "BBB-on-Chip" models.

Even though the first examples of such BBB-on-Chip models are functionally inferior to the conventional BBB models, they serve as proof-of-concept and as inspiration for further developments in the field.

There are numerous areas in which BBB-on-Chip models can be further developed, such as integration of complex extracellular matrices, co-cultures of cell types of the neurovascular unit, integration of shear stress and electrical stimulation, dynamic changes in microenvironmental factors, and the addition of more sophisticated real-time measurements. Proof-of-concept work for all these areas can already be found in literature and these examples have been highlighted in this chapter.

The versatility of the micro-engineered models, combined with their small footprint and their potential for increased throughput, makes them exciting new research tools in the field of brain research. As the "organs-on-chips"-approach is gaining momentum, BBB-on-Chip models will become more sophisticated and might even be integrated with other organs-on-chips to engineer more comprehensive models of human physiology.

Much of the technology behind BBB-on-Chip models is only just beginning to find its way into biomedical research settings. Within the foreseeable future, the integration of biomedical science and micro-engineering is set to generate exciting new results in the field of neurovascular biology.

Acknowledgements

Financial support of the European Research Council (ERC, grant no. 280281 MESOTAS) is gratefully acknowledged.

References

1. J. W. Mink, R. J. Blumenschine and D. B. Adams, *Am. J. Physiol.*, 1981, **241**, R203.
2. C.-H. Lai and K.-H. Kuo, *Brain Res. Rev.*, 2005, **50**, 258.
3. N. J. Abbott, L. Rönnbäck and E. Hansson, *Nat. Rev. Neurosci.*, 2006, **7**, 41.
4. G. Schiera, E. Bono, M. P. Raffa, A. Gallo, G. L. Pitarresi, I. Di Liegro and G. Savettieri, *J. Cell. Mol. Med.*, 2003, **7**, 165.
5. W. Löscher and H. Potschka, *Prog. Neurobiol.*, 2005, **76**, 22.
6. M. Demuele, A. Régina, J. Jodoin, A. Laplante, C. Dagenais, F. Berthelet, A. Moghrabi and R. Béliveau, *Vasc. Pharmacol.*, 2002, **38**, 339.
7. B. L. Urquhart and R. B. Kim, *Eur. J. Clin. Pharmacol.*, 2009, **65**, 1063.
8. E. Aronica, S. M. Sisodiya and J. A. Gorter, *Adv. Drug Deliv. Rev.*, 2012, **64**, 919.
9. H. Sun, H. Dai, N. Shaik and W. F. Elmquist, *Adv. Drug Deliv. Rev.*, 2003, **55**, 83.
10. P. L. Golden and G. M. Pollack, *J. Pharm. Sci.*, 2003, **92**, 1739.
11. A. Lazarowski, L. Czornyj, F. Lubienieki, E. Girardi, S. Vazquez and C. D'Giano, *Epilepsia*, 2007, **48**, 140.
12. Z. Guo, J. Zhu, L. Zhao, Q. Luo and X. Jin, *J. Exp. Clin. Cancer Res.*, 2010, **29**, 122.
13. A. M. Palmer, *J. Alzheimers Dis.*, 2011, **24**, 643.
14. P. Grammas, J. Martinez and B. Miller, *Exp. Rev. Mol. Med.*, 2011, **13**, DOI: 10.1017/S1462399411001918.
15. G. A. Rosenberg, *J. Cereb. Blood Flow Metab.*, 2012, **32**, 1139.
16. R. D. Bell, E. A. Winkler, A. P. Sagare, I. Singh, B. LaRue, R. Deane and B. V. Zlokovic, *Neuron*, 2010, **68**, 409.
17. P. M. Carvey, B. Hendey and A. J. Monahan, *J. Neurochem.*, 2009, **111**, 291.

18. B. V. Zlokovic, *Neuron*, 2008, **57**, 178.
19. A. P. Sagare, R. D. Bell and B. V. Zlokovic, *J. Alzheimers Dis.*, 2013, **33**, S87.
20. B. Jeynes and J. Provias, *J. Neurosci. Res.*, 2011, **89**, 22.
21. A. H. Abuznait and A. Kaddoumi, *ACS Chem. Neurosci.*, 2012, **3**, 820.
22. R. D. Bell and B. V. Zlokovic, *Acta Neuropathol.*, 2009, **118**, 103.
23. C. Larochelle, J. I. Alvarez and A. Prat, *FEBS Lett.*, 2011, **585**, 3770.
24. J. I. Alvarez, R. Cayrol and A. Prat, *Biochim. Biophys. Acta*, 2011, **1812**, 252.
25. A. Minagar and J. S. Alexander, *Mult. Scler.*, 2003, **9**, 540.
26. D. Xu, S. D. Miller and S. Koh, *Front Cell Neurosci.*, 2013, 7, DOI: 10.3389/fncel.2013.00195.
27. S. Y. Kim, M. Buckwalter, H. Soreq, A. Vezzani and D. Kaufer, *Epilepsia*, 2012, **53**, 37.
28. G. A. Grant, N. J. Abbott and D. Janigro, *News Physiol. Sci.*, 1998, **13**, 287.
29. A. Reichel, D. J. Begley and N. J. Abbott, *Methods Mol. Med.*, 2003, **89**, 307.
30. N. J. Abbott, *Drug Discov. Today Technol.*, 2004, **1**, 407.
31. J. Mensch, J. Oyarzabal, C. Mackie and P. Augustijns, *J. Pharm. Sci.*, 2009, **98**, 4429.
32. F. L. Cardoso, D. Brites and M. A. Brito, *Brain Res*, 2010, **64**, 328.
33. I. Wilhelm, C. Fazakas and I. A. Krizbai, *Acta Neurobiol. Exp.*, 2011, **71**, 113.
34. P. Naik and L. Cucullo, *J. Pharm. Sci.*, 2012, **101**, 1337.
35. A. D. van der Meer and A. van den Berg, *Integr. Biol.*, 2012, **4**, 461.
36. D. Huh, Y.-S. Torisawa, G. A. Hamilton, H. J. Kim and D. E. Ingber, *Lab Chip*, 2012, **12**, 2156.
37. D. Huh, B. D. Matthews, A. Mammoto, M. Montoya-Zavala, H. Y. Hsin and D. E. Ingber, *Science*, 2010, **328**, 1662.
38. D. Huh, D. C. Leslie, B. D. Matthews, J. P. Fraser, S. Jurek, G. A. Hamilton, K. S. Thorneloe, M. A. McAlexander and D. E. Ingber, *Sci. Transl. Med.*, 2012, **4**, 159.
39. R. Booth and H. Kim, *Lab Chip*, 2012, **12**, 1784.
40. L. M. Griep, F. Wolbers, B. de Wagenaar, P. M. ter Braak, B. B. Weksler, I. A. Romero, P. O. Couraud, I. Vermes, A. D. van der Meer and A. van den Berg, *Biomed. Microdevices*, 2013, **15**, 145.
41. A. K. H. Achyuta, A. J. Conway, R. B. Crouse, E. C. Bannister, R. N. Lee, C. P. Katnik, A. A. Behensky, J. Cuevas and S. S. Sundaram, *Lab Chip*, 2013, **13**, 542.
42. J. H. Yeon, D. Na, K. Choi, S. Ryu, C. Choi and J. Park, *Biomed. Microdevices*, 2012, **14**, 1141.
43. B. Prabhakarpandian, M. Shen, J. B. Nichols, I. R. Mills, M. Sidoryk-Wegrynowicz, M. Aschner and K. Pant, *Lab Chip*, 2013, **13**, 1093.
44. N. J. Abbott, A. A. K. Patabendige, D. E. M. Dolman, S. R. Yusof and D. J. Begley, *Neurobiol. Dis.*, 2010, **37**, 13.
45. E. S. Lippmann, S. M. Azarin, J. E. Kay, R. A. Nessler, H. K. Wilson, A. Al-Ahmad, S. P. Palecek and E. V. Shusta, *Nat. Biotechnol.*, 2012, **30**, 783.

46. E. S. Lippmann, A. Al-Ahmad, S. P. Palecek and E. V. Shusta, *Fluids Barriers CNS*, 2013, **10**, 2.
47. L. Cucullo, P. O. Couraud, B. Weksler, I. A. Romero, M. Hossain, E. Rapp and D. Janigro, *J. Cereb. Blood Flow Metabol.*, 2008, **28**, 312.
48. K. Hatherell, P. O. Couraud, I. A. Romero, B. Weksler and G. J. Pilkington, *J. Neurosci. Methods*, 2011, **199**, 223.
49. B. Weksler, I. A. Romero and P. O. Couraud, *Fluids Barriers CNS*, 2013, **10**, 16.
50. Y. Sano, F. Shimizu, M. Abe, T. Maeda, Y. Kashiwamura, S. Ohtsuki, T. Terasaki, M. Obinata, K. Kajiwara, M. Fujii, M. Suzuki and T. Kanda, *J. Cell Physiol.*, 2010, **225**, 519.
51. Y. C. Kuo and C. H. Lu, *Colloids Surf. B Biointerfaces*, 2011, **86**, 225.
52. M. Culot, S. Lundquist, D. Vanuxeem, S. Nion, C. Landry, Y. Delplace, M. Dehouck, V. Berezowski, L. Fenart and R. Cecchelli, *Toxicol. In Vitro*, 2008, **22**, 799.
53. W. Neuhaus, R. Lauer, S. Oelzant, U. P. Fringeli, G. F. Ecker and C. R. Noe, *J. Biotechnol.*, 2006, **125**, 127.
54. K. Cohen-Kashi Malina, I. Cooper and V. I. Teichberg, *Brain Res.*, 2009, **1284**, 12.
55. A. Patabendige, R. A. Skinner, L. Morgan and N. J. Abbott, *Brain Res.*, 2013, **1521**, 16–30.
56. S. Nakagawa, M. A. Deli, H. Kawaguchi, T. Shimizudani, T. Shimono, A. Kittel, K. Tanaka and M. Niwa, *Neurochem. Int.*, 2009, **54**, 253.
57. N. J. Abbott, D. E. M. Dolman, S. Drndarski and S. M. Fredriksson, *Methods Mol. Biol.*, 2012, **814**, 415.
58. Q. Xue, Y. Liu, H. Qi, Q. Ma, L. Xu, W. Chen and X. Xu, *Int. J. Biol. Sci.*, 2013, **9**, 174.
59. N. J. Abbott, *J. Anat.*, 2002, **200**, 629.
60. J. I. Alvarez, T. Katayama and A. Prat, *Glia*, 2013, **61**, 1939.
61. E. D. Hay, *Cell Biology of Extracellular Matrix*, ed. E. D. Hay, Plenum Press, New York, 2nd edn, 1991.
62. M. Jucker, M. Tian, D. D. Norton, C. Sherman and J. W. Kusiak, *Neuroscience*, 1996, **71**, 1153.
63. T. Tilling, D. Korte, D. Hoheisel and H.-J. Galla, *J. Neurochem.*, 1998, **71**, 1151.
64. L. S. Perlmutter and H. Chang Chui, *Brain Res. Bull.*, 1990, **24**, 677.
65. Y. Yamaguchi, *Cell. Mol. Life Sci.*, 2000, **57**, 276.
66. A. P. Wong, R. Perez-Castillejos, J. C. Love and G. M. Whitesides, *Biomaterials*, 2008, **29**, 1853.
67. A. P. Golden and J. Tien, *Lab Chip*, 2007, 7, 720.
68. L. L. Bischel, E. W. K. Young, B. R. Mader and D. J. Beebe, *Biomaterials*, 2013, **34**, 1471.
69. D. Hoheisel, T. Nitz, H. Franke, J. Wegener, A. Hakvoort, T. Tilling and H.-J. Galla, *Biochem. Biopys. Res. Comm.*, 1998, **244**, 312.
70. N. J. Abbott, *Cell. Mol. Neurobiol.*, 2000, **20**, 131.

71. F. Lin, W. Saadi, S. W. Rhee, S.-J. Wang, S. Mittal and N. L. Jeon, *Lab Chip*, 2004, **4**, 164.
72. Y.-S. J. Li, J. H. Haga and S. Chien, *J. Biomech.*, 2005, **38**, 1949.
73. A. D. van der Meer, A. A. Poot, M. H. G. Duits, J. Feijen and I. Vermes, *J. Biomed. Biotechnol.*, 2009, **2009**, 823148.
74. S. Duport, F. Robert, D. Muller, G. Grau, L. Parisi and L. Stoppini, *Proc. Natl Acad. Sci. USA*, 1998, **95**, 1840.
75. T. M. Pearce, J. A. Wilson, S. G. Oakes, S.-Y. Chiu and J. C. Williams, *Lab Chip*, 2005, **5**, 97.
76. K. Musick, D. Khatami and B. C. Wheeler, *Lab Chip*, 2009, **9**, 2036.
77. C. Hartmann, A. Zozulya, J. Wegener and H.-J. Galla, *Exp. Cell Res.*, 2007, **313**, 1318.
78. M. von Wedel-Parlow, S. Schrot, J. Lemmen, L. Treeratanapiboon, J. Wegener and H.-J. Galla, *Brain Res.*, 2011, **1367**, 62.
79. M. A. Deli, C. S. Ábrahám, Y. Kataoka and M. Niwa, *Cell. Mol. Neurobiol.*, 2005, **25**, 59.
80. G. Li, M. J. Simon, L. M. Cancel, Z.-D. Shi, X. Ji, J. M. Tarbell, B. Morrison III and B. M. Fu, *Ann. Biomed. Eng.*, 2010, **38**, 2499.

CHAPTER 4

The Use of Microfluidic-based Neuronal Cell Cultures to Study Alzheimer's Disease

ROBERT MEISSNER* AND PHILIPPE RENAUD

Microsystems laboratory (LMIS4), Ecole Polytechnique Fédérale de Lausanne, Station 17, CH-1015, Lausanne, Switzerland
*Email: robert.meissner@epfl.ch

4.1 Alzheimer's Disease – Increased Mortality Rates and Still Incurable

Alzheimer's disease (AD) is the most common cause of dementia and develops when nerve cells no longer function normally. This process results in memory loss and affects the individual's thinking and behavior. Basic human functions such as walking and swallowing become difficult to carry out. AD is a fatal disease.[1]

More than 35 million people worldwide are affected by AD and this number is expected to double every 20 years. Worldwide costs of dementia represented over US$600 billion in 2010.[2] Although death rates decreased significantly for 7 of the 15 leading causes of death in the USA,[3] those from AD experienced a rise of 66% between 2000 and 2008.[1] Despite much effort on the part of academic research and industry the causes of AD are not yet known and no treatment is currently available to slow or stop synaptic and neuronal decline in AD.[4] The need for a substantial change with regard to new testing strategies and tools seems to be indispensable. In fact, studies

RSC Nanoscience & Nanotechnology No. 36
Microfluidics for Medical Applications
Edited by Albert van den Berg and Loes Segerink
© The Royal Society of Chemistry 2015
Published by the Royal Society of Chemistry, www.rsc.org

on living animals are difficult to carry out due to a high number of uncontrollable parameters and the difficulty of *in situ* observation. Conventional cell culture, on the other hand, is limited by the inability to precisely control the environment of cells. A comprehensive study of AD requires the highly polarized neurons to be arranged in a complex but well-defined way, a condition that conventional cell culture is not able to fulfill.

To overcome limitations of conventional cell culture assays, microfluidic platforms have been developed for neuroscience applications. Microfluidic systems are capable of regulating fluid flow precisely owing to a laminar flow pattern, structuring cells, since many biological species are on the same length scale, and manipulating the extracellular environment. This chapter reviews recent developments in microfluidic culture platforms for structured neuronal cell culture and AD research. The first part describes the importance of neuronal networks in AD and how microsystems may be a key in understanding AD progression in the brain. The second part reviews microfluidic-based *in vitro* Alzheimer models and points out what specific questions may be addressed in future.

4.2   Unknowns of Alzheimer's Disease

4.2.1   Molecular Key Players of AD

AD is characterized by mainly two abnormalities; extracellular deposits of beta-amyloid (Aβ)[5] and intracellular aggregates of neurofibrillary tangles (NFTs)[6] (Figure 4.1a). Aβ is a fragment of the amyloid precursor protein (APP), a protein that has been shown to be implicated in synapse formation and transmission.[8] Arising from the proteolytique cleavage of APP (Figure 4.1b), Aβ is broken down and eliminated in healthy brains. In Alzheimer diseased brains, however, this process is disrupted and Aβ aggregates to plaques outside the cell. NFTs, on the other hand, are mainly comprised of the microtubule-binding Tau protein. Its major function is to stabilize the microtubules and hence to promote structural support and nutrient transport within neurons. The Tau protein can dissociate from the microtubules when being phosphorylated and reassociate when being dephosphorylated, a dynamic process that is mediated by enzymes called kinases (*e.g.* GSK) and phosphatases (*e.g.* PP2A) (Figure 4.1c). Possessing a multitude of phosphorylation sites, Tau can hence undergo conformational changes that are necessary to carry out its microtubule-supportive function. In diseased brains, however, the process of phosphorylation and dephosphorylation is unbalanced and numerous phosphorylation sites have been found to be hyperphosphorylated, leading to the complete dissociation of Tau from the microtubules.[9] As a result, hyperphosphorylated Tau aggregates to NFTs within the cell.

Although many studies have investigated the molecular mechanism of this disease, the reason behind the disrupted Aβ elimination and the increased Tau phosphorylation is still unclear.[10] Moreover, the sequential

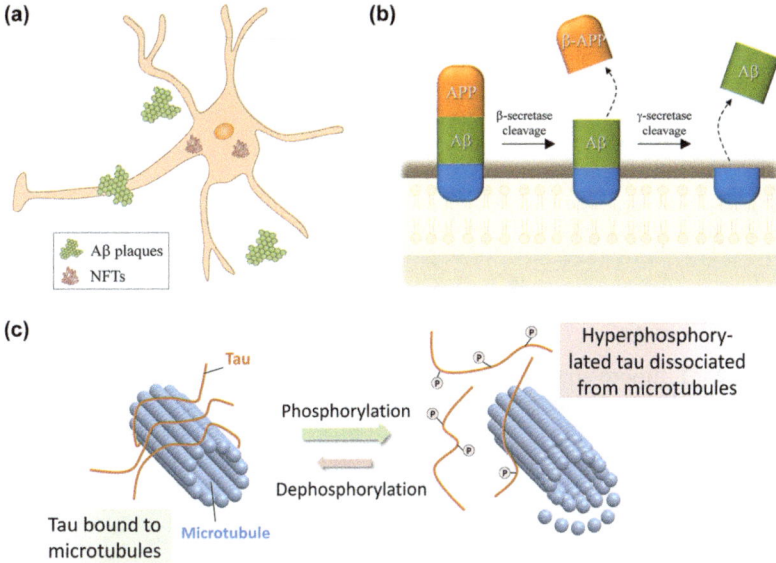

Figure 4.1   Major hallmarks in AD. a) Extracellular deposits of Aβ plaques and intracellular aggregation of pathological Tau to NFTs lead to neuronal decline in AD. b) Aβ plaque production is obtained through cleavage of the precursor protein APP by a β-secretase and the subsequent cleavage of the left membrane-bound fragment by a β-secretase. Aβ is released into the extracellular space. c) Balance of Tau-microtubule phosphorylation and dephosphorylation is disturbed in AD, leading to hyperphosphorylated Tau proteins that do not bind to the microtubules and that aggregate to NFTs in the cytoplasm (reprinted by permission from Wiley, *Biotechnology and Bioengineering*, copyright 2011 (ref. 7)).

order and the cause and effect, especially with regard to the interplay of Aβ and Tau, remain unsolved.[6]

4.2.2   From Molecules to Neuronal Networks

AD is a progressive disease that spreads throughout the brain according to a predictable pattern (Figure 4.2). It starts in the entorhinal cortex (EC), where the large projection neurons, which link the cerebral cortex with the hippocampus, are affected first.[11–14] It then propagates across the limbic system and association cortices.[15] Despite good knowledge about the hierarchical accumulation pattern of Aβ and Tau in the brain, little is known about the mechanism of the initiation and progression of these protein accumulations: why does it start in the EC and why is there a specific spreading pattern?[16] Recent findings have indicated that Tau aggregates can transfer to neighboring cells and to synaptically connected neurons in distant parts of the brain in a prion-like manner.[15] In fact, a transgenic mouse

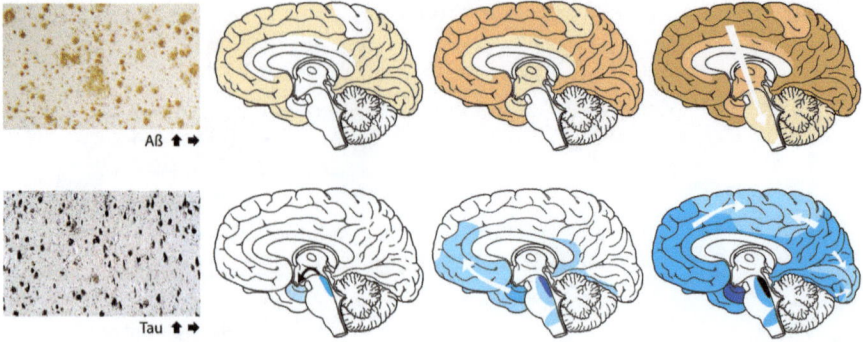

Figure 4.2   Temporo-spatial spreading of Tau-positive neurofibrillary lesions and Aβ
plaques in the process of Alzheimer's disease. Aβ-plaques (top) and NFTs
(bottom) spread through AD brains in specific patterns. Shading intensities are proportional to the clump density over time.
(Reprinted by permission from Wiley, *Annals of Neurology*, copyright
2011 (ref. 17).)

model with overexpression of human mutant Tau exclusively in layer II of the
EC has been established. Several months after lesions in the EC, the Tau
pathology appeared in other brain regions with neurons that exhibited no
detectable transgene expression; first to neighboring cells and then to
neurons downstream in the synaptic circuit. These experiments proved the
transfer and progression of misfolded Tau proteins from neuron to neuron
in a neuronal circuit-based manner. Moreover, the migrating pathological
Tau aggregates were shown to be endocytosed by healthy neurons[18] where
they act as seeds to induce the misfolding of intracellular native Tau proteins
and their conversion into NFTs.[19–21] In fact, this is a process that was previously reported in other cases where the accumulation of misfolded protein
aggregates is mediated by the intercellular spreading of oligomeric seeds.
Although the experimental proof of aggregated Tau-spreading may explain
the formation of NFTs in a stepwise characteristic pattern,[16] it still remains
unclear why only some of the interconnected neurons develop NFTs. There
may be conditions that favor or prevent pathological Tau proteins from
being transferred and/or from acting as seeds.

   In addition to the spreading of pathological Tau, the progression of
Aβ deposition was also reported. Experiments where Aβ oligomers were
microinjected into electrophysiologically defined primary hippocampal rat
neurons revealed their neuron-to-neuron transfer.[22] Moreover, a transgenic
mouse model with spatially restricted overexpression of mutant APP in the
EC, demonstrated that Aβ-induced molecular and functional impairments
can cross synapses.[23]

   All these studies have proven the characteristic spreading of NFTs and
Aβ deposits in the brain and have consequently shown the importance
of neuronal networks in the progression of AD. Additional elucidation of
the detailed mechanism behind the specific spatiotemporal pattern of

neuropathology in AD potentially may open new avenues to develop long-awaited strategies to stop the progression of this dreadful disease.[22]

4.3 Why Microsystems May Be a Key in Understanding the Propagation of AD

Although the latest important *in vivo* findings about the transfer of toxic species of Tau and Aβ from transfected brain parts to distinct regions have been a breakthrough in neurosciences, researchers cannot rule out that a low expression of the transgene Tau (below the level of detection) in areas other than the EC may have occurred[16] as observed previously.[24] More importantly, detailed investigation at the cellular level is limited: functional and live analysis such as signal recordings or live cell imaging can hardly be performed *in vivo*; stereology and immunocytochemistry are tedious to carry out; too many parameters that cannot be controlled tightly make it difficult to account for a specific phenomenon; and living organisms exhibit a great variability. This section describes how microtechnology-based platforms may provide the essential support for *in vitro* neuronal cell culture and its analysis that may fill the gap between current findings and a thorough understanding of disease progression in AD.

4.3.1 Requirements for *in Vitro* Studies on AD Progression

Experimental models such as animals (*e.g.* rats, mice) are used for brain studies owing to the preservation of anatomical structures. Different neuronal populations are connected with well-defined topologies and polarities. This directionality exists in developmental mechanisms (connection of different neuronal sub-types)[25] as well as in degeneration of neuronal networks in mature brains.[22,26] Conventional cell cultures, on the other hand, are missing the anatomical structure with its highly ordered connectivity of neurons and do not satisfy the requirements of directionality and orientation that are necessary for the study of neuropathology in AD. In fact, conventional cell cultures are characterized by random connections and intermingled networks that vary from one sample to another[26] (Figure 4.3). Therefore, in order to study network phenomena and disease progression, (1) different populations of cells need to be separated (spatial compartmentalization) while allowing them to interconnect (connectivity) in (2) a polarized and oriented way (directionality). Moreover, (3) to mimic the *in vivo* tissue structure closely, those interconnected but distinct cell populations need to be dispersed in three dimensions (3D cell culture). Advances in microfluidics and microfabrication have enabled such micro-structuring with the creation of microchips as tools for *in vivo*-like anatomically organized neuronal cell cultures on the one hand, and simultaneous cell culture analysis (*e.g.* signal transmission analysis and on-chip protein analysis) on the other.

(a)

*Conventional
cell culture*

(b)

*Compartmentalized
cell culture*

(c)

*Orientation /
directionality*

(d)

*Three-dimensional
cell culture*

Tissue sophistication

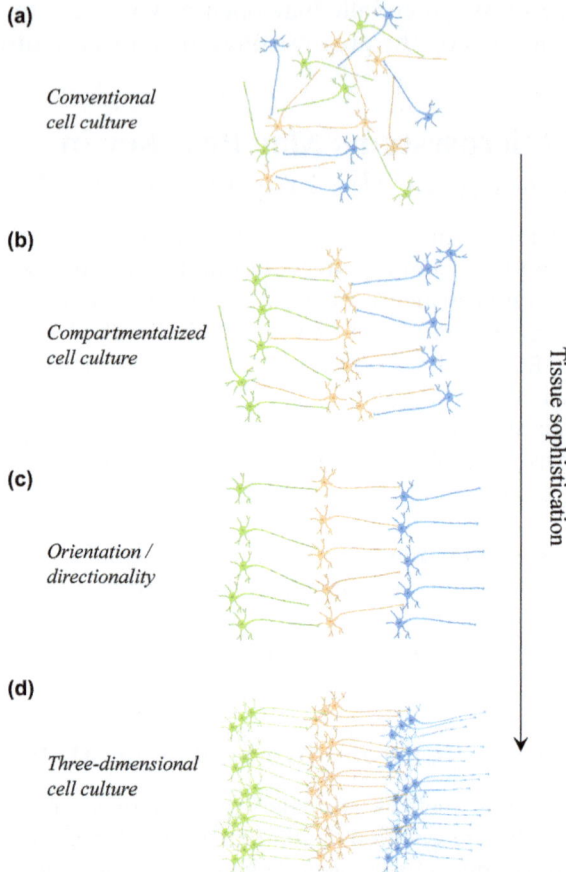

Figure 4.3   Requirements for *in vitro* studies on AD progression. a) Conventional Petri dish cell culture is characterized by intermingled networks with random connections. b) Compartmentalization of different cellular subtypes that are interconnected allows local treatment of specific subpopulations. c) Orientation of neuritic processes leads to well-defined topologies and polarization. d) Three-dimensional, structured and oriented neuronal cultures resemble the most *in vivo* brain tissues.

4.3.2   Establishing Ordered Neuronal Cultures with Microfluidics

4.3.2.1   *Compartmentalization of Neuronal Cell Populations*

While continuously striving for a better understanding of AD and simultaneously being limited in the direct observation of living organisms, researchers have made great advances in the last 40 years in reconstructing typical brain patterns *in vitro*. Microtechnology has played the essential role in establishing brain structures that may enable basic neuroscience research as well as the pharmaceutical industry to find strategies to counteract AD with its dreadful consequences on people's lives.

One of the first steps from random neuronal cell culture towards guided neurite outgrowth was achieved by Robert B. Campenot in 1977.[27] He was able to compartmentalize somas and their axons by having the processes grow from a soma-containing chamber to a side chamber that contained or did not contain nerve growth factor (NGF). Both chambers were separated by a Teflon divider that adhered on a collagen coated Petri dish by sealing with silicone grease (Figure 4.4). The seals could be penetrated by growing neurites, but not by the cell soma. He found out that neurites not only penetrated into the grease, but also grew the full distance under the grease and emerged into the side chamber if a high concentration of NGF was present, but they never appeared in the side chamber if no NGF was added. In this way, he was able to control the local environment of the distal part of the neurites and that of the somas separately. Interestingly, the method that was used to separate cell soma and distal part of neurites was relatively simple. It was based upon the fact that neurites squeezed themselves between the Teflon and the Petri dish surface through the grease (a gap in the micrometer range). Here, the microscale range was exploited to create a mechanical constraint for neuronal processes without the use of any of today's standard microfabrication techniques such as photolithography. Nonetheless, these devices were difficult to assemble and hard to image as well as prone to leakage.

A refinement of this mechanical-constraint strategy has been achieved by several following studies.[28,29] As an example, Taylor *et al.* fabricated a microfluidic culture platform consisting of a poly(dimethylsiloxane) (PDMS) mold placed against a coverslip (Figure 4.5). The design of the device incorporates two culture chambers that are separated by a physical barrier with embedded microgrooves. Those microgrooves are 10 µm wide and 3 µm

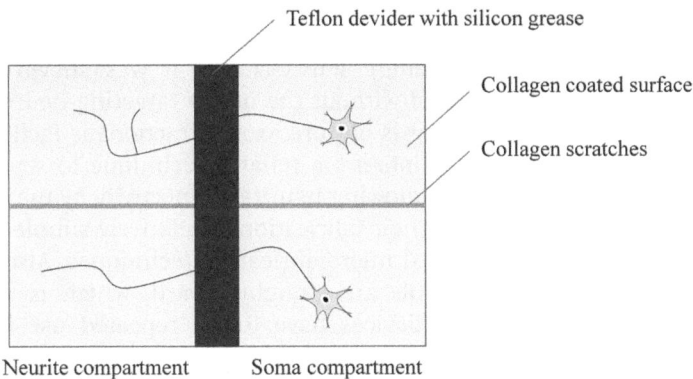

Figure 4.4   Schematic of Campenot chamber. Neurons are cultured in one compartment and neurites can grow through the grease under the Teflon divider into the second chamber allowing the separation of cell soma and neuronal processes. Hence, neurites and soma can be separately exposed to molecules. Scratches on the collagen-coated surface were made to hinder neurite outgrowth between neurons within the same compartment.

(a)

(c)

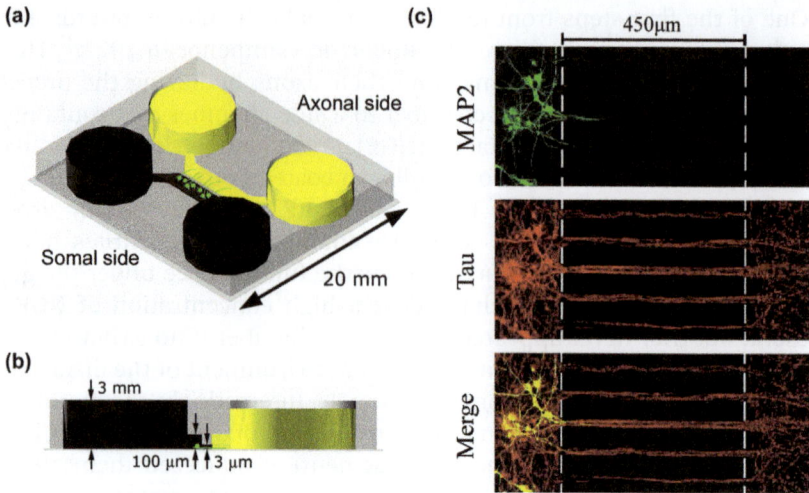

Figure 4.5 Microfluidic-based cell culture platform according to ref. 29. a) Perspective view of the PDMS culture chamber containing the relief pattern of the soma and axon compartment. Both are connected through microgrooves (10 μm wide and 3 μm high). b) Side view of the PDMS device. A volume difference between the reservoirs on each site allows chemical environments to be isolated for 20 h due to the high resistance of the microgrooves. c) Axons grew from one chamber to the other one (red: Tau). Dendrites on the other hand did not pass the microgrooves (green: MAP2). (Reprinted with permission from Macmillan Publishers Ltd, *Nature Methods*, copyright 2005 (ref. 29).)

high and allowed the neuritic processes, but not the cell bodies, to extent from one chamber to the other. Compared to its predecessor, the major advantages of this device such as reliability, reproducibility, and precision are due to its standardized fabrication process (photolithography, soft-lithography, replica molding). Furthermore, it was shown that the axonal growth can be polarized without the use of targeting neurotrophins and that the biochemical analysis of pure axonal fractions is facilitated.

Overall, mechanical constraints are a reliable technique to separate cell bodies from each other while allowing neurite connectivity by the means of micro-channels for example. Their fabrication is relatively simple and consistent, especially with standard microfabrication techniques. Also, the cell loading and culturing protocols are straightforward, which is why such microfabricated microfluidic devices have found repeated use in neuroscientific studies.[28–34]

A different technique towards reestablishing an ordered arrangement of neuronal cells *in vitro* can be achieved through surface patterning. Commonly used procedures include a combination of surface chemistry and photolithographic techniques[35] or soft-lithography. Small organic molecules bind to a surface such as glass or silicon, whose presence promotes (laminin, fibronectin) or inhibits (alkane,[35] untreated silicon[36] or glass,[37] albumine) neuronal cell attachment. Molecules that promote cell attachment generally have amino acid fragments that interact with specific cell membrane receptors such as integrins. Receptor binding to these sequences can initiate second messenger systems within the cell that promote growth, gene expression, and differentiation.[38] Adhesion promoters or inhibitors can be immobilized through simple adsorption. However, the coatings are easily washed off. To avoid this problem, covalent binding may be employed and can be virtually achieved for any molecule. For example, molecules that cannot be formed into silane derivative, such as the glycoproteins fibronectin[39] or laminin,[40] can be patterned by first binding an amine derivative in the desired pattern and then binding the glycoprotein to the amine with chemical cross-linking reagents.[35] Surface patterning with photolithography, for example, is attained with a photoresist (*e.g.* lift-off resist) that protects specific regions from being surface treated. The desired patterning structure on the surface is obtained after the removal of the photoresist (Figure 4.6a).

Surface patterning through soft-lithography, on the other hand, is a set of techniques that uses elastomeric stamps made of PDMS with patterned relief features in order to immobilize the adhesion molecules onto a surface (Figure 4.6b).[41,42] Early studies demonstrated the applicability of surface patterning techniques to the patterning of neuronal cell cultures[42–45], however, both the viability and the pattern of outgrowth were not maintained long enough for developing neurons that attain electrical excitability. Later studies showed extended maintenance of viability as well as higher resolution patterns with cells being maintained on lines as fine as 5 μm[35]

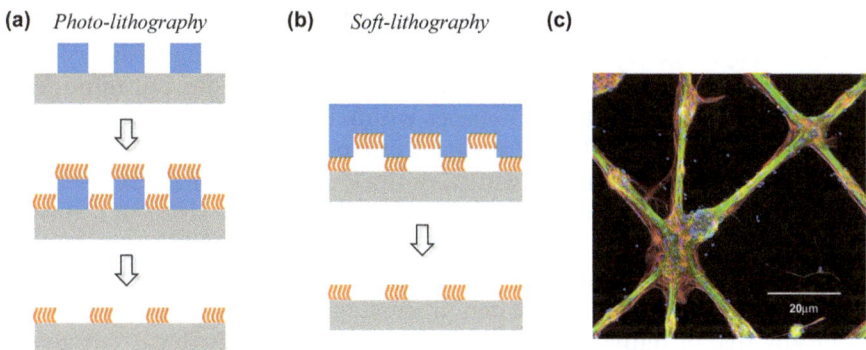

(a) *Photo-lithography*   **(b)** *Soft-lithography*   **(c)**

Figure 4.6   Monolayer patterning through surface functionalization. a) Schematic of photolithography-based patterning. Regions protected by the photoresist retain their initial surface when being exposed to the functionalization molecules and subsequently stripped. b) Soft-lithography-based patterning. An elastomeric stamp with patterned relief features is used to print molecules on a surface. First, the stamp is dipped into a solution containing the functionalization molecules and then it is brought into contact with the surface, thus leaving molecules only on the contact area. c) Patterned growth of neurons on a patterned substrate (reprinted with permission from E. Ujhelyi).

(see Figure 4.6c). More recent studies generated more precise micro- and even nanoscale patterns.[46] Overall, surface patterning allows the separation of cell bodies and hinders cell migration outside the surface patterning area (which results in a preserved ordered culture). On the other hand, localized chemical treatments of cell bodies and axons separately, for example, is not possible and would require additional mechanical constraints (*e.g.* for the formation of chemical gradients).

A more recent technique to create patterned neuronal cell cultures is by using hydrogels. Samples of hydrogels are first premixed with different cell populations and are then polymerized in layers that are located in direct contact to each other. Hence cells from different layers can interact through the hydrogel by the diffusion of cellular factors, by neurite outgrowth from one hydrogel layer to the other, or by cell migration. Hydrogel layers can be stacked on top of each other[47] or horizontally side-by-side.[48] As an example, Kunze *et al.* exploited the laminar flow at microscale to pattern up to four agarose/alginate hydrogel layers (Figure 4.7a). The device was comprised of four micro-channels that were connected to a common main channel. The laminar flow pattern of the hydrogel mixtures resulted in parallel hydrogel layers flowing side-by-side in the main channel (Figure 4.7b, left). After having thermally gelled the agarose-alginate hydrogels, their parallel pattern within the main channel was maintained (Figure 4.7b, right). The different hydrogel layers were premixed or not with primary cortical neurons and cultured for up to three weeks. Within that time, cell populations from different layers connected through the outgrowth of neuritic processes (Figure 4.7c). Furthermore, pre-mixing specific hydrogel layers with B27 and the subsequent establishment of a timely limited gradient allowed the enhancement of neurite outgrowth into a specific direction.

Overall, hydrogel layers are a very elegant way of patterning cells in 3D (as will be discussed in more detail in Section 4.3.2.3) and thus of mimicking more closely the native state of neurons within the brain. For this, patterning hydrogels horizontally allows for easier microscopic observation compared to vertically stacked hydrogel with very thick samples. On the other hand, cell loading protocols are more complicated and delicate since all hydrogel-cell mixtures need to be loaded at the same time while establishing more or less identical flows in all micro-channels; this technique is still suffering from reproducibility. An automated processing with a cell loader-device that is mostly free of human interaction would potentially enable a very reproducible fabrication of such three-dimensional patterned neuronal cell cultures.

4.3.2.2   *Directionality of Neurite Outgrowth*

It has been shown that relatively simple microfabrication tools can be used to spatially isolate the soma of a neuron from its axon. It has furthermore been shown that different cell populations can be separated from each other while maintaining their connectivity. In the brain, however, different

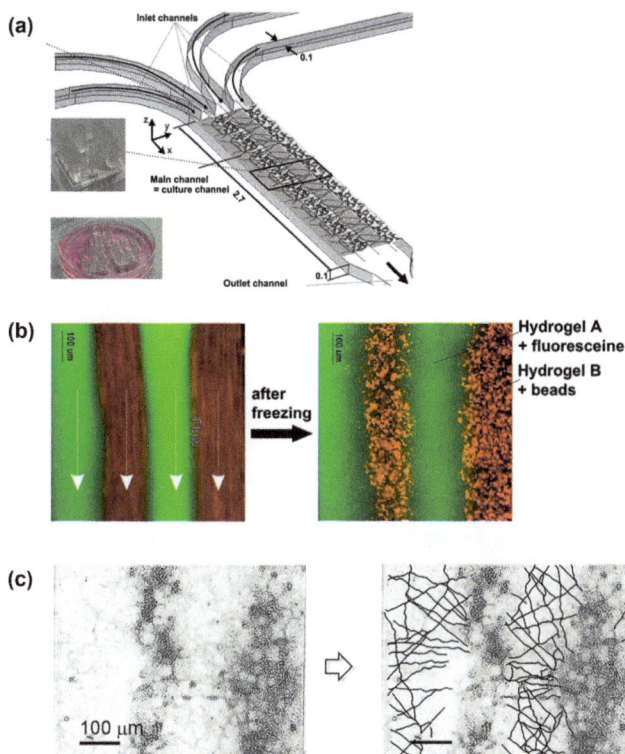

Figure 4.7 Patterned 3D neuronal cell culture in hydrogels. a) Illustration of the layered structure of the 3D neural cell culture. The device is composed of four inlet channels that lead into a main channel with one outlet channel. The two inserts show the final microfluidic device fabricated using PDMS and three devices placed in a Petri dish for incubation during cell culture. b) Fluorescence images taken in the main channel area present the four laminar hydrogel microstructures before (left) and after (right) gelling of the agarose components by cooling down under 26 °C. c) DIC images of the main channel at 0 (left) and 5 DIV (right). Single neurites were traced with NeuronJ.
(Reprinted from ref. 48, Copyright (2010), with permission from Elsevier.)

neuronal populations are connected with well-defined polarities,[26] a crucial condition of neuronal networks to guarantee the proper information flow. Hence, the *in vitro* study of polarized neurite outgrowth is essential for the investigation of cellular mechanisms that influence axonal plasticity and response to injury as well as etiology of neurodegenerative diseases. While the study of these phenomena with traditional technologies on a conventional scale is very difficult, axon growth guidance based on microfluidics creates new opportunities for research.[49]

Peyrin *et al.* established a microfluidic platform comprising independent cell culture chambers that are connected by an array of asymmetrical microchannels, so-called ''axonal diodes'' (Figure 4.8a and b). The asymmetrical

(a)

(b)

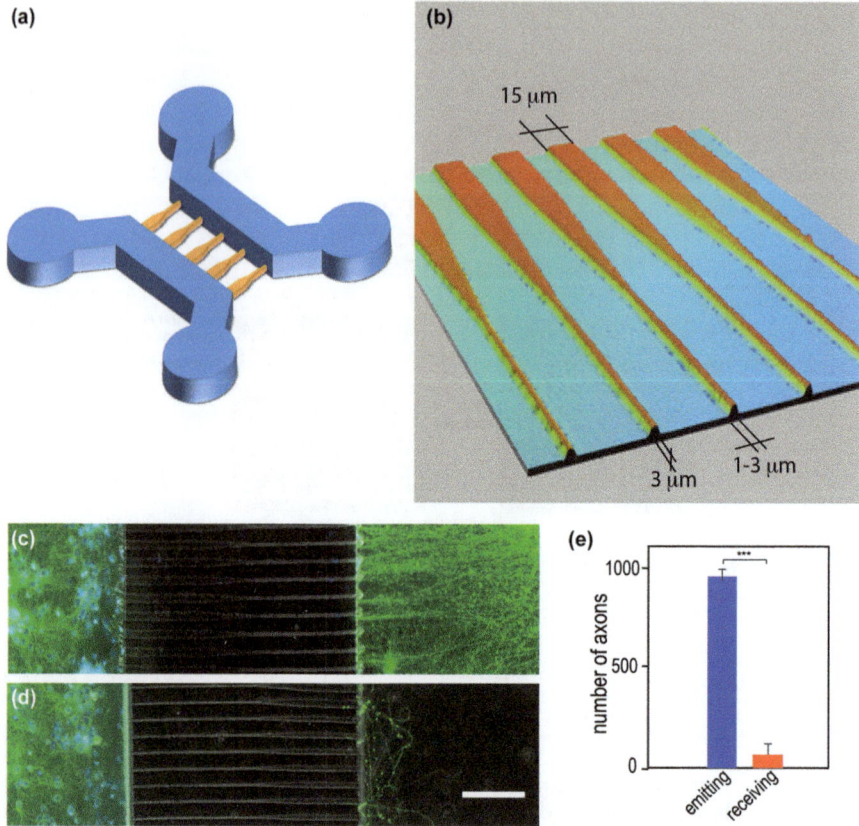

Figure 4.8 Orienting neurite outgrowth with physical constraints. a) 3D view of the microfluidic device used by Peyrin *et al.*[26] b) 3D view of the funnel-shaped micro-channels as observed by white-light optical profiling. c) Immunofluorescent images of microfluidic cultures (green: α-tubulin, blue: Hoechst) in which cortical neurons were seeded either on the c) wide (15 μm) or the d) narrow (3 μm) side (bar = 50 μm). e) Neurite outgrowth to opposite chamber is much higher when seeded on the wide side, bar = 50 μm (reproduced from ref. 26).

geometry acts as a directional selective filter for axons due to two effects. First, asymmetry imposes directionality by a probabilistic effect easing the entrance of axons on the wide side of the diode compared to the narrow side. Second, axons from neurons, seeded in the receiving chamber, once encountering the diode wall instead of the micro-channels entrance tend to grow perpendicularly to the diode along the chamber wall, without entering to a great extent the narrow side of the diodes, an effect that was correlated to the axonal stiffness that prevents frequent axonal turning. Accordingly, a cortico-striatal oriented network was established, cultured for three weeks, and proven active through structural (immunostaining of synaptic contacts) and functional analysis (KCl depolarization and electrical recording). The unidirectional axon connectivity was shown to have a very high selectivity of 97% (Figure 4.8e).[26]

Another very frequently used method for directing neurite outgrowth is by using axon guidance molecules (growth factors, neurotransmitters, netrins, adhesion molecules) or repellents (semaphorins, slits). For this, neuronal cells need to be exposed to biomolecule gradients that are controllable and that mimic those that are present *in vivo*.[50] Biomolecule gradients can be established either by substrate-bound molecules[51] or by a constant supply and removal of molecules at precise locations.[52] Kunze *et al.* reported the use of a microfluidic device to expose a patterned three-dimensional culture of primary neurons to gradients of B27 and NGF. The hydrogel-based patterned 3D culture region was separated from two lateral perfusion channels (that acted as a "source" and "sink") by small micro-channels that delivered the biomolecules on one side of the hydrogel culture and removed them on the other side (Figure 4.9a). By this means, it was shown that cortical neurons responded only to synergistic NGF-B27 gradients and that the synaptic density increased proportionally to these gradients (Figure 4.9b).

Unlike physical constraints as presented by Peyrin *et al.*, directionality obtained through biomolecular gradients enables the creation of directed neurite networks through three-dimensional neuronal cultures. On the other hand, neurite guidance by "axonal diodes" attracts specific attention owing to its ease of fabrication and use as well as to its high efficiency.

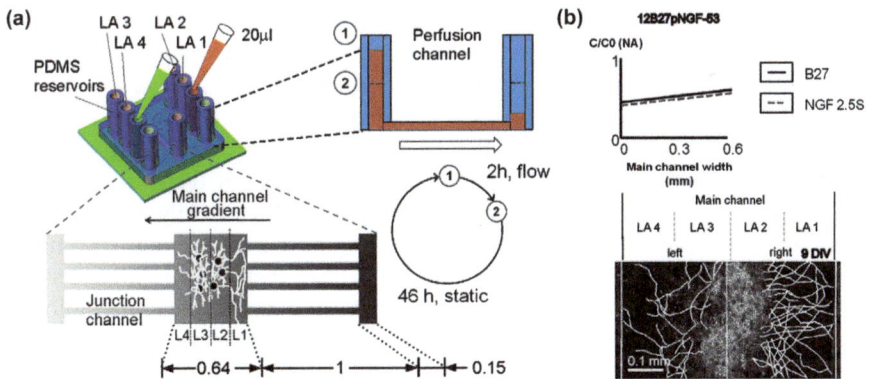

Figure 4.9　Orienting neurite outgrowth using guidance molecules. a) Illustration of experimental setup. Empty polydimethylsiloxane (PDMS) reservoirs are selectively filled with medium (red: enriched NGF/B27 medium, green: pure medium). A 2 h perfusion flow due to liquid height differences in the reservoirs establishes a linear gradient of NGF/B27. The long perfusion channels maintain the gradient in the main channel once the perfusion flow stops. Every other day, refilling was repeated. b) Schematic view of synergistic gradient of B27 and NGF and DIC image of neuronal culture with traced neurites at 9DIV, bar = 0.1 mm (reproduced from ref. 52).

No complicated setup with a pumping system for long-term maintenance of biomolecule gradients is necessary. Moreover, directional efficiency does not depend on the neuronal type that is used since different neuronal cell types may have different responses to various biomolecules (receptor-dependence). Also, the selectivity of physical constraints seems to be higher than the one obtained by the means of biomolecular cues, apart from the fact that an additional culturing dimension renders high selectivity-efficiency more difficult to achieve. Finally, in view of the study of higher order networks, *e.g.* to study inter-neuronal transport of pathological Tau species from one neuron to another that is located multiple neuronal connections further away, the axonal diode-technique seems to be a highly promising approach.

4.3.2.3   Three-dimensional Neuronal Culture

In vivo tissues have a very complex but well-organized 3D structure. Culturing cells in 3D provides another dimension for external mechanical inputs and for cell adhesion, which dramatically affects integrin ligation, cell contraction, and associated intracellular signaling.[53-55] Moreover, the diffusion of soluble factors changes dramatically compared to 2D cultures. Hence, culturing cells in a 3D environment seems highly promising since the effect of mechanical and chemical stimuli on 3D neuronal cultures as well as its propagation from one population to the other can be studied.

Establishing patterned 3D cell cultures is a challenging task. It requires a set of biomaterials, scaffold, and devices that can support the formation and maintenance of 3D tissue structures and that can be used for specific applications. As already described in Section 4.3.2.1, Kunze *et al.* showed that an agarose-alginate mixture can be gelled thermally (Figure 4.7). It is therefore an excellent candidate for forming horizontal multilayered scaffolds for micropatterning embedded cells. A further technique, named layer-by-layer deposition, enabled the creation of vertically stacked cell-matrix assemblies.[47] This method involved the stacking of cell-matrix assemblies (collagen, matrigel). The first layer was immobilized on a pretreated surface using a PDMS stamp; it then contracted and allowed the subsequent overlaying with another cell-matrix assembly. However, vertically stacked cell-hydrogel mixtures make optical observation difficult. Using confocal microscopy, for example, the thickness of the samples may exceed the working distance of the microscope. In addition, images of a specific focal plan observed under standard transmission microscope comprise a wide range of planes that are out of focus, resulting in lower image quality.

A hydrogel-free method to establish 3D neuronal networks was reported by Pautot *et al.*[56] This technique involves the culture on silica beads that provide a growth surface for neuronal cells (Figure 4.10). The use of beads allows moving them without disrupting the neuronal adhesion. They can be

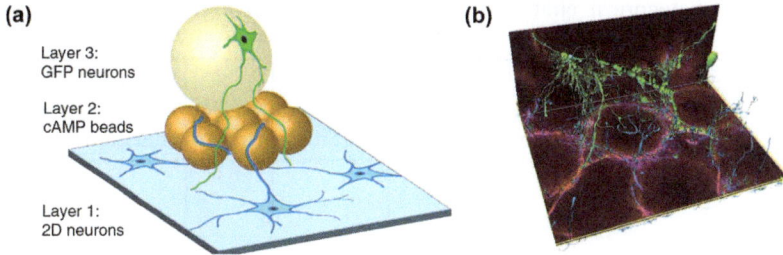

Figure 4.10   Neuronal culture on silica beads. (a) Schematic of the silica beads culture: a 2D culture of neurons was overlaid by cAMP (attractant signaling molecule) coated 45 mm beads and a third layer of beads containing GFP-expressing neurons. (b) A 3D reconstruction of axons (blue) from neurons on the coverslip (layer 1) growing up into the intermediate layer of cAMP beads (layer 2) and encountering there descending dendrites from bead layer (layer 3) of GFP-expressing cells (green).
(Reprinted by permission from Macmillan Publishers Ltd, *Nature Methods*, copyright 2008 (ref. 56).)

assembled to form 3D layered arrays containing distinct subsets of neurons in different layers with constrained connectivity between neurons on different beads. This system could prove very useful for future applications, especially because cell manipulation and culturing is simpler than cell-hydrogel mixture-based protocols. Their future integration into a microfluidic system where different beads (with different neuronal cells) can be patterned and used to build up an oriented neuronal network thanks to biomolecular gradients or physical constraints may allow further refining this technique and obtaining *in vivo*-like orientations of neuronal networks.

4.4   Micro-devices-based *in Vitro* Alzheimer Models

Microtechnology-based devices are the key to structuring neuronal cells *in vitro* such that *in vivo*-like neuronal networks can be reconstructed; *i.e.* in terms of connected but distinct cell layers, directionality, and three-dimensional structures. Although all these prerequisites for establishing an environment similar to the human brain have been achieved individually, no integrated device or technology that is fulfilling all conditions was reported yet (Table 4.1). However, avenues for preliminary studies of AD with simplified models are open already and should be pursued in order to obtain important insights into how AD is progressing in the brain and how it can be stopped.

4.4.1   First Microtechnology-based Experimental Models

A first microfluidic approach that allows for the study of axonal biology in AD including transport within neurons was carried out by Poon *et al.*[57] Transgenic mouse neurons that overexpressed Aβ (Tg2576) were cultured

Table 4.1 Categorical chart of prerequisites for studying disease progression in microfluidic devices and how recent studies fit into these categories.

Studies	Compartmentalization of cell somas				3D cell culture	Directionality	
	Surface patterning	Mechanical constraints	Gels	Cell migration		Mechanical constraints	Chemical gradients
Campenot 1977		a					+
Kleinfeld 1988	+						
Taylor 2005		+a					
Kunze 2010			+	+	+		
Pautot 2006		+		+	+		+
Peyrin 2011		+				+	
Ideal		+		−	+	+ (high efficiency)	

aOnly cell soma and axons were compartmentalized.

in a microfluidic device (similar to the one reported by ref. 29) where cell somas and axons were compartmentalized. Neuritic processes grew from the seeding chamber through >150 μm long microgrooves to another chamber that consequently allowed the selective application of a neurotrophin (brain-derived neurotrophic factor (BDNF)) to the axonal terminals. In healthy neurons, BDNF binds to its receptor tropomyosin-related kinase B (TrkB) at the axon terminal; the signal is propagated along the axon to the cell body, where it regulates gene expression and neuronal function. By locally applying BDNF to the axonal terminals, it was shown that the TrkB processing was impaired in transgenic Aβ neurons, but not in non-transgenic mouse neurons (WT). Also, the BDNF vesicle retrograde transport was slower in Tg2576 when compared to WT. Furthermore, it was reported that WT neurons that were exposed to Aβ oligomers prepared *in vitro* exhibited an impaired retrograde transport of axonally applied BDNF and demonstrated a three-fold reduction in the accumulation of somal BDNF. Altogether, these findings strongly supported the hypothesis that the presence of either intracellular overexpressed Aβ or extracellular oligomers of Aβ impaired BDNF retrograde transport. More importantly, this study essentially foretasted the potential that underlies studies of compartmentalized neuronal cells and their axons.

A further study on the impact of Aβ was carried out by Kim and Choi et al.,[58] where it was shown that Aβ induced rupture of axonal transport can be restored by manipulating the level of α-tubulin acetylation. Axons were separated from their soma by using the Taylor device. Accordingly, the authors were able to examine the mitochondrial movement velocity within axons thanks to their spatial isolation from cell somas in a precise manner.

This study demonstrates the utility of structured neuronal cultures for the investigation of transport kinetics within axons after local treatments of cell somas and can therefore provide essential clues for the discovery of potential drug targets.

Kunze and Meissner *et al.*[59] reported the culture of two spatially distinct, but interconnected cell populations where one of them was set into a Tau-pathological state. This model represents the first approach for studying disease progression from one cell population to another in a microfluidic device. For this, a microfluidic chip with three compartments was used: two lateral channels where primary cortical neurons were cultured and one main channel where the neuritic processes could interconnect (Figure 4.11a). Neurites started to grow out after 2 days *in vitro* and neurite density in the main channel saturated after 8 DIV (Figure 4.11b, top). Subsequently,

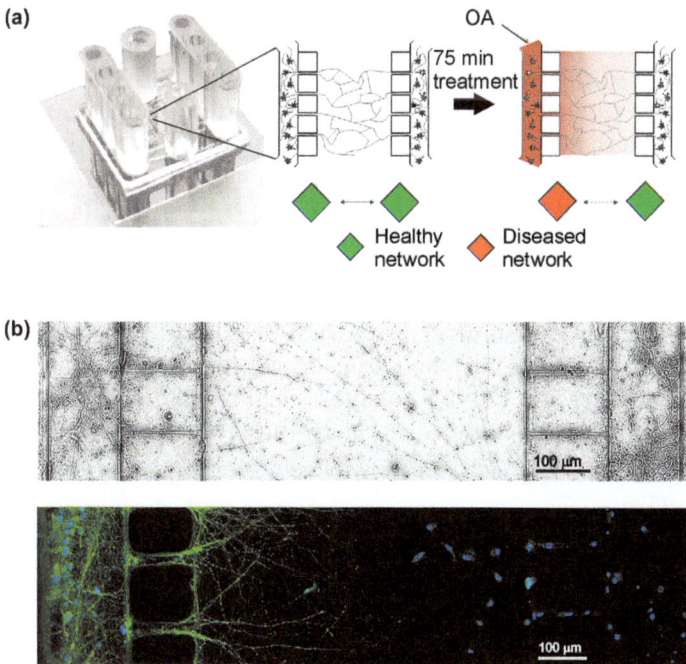

Figure 4.11   Generation of a co-pathological neural cell culture. a) Concept of establishing a co-pathological culture: two cell populations are spatially separated while being interconnected. Only one of them is treated with okadaic acid (OA), leading to a local hyperphosphorylation of Tau and thus a co-culture of "healthy" and "diseased" neurons. b) Top: DIC image of neuronal culture with two lateral cell soma compartments and a central neurite interconnection chamber. Bottom: Tau phosphorylation imbalance within the entire cell culture is visualized (green: Tau Ser262 – Cy2, blue: DAPI).
(Reprinted by permission from Wiley, *Biotechnology and Bioengineering*, copyright 2011 (ref. 59).)

okadaic acid (OA), a protein phosphatase 2A inhibitor, was perfused over one cell culture compartment leading to a linear gradient in the main channel within one minute and a quasi-zero concentration in the second cell compartment. The OA treatment of only one cell population resulted in Tau protein hyperphosphorylation of these cells and hence generated an imbalance of the Tau phosphorylation states between the two connected cell populations (Figure 4.11b, bottom). This controlled generation of two different phosphorylation states within one cell population was referred to as a "co-pathological state". The chemical gradient of okadaic acid induced a Tau phosphorylation gradient throughout the whole cell culture. This metabolic gradient was visualized through immunostaining of the Tau phosphorylation site Ser262. Okadaic acid-exposed cells displayed high intensity Tau aggregates within the cell soma and neurites whereas non-exposed cells exhibited a much less intense and homogeneous distribution of Tau. In conclusion, the established model included (1) gradient controlled Tau phosphorylation states and (2) connected diseased and healthy neuronal cell populations and thus illustrated the feasibility of reconstructing typical disease patterns *in vitro*. In addition, Kunze and Meissner *et al.* demonstrated the application of this concept to a 3D neuronal cell culture.[7] For this, neuronal cells were micropatterned as described in Section 4.3.2.1[48] before being exposed to an okadaic acid gradient. Altogether, this new strategy of artificially creating disease states of only precise parts of a neuronal tissue may be an important step towards a thorough understanding of how pathological states of Tau proteins propagate to some cells and not to others. No less important is the question of what effect potential drugs, that counteract the disease progression, have on non-pathological healthy neurons (potential side effects). This is an issue that is especially important in terms of prevention treatment.

4.4.2   Requirements of Future Micro-device-based Studies

Ideally, for reliable *in vitro* studies, experiments should be carried out on tissues that resemble entirely the one of the human brain, but this strategy is thwarted for three reasons. (1) The high complexity of the human brain's macro- and micro-structure makes it impossible to be artificially reproduced. In addition, reported models that strive for more sophistication are often difficult to fabricate and to use (complicated protocols) and they are lacking in reproducibility, an aspect that is especially important when the application is for drug discovery. (2) Primary cell sources are very limited, even unavailable for human origin. Most of the *in vitro* studies for AD use primary animal cells (*e.g.* from rats). These sources require costly and time-consuming brain tissue extraction protocols and they are controversial due to species differences between animals and humans. Immortal human cell lines (*e.g.* SHSY-5Y), on the other hand, are abundantly available, but they

(a) *Network functionality test*   **(b)** *Gentle molecule delivery*

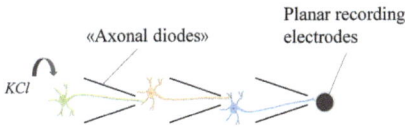

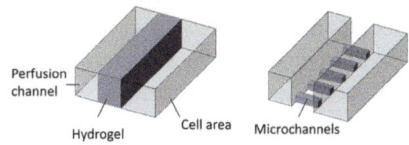

Figure 4.12   Example requirements for future microfluidic-based neuroscience applications. a) Functionality of an oriented neuronal network can be probed as a function of disease progression. b) Molecules can be delivered to the cells in a gentle way, either through a porous hydrogel wall or through small high-resistance micro-channels.

have undergone genotypic and phenotypic changes and are thus metabolically different from the original primary cells. (3) The observation of disease-related phenomena within the tissue is a big challenge in complex tissue samples (especially 3D cell cultures). In fact, visual characterization with fluorescence labeling and confocal microscopy for example is expensive and time consuming. Label-free methods such as impedance spectroscopy, on the other hand, may not provide the deepness of information that is necessary to understand molecular processes. In addition, neuronal signals upon chemical stimulation are difficult to record due to the larger distance between neuronal cells and planar recording electrodes.

Finally, the points mentioned above imply that compromises specific to certain applications may be the key to elucidating disease-related questions. Simple detection methods can be used at the expense of decreased tissue sophistication or *vice versa*. For example, reconstructing complex oriented networks with multiple cellular compartments combined with signal recording electrodes and simple microscopic characterization may be acceptable at the expense of three-dimensionality (Figure 4.12a). Also, the study of disease propagation in a 3D context under confocal microscopy analysis may be tolerable at the cost of decreased neurite orientation efficiency. Furthermore, cell lines may be turned into more neuron-like cells in terms of morphology and biochemistry of human mature neurons by treatment with differentiation promoting chemicals such as retinoic acid[60] and staurosporine.[61] Moreover, upcoming suppliers of stem cell- or induced pluripotent stem cell (iPS)-derived neurons that are reliable and low-priced may represent a further alternative to primary cells.

The possibility to analyze the biological specimen is a fundamental condition for studying disease propagation in microsystems. Representing the ideal case, *in situ* observation of specific phenomena within the neuronal network may not always be possible depending on the technique that is used. Although fluorescence microscopy and electrical signal recording techniques can be carried out exactly at their experimental location, other methods for protein or RNA quantification require the cells to be isolated from the microsystem. Western blot analysis, for example, may be used to address if pathological Tau or Aβ proteins are spread to downstream synaptically connected neurons. For this, cells from specific compartments or locations need to be removed selectively and efficiently. Following that, protein lysates are separated on a polyacrylamide gel and visualized on a nitrocellulose membrane with protein-specific antibodies. Due to a decreased number of cells in microfluidic channels in general and thus less total protein amount, it is necessary (1) to extract a maximum number of cells from the microsystem and (2) to guarantee a highly concentrated sample (thus avoiding dilution) in order to obtain a strong signal (coloration).

Moreover, local treatment of specific parts of the neuronal network requires the supply of molecules to each cell sub-population separately. The molecule delivery process needs to be carried out such that the treated cells are not disturbed and do not respond to unwanted stress (flow stress, vibration). For this, special attention is to be paid during pipetting. Alternatively, a molecule injection chamber may be spatially separated from the cell compartment by either high-resistance microgrooves or by a permeable hydrogel wall, both allowing the molecule penetration while hindering fluid flow (Figure 4.12b).

Altogether, complexity and sophistication of artificial neuronal tissues most often stands in contrast to the ease with which the sample is examined and also exposed to stimuli (*e.g.* molecule delivery). Therefore a compromise for each application specific to a biological question needs to be found. Similarly, metabolically relevant cell models are rare and costly (primary cells, stem cell-derived neuronal cells), while easily available cells lines may be physiologically different from their original cells. Again, the tolerable type of cells used in an experiment strongly depends on the biological question.

4.5 Questions that May Be Addressed by Micro-controlled Cultures

Microsystems have shown to be an important tool for studying cell interactions because they allow both an *in vivo*-like reconstitution of tissue structures and a more precise functional analysis of neuronal networks. The high potential that underlies microtechnology implies its indispensable use in future neuroscience research. Therefore an interdisciplinary exchange between neuroscientists and microtechnologists seems inevitable.

In fact, local treatment of specific sub-populations of cells within a larger neuronal network requires an accurate control of the cells microenvironment that is reliant on microfluidics. In addition, *in situ* and live observation of the sample under investigation necessitates experiments to be carried out on artificial tissues that are reconstituted on thin glass slides. Also, the analysis of electrical functioning of the entire neuronal network depends on the use of signal recording electrodes that are the same size as neuronal cells (*i.e.* in the micrometer range). Overall, microfabrication techniques can help to elucidate a large number of still open questions:

What factors and mechanisms are responsible for the formation of the first misfolded Tau seeds? This may be due to spontaneous triggers (mutations, transcriptional or translational errors, tissue injury) or exogenous triggers (exposure to preformed aggregates). Misfolded aggregates of an additional protein may promote Tau aggregation (cross-seeding).[62] To find an answer to this question, tests may involve the examination of Tau aggregation upon cell exposure to different protein aggregates in order to find out what conditions make healthy cells turn into pathological ones.

How do different stages of Tau aggregation (kinetics, extent) influence the communication between neuronal cells? Following studies may involve the examination of pathological protein transfer to neighboring cells as well as its mechanisms and kinetics. Further questions include how protein aggregates are transferred from one cell to another (secretion, damage of host cell – cell injury)[20,21] and what are the mechanisms that lead to the release of those pathological proteins (freely emerged Tau, sequestered within exosomes)? Which species of Tau is specifically responsible for the propagation of NFTs? Also, how are synaptic destruction and neurodegeneration related in time (dying back terminal degeneration, release of Tau is toxic)? Moreover, only some of the interconnected neurons develop NFTs; what conditions prevent the Tau proteins from being transferred? This issue can be investigated by using microsystems where pathologic protein aggregation may be achieved by locally applying exogenous factors or by compartmentalized co-culture of a variety of transgenic diseased cells with healthy cells.

The answers to these questions are of enormous therapeutic interest. Effective drug targets may be developed to counteract not only the symptoms, but the fatal consequences as close as possible at the origin of the pathology. In addition, knowledge about the reason behind individual-dependent progression rates may give rise to personalized treatments (type of medication, doses). Altogether, there is a gap of understanding that may be filled by the means of microtechnology, a tool that has hardly been exploited in AD research with regard to the various studies that can and have to be carried out in order to make headway against AD.

References

1. Alzheimer's Association, *2012 Alzheimer's Disease Facts and Figures*, 2012.
2. A. Wimo and M. Prince, *Alzheimer's Disease International*, 2010.
3. S. L. Murphy, J. Xu and K. D. Kochanek, *National Vital Statistics Reports*, 2012, **60**, 1–51.

4. Alzheimer s Association, *Alzheimer's Disease Facts and Figures*, 2010.
5. J. Hardy and D. J. Selkoe, *Science (New York, NY)*, 2002, **297**, 353–356.
6. D. Geschwind, *Neuron*, 2003, **40**, 457–460.
7. A. Kunze, R. Meissner, S. Brando and P. Renaud, *15th International Conference on Miniaturized Systems for Chemistry and Life Sciences*, 2011, **1**, 696–698.
8. C. Priller, T. Bauer, G. Mitteregger, B. Krebs, H. A. Kretzschmar and J. Herms, *J. Neurosci.*, 2006, **26**, 7212–7221.
9. J. Bulinski and G. Gundersen, *Bioessays*, 2005, **13**, 285–293.
10. L. M. Ittner and J. Götz, *Nat. Rev. Neurosci.*, 2011, **12**, 65–72.
11. H. Braak and E. Braak, *Brain Pathol.*, 1991, **1**, 213–216.
12. B. T. Hyman, G. W. Van Hoesen, A. R. Damasio and C. L. Barnes, *Science*, 1984, **225**, 1168–1170.
13. T. Gomez Isla and H. West, *Ann. Neurol.*, 1996, **39**, 62–70.
14. B. Hyman and G. Van Hoesen, *Neurobiol. Aging*, 1987, **8**, 555–556.
15. A. de Calignon, M. Polydoro, M. Suárez-Calvet, C. William, D. H. Adamowicz, K. J. Kopeikina, R. Pitstick, N. Sahara, K. H. Ashe, G. A. Carlson, T. L. Spires-Jones and B. T. Hyman, *Neuron*, 2012, **73**, 685–697.
16. C. Soto, *Neuron*, 2012, **73**, 621–623.
17. M. Jucker and L. C. Walker, *Ann. Neurol.*, 2011, **70**, 532–540.
18. F. Clavaguera and T. Bolmont, *Nat. Cell Biol.*, 2009, **11**, 909–913.
19. B. Frost, R. Jacks and M. Diamond, *J. Biol. Chem.*, 2009, **284**, 12845–12852.
20. J. L. Guo and V. M.-Y. Lee, *J. Biol. Chem.*, 2011, **286**, 15317–15331.
21. T. Nonaka, S. T. Watanabe, T. Iwatsubo and M. Hasegawa, *J. Biol. Chem.*, 2010, **285**, 34885–34898.
22. S. Nath, L. Agholme, F. R. Kurudenkandy, B. Granseth, J. Marcusson and M. Hallbeck, *J. Neurosci.*, 2012, **32**, 8767–8777.
23. J. A. Harris, N. Devidze, L. Verret, K. Ho, B. Halabisky, M. T. Thwin, D. Kim, P. Hamto, I. Lo, G. Yu, J. J. Palop, E. Masliah and L. Mucke, *Neuron*, 2010, **68**, 428–441.
24. K. Santacruz, J. Lewis, T. Spires and J. Paulson, *Science*, 2005, **309**, 476–481.
25. J. Stiles and T. L. Jernigan, *Neuropsychol. Rev.*, 2010, **20**, 327–348.
26. J.-M. Peyrin, B. Deleglise, L. Saias, M. Vignes, P. Gougis, S. Magnifico, S. Betuing, M. Pietri, J. Caboche, P. Vanhoutte, J.-L. Viovy and B. Brugg, *Lab on a Chip*, 2011, **11**, 3663–3673.
27. R. Campenot, *Proc. Natl Acad. Sci. USA*, 1977, **74**, 4516–4519.
28. J. Park, H. Koito, J. Li and A. Han, *Biomedical Microdevices*, 2009, **11**, 1145–1153.
29. A. Taylor, M. Blurton-Jones and S. Rhee, *Nat. Methods*, 2005, **2**, 599–605.
30. I. H. Yang, R. Siddique, S. Hosmane, N. Thakor and A. Höke, *Exp. Neurol.*, 2009, **218**, 124–128.

31. H. Hayashi, R. B. Campenot, D. E. Vance and J. E. Vance, *J. Biol. Chem.*, 2004, **279**, 14009–14015.
32. J. Bertrand, M. J. Winton, N. Rodriguez-Hernandez, R. B. Campenot and L. McKerracher, *J. Neurosci.*, 2005, **25**, 1113–1121.
33. T. Ishibashi, K. A. Dakin, B. Stevens, P. R. Lee, S. V Kozlov, C. L. Stewart and R. D. Fields, *Neuron*, 2006, **49**, 823–832.
34. R. Campenot, *Dev. Biol.*, 1982, **93**, 13–21.
35. D. Kleinfeld, *J. Neurosci.*, 1988, **8**, 4098–4120.
36. M. Grattarola, M. Tedesco, A. Cambiaso, G. Perlo, G. Giannetti and A. Sanguineti, *Biomaterials*, 1988, **9**, 101–106.
37. S. Schacher and E. Proshansky, *J. Neurosci.*, 1983, **3**, 2403–2413.
38. J. P. Ranieri, R. Bellamkonda, E. J. Bekos, J. A. Gardella, H. J. Mathieu, L. Ruiz and P. Aebischer, *Int. J. Dev. Neurosci.*, 1994, **12**, 725–735.
39. H. Kleinman, R. Klebe and G. Martin, *J. Cell Biol.*, 1981, **88**, 473–485.
40. R. Timpl, H. Rohde and P. Robey, *J. Biol. Chem.*, 1979, **254**, 9933–9937.
41. H. Kaji, G. Camci-Unal, R. Langer and A. Khademhosseini, *Biochim. Biophys. Acta*, 2011, **1810**, 239–250.
42. D. Heller, V. Garga, K. Kelleher and T. Lee, *Biomaterials*, 2005, **26**, 883889.
43. O. Ivanova and L. Margolis, *Nature*, 1973, **242**, 200–201.
44. P. Letourneau, *Dev. Biol.*, 1975, **44**, 92–101.
45. A. Cooper, H. Munden and G. Brown, *Exp. Cell Res.*, 1976, **103**, 435–439.
46. R. S. Kane, S. Takayama, E. Ostuni, D. E. Ingber and G. M. Whitesides, *Biomaterials*, 1999, **20**, 2363–2376.
47. W. Tan and T. A. Desai, *Biomaterials*, 2004, **25**, 1355–1364.
48. A. Kunze, M. Giugliano, A. Valero and P. Renaud, *Biomaterials*, 2011, **32**, 2088–2098.
49. J. Wang, L. Ren, L. Li, W. Liu, J. Zhou, W. Yu, D. Tong and S. Chen, *Lab on a Chip*, 2009, **9**, 644–652.
50. T. M. Keenan and A. Folch, *Lab Chip*, 2008, **8**, 34–57.
51. N. L. Jeon and H. Baskaran, *Nat. Biotechnol.*, 2002, **20**, 826–830.
52. A. Kunze, A. Valero, D. Zosso and P. Renaud, *PloS One*, 2011, **6**, e26187.
53. L. G. Griffith and M. A. Swartz, *Nat. Rev. Mol. Cell Biol.*, 2006, 7, 211–224.
54. B. Knight, C. Laukaitis, N. Akhtar and N. Hotchin, *Curr. Biol.*, 2000, **10**, 576–585.
55. C. Roskelley, *Proc. Natl Acad. Sci. USA*, 1994, **91**, 12378–12382.
56. S. Pautot, C. Wyart and E. Isacoff, *Nat. Methods*, 2008, **5**, 735–740.
57. W. W. Poon, M. Blurton-Jones, C. H. Tu, L. M. Feinberg, M. A. Chabrier, J. W. Harris, N. L. Jeon and C. W. Cotman, *Neurobiol. Aging*, 2011, **32**, 821–833.

58. C. Kim, H. Choi, E. S. Jung, W. Lee, S. Oh, N. L. Jeon and I. Mook-Jung, *PloS One*, 2012, 7, e42983.
59. A. Kunze, R. Meissner, S. Brando and P. Renaud, *Biotechnol. Bioeng.*, 2011, **108**, 2241–2245.
60. E. Jones-Villeneuve, *Mol. Cell. Biol.*, 1983, **3**, 2271–2279.
61. T. Shea and M. Beermann, *Cell Biol. Int. Rep.*, 1991, **15**, 161–168.
62. R. Morales, K. Green and C. Soto, *CNS Neurol. Disord. Drug Targets*, 2009, **8**, 363–371.

CHAPTER 5

Microbubbles for Medical Applications

TIM SEGERS,* NICO DE JONG, DETLEF LOHSE AND
MICHEL VERSLUIS

Physics of Fluids Group, MIRA Institute for Biomedical Technology and
Technical Medicine, MESA+ Institute for Nanotechnology, University of
Twente, PO Box 217, 7500 AE Enschede, The Netherlands
*Email: t.j.segers@utwente.nl

5.1 Introduction

Ultrasound is the most widely used medical imaging modality. It offers low-risk and portable imaging, it produces real time images, which can be taken at bedside, and it is inexpensive as compared to computed tomography (CT) and magnetic resonance imaging (MRI). The working principle is based on the scattering and reflection of transmitted ultrasound waves at interfaces and tissue inhomogeneities with different acoustic impedances. An ultrasound image can be constructed from the time of flight and the intensity of the received echoes. Typical ultrasound frequencies used are in the range of 1 to 50 MHz, where the highest frequencies result in the highest spatial resolution due to its shorter wavelength. For higher frequencies, however, the penetration depth is reduced as a result of attenuation, which increases linearly with frequency. Therefore the optimum imaging frequency is always a trade-off between resolution and imaging depth.

Tissues in the body contain acoustical inhomogeneities, which scatter ultrasound and light up in ultrasound images. On the other hand, blood is a poor ultrasound scatterer and the visibility of the blood pool can be

RSC Nanoscience & Nanotechnology No. 36
Microfluidics for Medical Applications
Edited by Albert van den Berg and Loes Segerink
© The Royal Society of Chemistry 2015
Published by the Royal Society of Chemistry, www.rsc.org

enhanced by the use of stabilized microbubbles as an ultrasound contrast agent (UCA). Ultrasound contrast agents consist of a suspension of stabilized microbubbles with a size between 0.5 and 10 μm in diameter, so they can safely pass even the smallest vascular beds. Bubbles are highly echogenic due to the acoustic impedance mismatch with their liquid surrounding and due to the resonance behavior of the bubble vibrations. The interaction of bubbles and ultrasound and the influence of the stabilizing shell is the subject of extensive study, summarized in this chapter, and our improved insight into the underlying physical mechanisms gives rise to increased performance of ultrasound contrast agents and their use in medical imaging and therapy.

This chapter will give an overview of the applications and design considerations of microbubbles for medical applications. First, the use of microbubbles for imaging and therapeutic purposes is introduced. The physical characteristics of microbubbles in a sound field are then described in order to understand the processes by which microbubbles enhance ultrasound imaging contrast. The third section explains that microbubbles may quickly dissolve in the surrounding liquid and that they need to be stabilized by a coating. The standard contrast agent production methods are given in the fourth section as well as more recent methods to produce monodisperse bubble suspensions using lab-on-a-chip devices. Finally, the dynamics of coated bubbles is discussed and methods to characterize the dynamics are presented.

5.1.1   Microbubbles for Imaging

The mechanism by which the microbubbles enhance the contrast is twofold. First, acoustic waves are scattered by the bubbles due to the large difference in acoustic impedance between the gas and the surrounding liquid. Second, the large compressibility of the gas bubbles results in radial oscillations of the bubbles in response to the ultrasound pressure waves. The radial oscillations are highly non-linear and produce non-linear sound waves, which are exploited in contrast-enhanced ultrasound imaging. The strong scattering of the microbubbles allows for the visualization and quantification of blood perfusion in organs, *e.g.* heart, liver, or kidney.[1] Figure 5.1 shows an ultrasound image of a rabbit kidney before (a) and after (b) arrival of a bolus injection of an ultrasound contrast agent.

The sensitivity of the bubble detection, down to a single bubble *in vivo*, facilitates targeted molecular imaging applications for the diagnosis of disease at the molecular level. Targeting ligands that bind specifically to selective biomarkers on the blood vessel wall can be labeled to the microbubble shell,[2] see Figure 5.2. The approach here is to inject targeted bubbles intravenously, then to wait 5 to 10 minutes for the freely flowing bubbles to be washed out by the lungs and the liver and then to image the adherent bubbles using ultrasound. One other approach is to discriminate acoustically between freely flowing and adherent bubbles through spectral

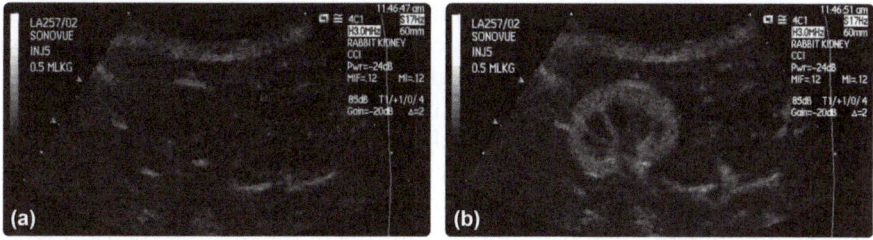

Figure 5.1 Ultrasound contrast microbubble imaging of the kidney.
Ultrasound echo image of a rabbit kidney a) before arrival of the ultrasound contrast agent and b) contrast enhancement due to presence of the ultrasound contrast agent.[83]

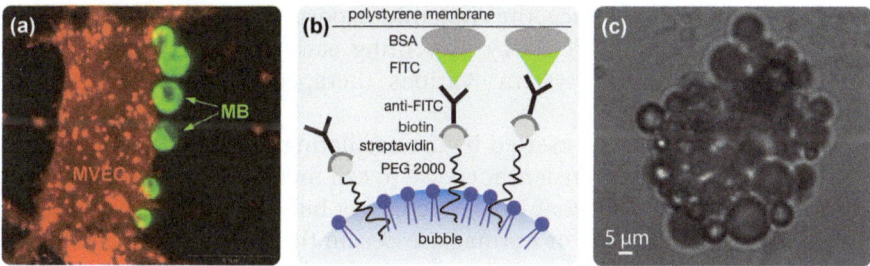

Figure 5.2 a) Fluorescent microscopy image of microbubbles (MB) targeted to a microvascular endothelial cell (MVEC).[1] b) Microbubbles can be targeted to a membrane in an experimental setting by labeling the bubbles with a PEG2000 spacer connected to streptavidin. A strong biotin-streptavidin bond between the bubble and the membrane is utilized when the membrane is labeled with BSA connected to FITC, anti-FITC, and biotin.[4] c) A complex of microbubbles targeted to a stem cell.[5] These complexes are echogenic and they can therefore be directed using acoustic radiation forces towards diseased tissue to up-regulate the therapeutic targeting efficiency.

differences through a resonance shift of the adherent bubbles due to the interaction between the bubble and the vessel wall.[3,4] This approach would require that all bubbles have the same response to the ultrasound driving signal, which is up to now not possible because of the large size distribution of the commercial agents. Recently microbubbles were targeted to stem cells[5] to produce echogenic complexes, which can be directed towards diseased tissue using acoustic radiation forces to up-regulate the therapeutic targeting efficiency (Figure 5.2c).

5.1.2 Microbubbles for Therapy

Contrast agent microbubbles can themselves serve as therapeutic agents. Several configurations are possible here. First, a drug can be co-administered with the bubbles while ultrasound-induced bubble oscillations close to cells promote local drug uptake through mechanical stress to the adjacent cell membranes, a process called sonoporation. Second, UCAs can be loaded with drugs, *e.g.* for the local delivery of chemotherapeutic drugs[1,2] with a narrow therapeutic index, or for the delivery of genes, such as siRNAs.[6–8] The payload can be directly incorporated in the bubble coating,[9] or in the core of the bubble,[10,11] or in liposomes attached to the bubble shell.[12] Drug-loaded microbubbles can be imaged at low acoustic pressures aiding the guidance and real-time monitoring of the therapy, then at higher acoustic pressures rupture of the bubble shell triggers drug release. Ideally, the release of such a payload follows a step response after passing an insonation pressure threshold for the controlled release of the payload in a confined and highly localized region determined by the position of the acoustic focus and the position of the bubbles. The prime focus of such a local delivery is to reduce the systemic exposure to toxic drugs and to increase the delivery efficacy by preventing early capture of the drugs or genes by the systemic system. Various therapeutic bubble systems are displayed in Figure 5.3.

The mechanical stress exerted by the oscillating microbubbles leading to sonoporation, *i.e.* the transient increase in cell membrane permeability, can be caused by several mechanisms. Oscillating bubbles in close proximity to cells will cause palpation or normal stresses on the cell membranes through the attractive secondary radiation forces between the bubbles and the cells.[13,14] Furthermore, asymmetric bubble oscillation and oscillatory translations may lead to the build-up of acoustic streaming around the bubble,[15] which will induce a shear stress on the cell membrane. Moreover, acoustic streaming will enhance the influx of fresh therapeutic agent through local mixing. Driving the bubbles at higher acoustic pressures can lead to an asymmetric collapse of the bubble and the formation of a liquid jet, a violent process called inertial cavitation.[16–18] The liquid jet can reach a speed ranging from 10–100 m s^{-1},[19,20] resulting in pore formation in the cell membrane. Typically, short ultrasound pulses are used to avoid irreversible membrane poration, leading to cell death. The interaction between oscillating bubbles and the blood-brain barrier (BBB) has shown the ability to transiently disrupt the tight junctional complexes, which normally prevent the entry of therapeutic drugs into the brain.[21–23]

5.1.3   Microbubbles for Cleaning

High-intensity focused ultrasound, or HIFU, utilizes long pulse ultrasound to create thermal ablation for deep tissue surgery and is promoted by the use of microbubbles.[24–26] In extracorporeal shockwave lithotripsy, cavitating clouds of microbubbles created in the focus of a shockwave lead to mechanical ablation to break kidney stones.[27,28] Each year many people suffer from a stroke. Ischemic stroke results from an occlusion of a major blood vessel in the brain by a blood clot, which decreases or suppresses blood

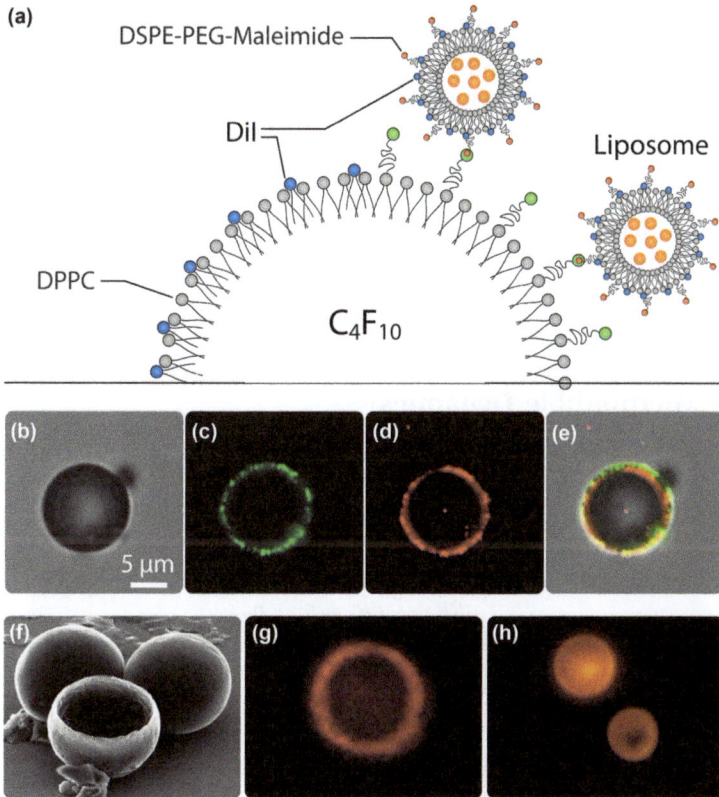

Figure 5.3 a) Schematic representation of a typical drug-loaded fluorescently labeled phospholipid coated microbubble.[73] The drug is loaded in liposomes attached to the bubble shell. b) A brightfield image of such a drug-loaded microbubble[73] and fluorescent images of c) bodipy-labeled liposomes and d) the DiI-labeled phospholipid shell. e) An overlay of the images shows that the loaded liposomes are situated outside the bubble shell. f) A SEM image of a hard-shelled microcapsule.[51] g) These capsules can be filled with air or h) they can be filled with an oil containing hydrophobic drugs.

supply downstream which may lead to long-term disability or death. Contrast agent microbubbles can be used in combination with ultrasound to lyse thrombi in the blood vessel, restoring blood flow, a process termed sonothrombolysis.[29] The microbubbles can be injected next to the blood clot while being insonified with ultrasound or, alternatively and in a more controlled manner, injected microbubbles targeted to the blood clot bind to the target and are subsequently insonified with ultrasound. The microbubble oscillations may lead to microstreaming and cavitation (microjetting) and the combined action of mechanical stress exerted, mixing of an anti-coagulant such as rTPA, and flow may lead to a controlled disintegration of the blood clot in such a high-risk area.

5.2 Microbubble Basics

Contrast microbubbles produce a very strong echo, which can be 1 billion times stronger than the echo of solid particles of the same size,[30] owing to the large compressibility of the gas core of the bubbles, see Figure 5.4. In addition, the bubbles resonate to the driving ultrasound field, adding another two orders of magnitude to its scattering cross-section. Finally, the bubbles oscillate non-linearly, generating harmonics of the fundamental driving frequency, which allows efficient discrimination of the linear tissue echoes from the harmonic bubble echoes, boosting its contrast ability, which is expressed in the contrast-to-tissue ratio (CTR).

5.2.1 Microbubble Dynamics

The bubble dynamics is governed by the equation of motion for a spherical bubble, known as the Rayleigh–Plesset (RP) equation.[31] The RP-equation follows directly from Bernoulli's principle (Newton's second law) and the continuity equation:

$$\rho R \ddot{R} + \frac{3}{2} \rho \dot{R}^2 = p_i - p_e \qquad (5.1)$$

where ρ is the liquid density and R, $\dot{R}$, and $\ddot{R}$ represent the radius of the bubble, the velocity of the bubble wall, and the acceleration of the bubble wall, respectively. p_i is the bubble's internal pressure and it depends on the properties of the gas. p_e is the external pressure and it includes the pressure contributions of the ambient pressure, the interfacial pressure, and the driving pressure $p_A(t)$. Equation (5.1) is an ordinary differential equation and its solution, given any driving pressure pulse $p_A(t)$, is the radius of the bubble

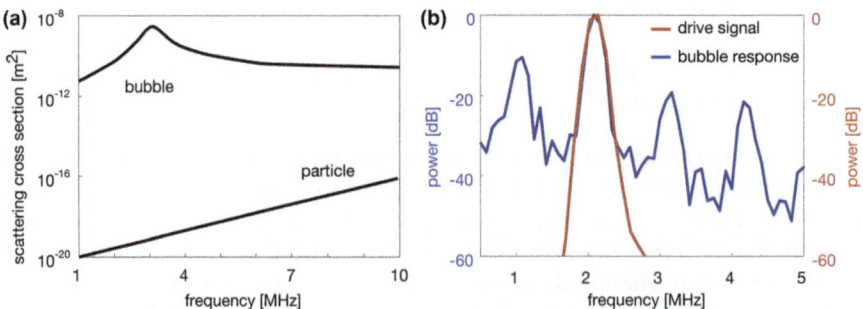

Figure 5.4 a) The scattering cross-section of a bubble can be 1 billion times larger than that of a solid particle of the same size.[30] In addition, the scattering cross-section of a bubble gains another two orders of magnitude if it is driven at its resonance frequency. b) The bubble oscillations are highly non-linear, generating harmonics of the fundamental driving frequency exploited to discriminate between linear tissue echoes and harmonic bubble echoes.[84]

as a function of time, $R(t)$. The non-linear features of the bubble response, or echo, originate directly from the non-linearities in the above equation. Once the radial dynamics is known, the pressure emitted by the bubble, $p_s(t)$ (in medical ultrasound literature termed the scattered pressure), can be calculated from the conservation of mass and momentum.[32]

$$p_s = \frac{\rho R}{r}\left(2\dot{R}^2 + R\ddot{R}\right) \tag{5.2}$$

where r is the distance to the bubble. Thus, for any given driving pressure pulse $p_A(t)$ the non-linear echo $p_s(t)$ can be calculated for an oscillating spherical bubble.

To understand the resonance behavior of microbubbles it is insightful to linearize the bubble dynamics equation. Assuming a sinusoidal pressure pulse in addition to the ambient pressure, $p_e = p_0 + p_A \sin\omega t$, with angular frequency $\omega = 2\pi f$ with f the ultrasound frequency, and assuming small amplitude oscillations $R = R_0(1 + x)$, where R_0 is the equilibrium bubble radius and $x \ll R_0$. Neglecting higher-order terms in x, eqn (5.1) then reduces to the classical differential equation of a simple driven harmonic oscillator:[33]

$$\ddot{x} + \omega_0^2 x = \frac{p_A}{\rho R_0}\sin\omega t \tag{5.3}$$

with a characteristic bubble oscillation eigenfrequency:

$$\omega_0 = \sqrt{\frac{3\kappa p_0}{\rho R_0^2}}, \tag{5.4}$$

with κ the polytropic exponent of the gas. The analogy to a mass-spring system is that the liquid surrounding the bubble acts as a mass, while the gas acts as a spring. Thus, the bubble resonates to the driving pressure field, and it has a resonance frequency f_R where a maximum amplitude is observed, see the resonance curve displayed in Figure 5.10. Figure 5.10 will be discussed in more detail later in this chapter. The system experiences damping, which can be appreciated from the width of the resonance curve. The energy loss mechanisms associated with damping can be attributed to acoustic reradiation and following a more detailed analysis also from viscous damping through the liquid viscosity and thermal damping due to the build-up of a thermal boundary layer around the bubble.[34] The total dimensionless damping constant, *i.e.* the sum of all the damping contributions, for microbubbles of a typical size of 5 μm is 0.1, therefore the resonance frequency is very close to the eigenfrequency of the bubble,[33] $f_R \approx f_0$. Note that the eigenfrequency is inversely proportional to the radius of the bubble, through eqn (5.4). For an air bubble in water the relation $f_0 \cdot R_0 \approx 3$ μm · MHz applies, as was derived earlier by Minnaert.[35] Thus, a 3 μm diameter bubble will have an eigenfrequency near 2 MHz, which is right at the heart of clinical medical ultrasound imaging.

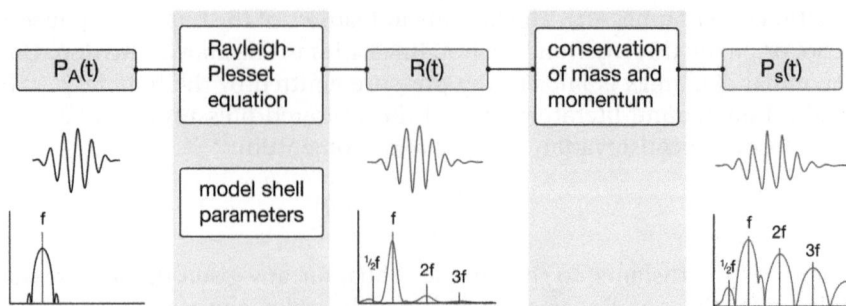

Figure 5.5   Microbubble modeling after Sijl *et al.*[85] An ultrasound pulse $P_A(t)$ of several cycles at a single frequency f drives radial bubble oscillations, $R(t)$. These can be modeled with a Rayleigh–Plesset-type model with the inclusion of shell pressure terms resulting from the shell viscosity and shell elasticity parameters.[66] The power spectrum of the radial bubble oscillations shows harmonics of the fundamental driving frequency. The bubble oscillations lead to a sound wave in the far field containing harmonics of the driving ultrasound frequency.

The way the bubbles are driven by ultrasound leading to an echo is depicted in Figure 5.5. A typical transmit pulse consists of a number of cycles of ultrasound at a single frequency and propagates in the direction of the focal region where the bubble is contained. The bubble is driven into oscillation by the successive wave of high and low pressure. The radial response of the bubble and the subsequent movement of the liquid surrounding the bubble leads to a sound emission that is propagating spherically in all directions.[32] The transducer will pick up the echo emitted within the receiving cone angle of the transducer. Given the non-linear response of the radial bubble dynamics the echo will also contain this particular non-linear signature and the fundamental frequency and its harmonics (second and third harmonic and subharmonic) will be picked up by the receiving unit within the available bandwidth of the transducer.

Harmonic imaging techniques make use of the non-linear bubble echo to cancel the linear echoes originating from tissue.[36] The pulse inversion technique[37] makes use of two ultrasound pulses with a 180° phase shift and adds the two echoes to cancel out the linear tissue response. Power modulation imaging[38] is a pulse-echo scheme in which two pulses with different acoustic pressure are emitted. The echoes acquired in receive mode are then scaled to the transmit pressure and subtracted to cancel out the linear signal leaving the non-linear signal originating from the bubbles. Finally, subharmonics of the bubble echo[39] are of interest to build up a contrast image, even though it has a lower resolution due to its lower frequency. The reason is that the transmitted ultrasound pulse may deform through non-linear propagation, forming higher harmonics, and tissue signal leaks into the harmonic bubble signal through linear scattering. On the other hand this exact feature is exploited in tissue harmonic imaging, without the use of contrast bubbles.

5.3 Microbubble Stability

Microbubbles are inherently unstable as a result of the capillary force and the resulting diffusion. The internal pressure of a bubble P_i is given by the ambient pressure P_0 plus the interfacial pressure contribution $2\sigma/R$, with σ the surface tension and R the radius of the bubble. For a 2 μm air bubble in water the interfacial pressure amounts to 1 bar. Hence the gas is rapidly squeezed out of the bubble and the bubble dissolves. The rate at which dissolution occurs comes from a diffusive balance:[40]

$$\dot{R} = \frac{D(c_i - c_s)R_g T}{M\left(P_0 + \dfrac{4\sigma}{3R}\right)}\left(\frac{1}{R} + \frac{1}{\sqrt{\pi Dt}}\right) \tag{5.5}$$

where R is the bubble radius, $\dot{R}$ the time derivative of the bubble radius, D is the diffusivity constant for the interface at a given temperature and pressure, c_i the initial dissolved gas concentration in the liquid, c_s is the gas concentration at the bubble surface, R_g is the universal gas constant, T the absolute temperature, σ the interfacial surface tension, R_0 the initial bubble radius, P_0 the ambient pressure, M the molecular weight of the gas, and t is time. Figure 5.6 shows the fate of microbubbles. In this example it takes a 5 μm diameter bubble approximately 90 ms to dissolve. Microbubbles of a size of 3 μm and 1 μm in diameter take a considerably shorter time to dissolve, 25 ms and 2 ms, respectively. Note that the bubbles follow the exact same dissolution curve once they reach the same size. In practice, the dissolution rate is strongly dependent on the local gas fraction of the surrounding liquid, c_i, which may change in the medium surrounding the bubble once the bubble dissolves and, in the presence of flow, other bubbles and solid and free interfaces. Thus, for the use of microbubbles for the aforementioned medical applications, dissolution of the microbubble must be prevented.

There are two ways to approach the dissolution problem: to add a coating to the interface of the bubble and to change the gas in the core of the bubble. The first approach has two main effects. First, the addition of a surfactant coating leads to a reduction of the surface tension, which will decrease directly the dissolution rate. Second, the presence of a surfactant layer at the interface will hinder dissolution through that interface. By changing the gas in the core to an inert high-molecular-weight gas with low solubility, *e.g.* sulfur hexafluoride SF_6 or perfluorocarbons, *e.g.* octafluoropropane C_3F_8 or perfluorobutane C_4F_{10}, the lifetime of the bubble can be extended. Figure 5.6 shows the four-fold increase of the lifetime of the 3 μm bubble by exchanging the gas core of the bubble. With the above approaches combined and implemented the total lifetime of the bubble can be extended to 5–10 minutes following intravenous injection. It should also be noted that typically long-chain poly-ethylene glycol (PEG) spacers are added to the bubble surface to improve circulation lifetime *in vivo*. Perfluorocarbon gas-filled microbubbles can be stable for several days when kept in a vial

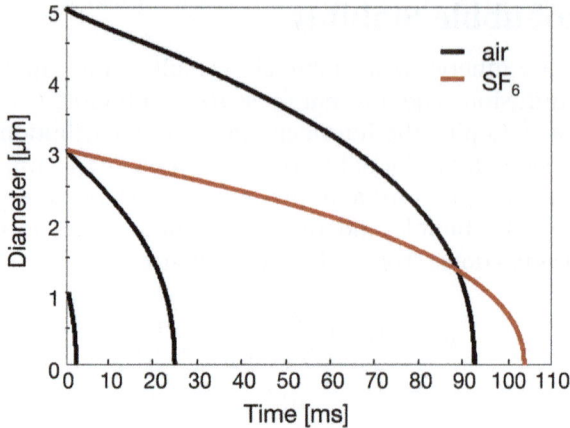

Figure 5.6 Microbubble dissolution curves.[40] A 5 μm air bubble in water dissolves in approximately 90 ms. A 3 μm and a 1 μm air bubble dissolve in 25 ms and 2 ms, respectively. Exchanging the gas core of the 3 μm bubble by a high-molecular-weight gas (SF_6) shows a four-fold increase of its lifetime.

under protective gas cover and freeze-dried bubbles can be stored for several years.

The very first class of stabilized microbubbles were coated by denatured human serum albumin. Agents in this category include air-filled Quantison (Upperton Limited, Nottingham, UK) (Figure 5.7a)[41] and octafluoropropane-filled Optison (GE Healthcare).[42] The albumin coating results in a relatively stiff shell[43] and the stiffness of the shell can be tuned by the use of coatings of different thickness. While the oscillation amplitude of hard-shelled microbubbles is dramatically reduced, the mechanism of contrast echo generation lies more in the bursting of the shell, leading to the release of a free gas bubble, with limited lifetime, but long enough to generate an echo above the noise level.[44]

Microbubbles can also be stabilized by the use of surfactants containing a hydrophobic tail and a hydrophilic head, which desorb and arrange at the free interface of the microbubble. Soap detergent surfactants, such as Tween and Dreft, are popular microbubble stabilizers for *in-vitro* use. Clinically approved microbubble contrast agents use purified phospholipids such as DPPC, also well-known pulmonary surfactants found in the alveoli, to cover and stabilize the interface. Unlike biological membranes and liposomes, where these lipids form a lipid bilayer, these long-chain lipids form a monolayer, which may change the bubble dynamics upon insonation through its viscoelastic properties. The flexible nature of the monolayer however retains much of its ability to oscillate. Interestingly, the non-linear effects introduced by the viscoelastic monolayer promote the non-linear echo and therefore boost the contrast-to-tissue ratio. Commercial agents in this category include Sonovue (Bracco Imaging) (Figure 5.7b),[45] Definity (Lantheus Medical Imaging)[46] and Sonazoid (GE Healthcare).[42]

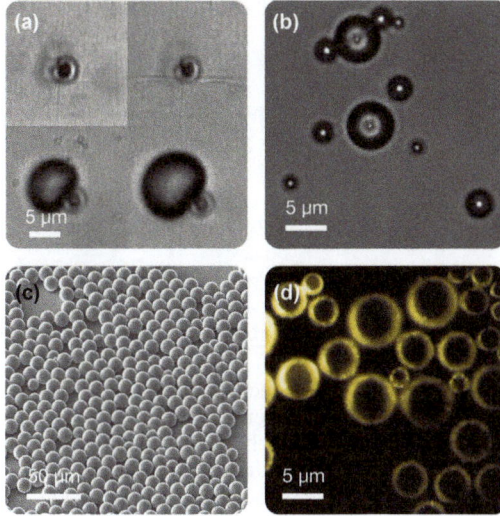

Figure 5.7 Various types of encapsulated agents for imaging and drug-delivery applications. a) Albumin-coated bubbles (Quantison) showing release of a free gas bubble upon rupture of the shell.[41] b) Soft-shelled lipid-coated microbubbles (BR-14 Bracco Suisse S.A.). c) Monodisperse polymeric microbubbles[52] (image courtesy of Philips Research Laboratories Europe). d) DiI fluorescently labeled lipid bubbles.[74]

For several reasons the use of serum albumin has been less desirable and the coatings of the newer class of agents have now been replaced by biocompatible polymers, ranging from polyvinyl alcohol (PVA), to polylactic acid (PLA) (see Figure 5.7c) and polylactic-*co*-glycolic acid (PLGA). This type of polymeric microbubble is also popular as a drug-delivery vehicle in the form of acoustically triggerable capsules. The *in-vitro* use of thermoplastic-coated microbubbles, *e.g.* with a PVC-AN shell,[47] remains popular for scientific purposes. Agents of this type include nitrogen-filled Cardiosphere (Point Biomedical) and perfluorobutane-filled Imagify (Acusphere).[48]

5.4 Microbubble Formation

Ultrasound contrast agents for medical use are widely available. Most of them come in a vial and have to be resuspended to form a bubble suspension. Typically one to ten billion bubbles in a volume of 1–2 mL are injected in a human perfusion study. The traditional contrast agent production methods are sonication and shaking of a solution of water-soluble surfactants or polymers. Agitation of the fluid results in the inclusion of small air bubbles in the fluid and subsequent diffusion of the coating material to the gas-water interface results in stable microbubbles by a self-assembly of the coating material.[12] The conventional processing techniques offer high yield and low production cost but poor control over microbubble size and uniformity.

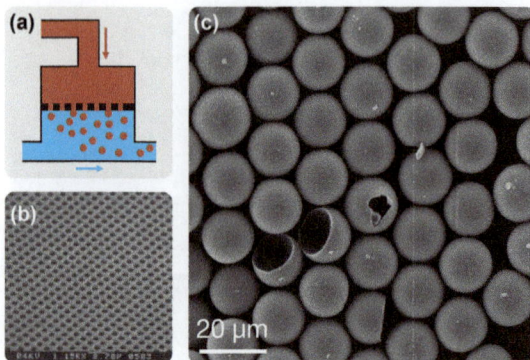

Figure 5.8   Membrane emulsification technology to make monodisperse polymeric microbubbles.[86] a) A polymer dissolved in a solvent is emulsified by a Nanosieve membrane (b) with mono-sized pores to make monodisperse polymeric emulsion microdroplets. c) SEM image of microcapsules formed after freeze drying the emulsion droplets.
Images courtesy of Nanomi monosphere technology, The Netherlands.

Emulsification processes were also shown to lead to high-yield bubble production methods. A polymer is dissolved in a solvent and mixed with water by high shear emulsification[49] resulting in micrometer droplets of dissolved polymer in water. Hard-shelled water-filled microspheres are formed when the solvent is evaporated. The capsules are then washed to remove excess solvent, then freeze-dried to produce gas-filled capsules.[50] A narrow size distribution can be obtained by membrane emulsification of a premix of polymer, polymer solvent and water through a mono-sized porous membrane, see Figure 5.8. The capsules can also be partly or fully filled with oil by these emulsification techniques to produce acoustically active hydrophobic drug carriers.[51]

For therapeutic applications of microbubbles it becomes increasingly important to be able to predict the acoustic response of the microbubble or of loaded microbubble suspensions which resulted in the development of new production methods for even narrower size distributions. Inkjet printing has been employed to improve microbubble uniformity.[52] Monodisperse droplets of a polymer solution are printed in water from an ink-jet printhead. Polymer-shelled microbubbles are then produced by evaporation of the solvent in a similar way to the membrane emulsification method. Although it has a lower production rate, the ink-jet printing technique has the advantage of being able to change the droplet size rather easily by changing the droplet formation parameters online in the printhead.

Highly monodisperse droplets and bubbles can be produced using flow focusing techniques.[53–56] Figure 5.9 shows how a central gas or a liquid thread is focused in between two external liquid flows through an orifice, where the thread is pinched off to form monodisperse bubbles or droplets. This microfluidic technique offers advantages over methods such as piezo ink-jet printing in that bubbles can be formed in a single step without the

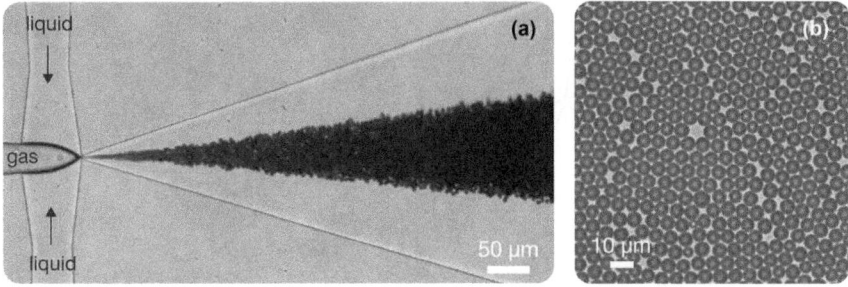

Figure 5.9 a) Flow-focusing geometry for monodisperse microbubble formation by focusing a gas thread between two liquid co-flows through a narrow orifice[87] at a production rate of $\sim 300\ 000$ bubbles per second. b) Harvested stable monodisperse bubbles produced in a flow focusing device.[88]

need for further processing steps, *e.g.* to remove solvents. Flow focusing techniques have also shown to be able to produce concentric bubbles with an outer oil layer facilitating drug loading.[57] A challenging aspect of this approach is to understand how the monodisperse bubbles can be encapsulated with a biocompatible coating to stabilize them,[58–60] to investigate how to maintain the monodispersity of bubbles produced at high production rates over time,[57] and to investigate the dynamics of different coating materials for *in-vivo* and clinical use.[61]

5.5 Microbubble Modeling and Characterization

The coating of the microbubble changes the bubble dynamics.[62–65] The addition of a surfactant layer and resulting pressure contributions from the viscoelastic shell must be incorporated into the RP-equation, eqn (5.1). The elasticity of the shell increases the stiffness of the system, hence an increase of the resonance frequency is expected. Similarly the viscosity of the shell leads to energy loss and increased damping. The dynamical behaviour can be modeled as was shown before in Figure 5.5 by including the shell parameters in the Rayleigh–Plesset model. Literature reports a shell viscous damping contribution of 75% of the total damping, so governing the resulting oscillation amplitudes and scattered pressures. Figure 5.10a shows the normalized resonance curve of a coated bubble. The effect of the viscoelastic shell is evident with an increase of the resonance frequency of nearly 40% and broadening of the resonance curve as a result of increased damping.

Recently it was discovered that non-linear properties of the bubble shell are important at small oscillation amplitudes.[66] A high concentration of lipids for example may lead to buckling of the lipid shell upon compression, while the bubble may rupture upon expansion, forming lipid islands on the otherwise free interface, thereby exposing the gas directly to

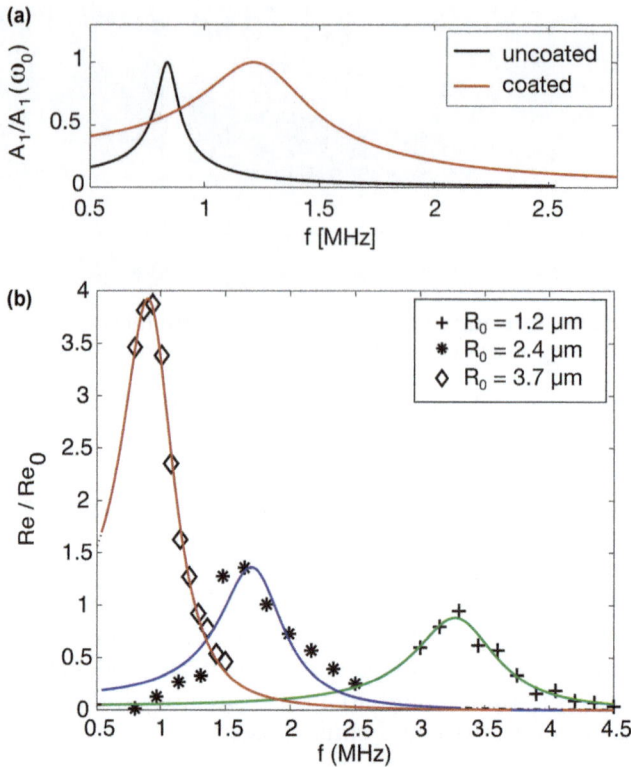

Figure 5.10 Resonance curves. a) Resonance curve of an uncoated 3.8 μm bubble and of a bubble of the same size with a phospholipid coating. The coating increases the stiffness of the system resulting in an increased resonance frequency. The viscosity of the coating leads to an increased energy dissipation, or damping, resulting in a broadened resonance curve. b) A scan of the relative amplitudes of oscillation, of three bubbles of different size, as a function of the insonation frequency. A Rayleigh–Plesset type model accounting for the bubble shell can be fitted to the measured resonance curves (solid lines) to extract information on the bubble shell viscoelastic parameters.

the liquid. In addition gas loss through diffusion leads to a higher concentration of lipids at the bubble surface, which may lead to folding or budding of the lipid layer. This may eventually result in expulsion of the excess lipids, which in turn leads to a higher surface tension, promoting gas loss, *etc.* In addition, the interfacial rheology is important to fully describe the origin and the details of the non-linear echo, and details can be found in refs 60 and 67. One important conclusion of the recent work on non-linear properties of the shell is that the frequency of maximum response is invariably pressure-dependent,[67] and a breakaway from the classical Minnaert theory.

5.5.1 Optical Characterization

Microbubble dynamics models are validated through single bubble optical characterization. The optical visualization of a sample of microbubbles under the microscope is relatively straightforward, but the real-time visualization of the bubble dynamics upon insonation with a 2-MHz ultrasound pulse is quite challenging and it requires high-speed imaging at a frame rate near 10–20 million frames per second. Several camera systems[68-70] are capable of reaching such high frame rates and in addition acquire a high number of frames of 100 or more. A higher number of frames allows for the measurements of the bubble response at various acoustic parameters, such as pressure[71] and frequency,[72] or both.[67] A scan of the bubble response with varying frequency leads to the recording of the resonance curve,[72] see Figure 5.10b, and provides valuable information on the shell viscoelastic properties. Recent advances in ultra-high-speed fluorescence microscopy, some performed at frame rates exceeding 20 million frames per second,[69,70] have revealed new insight in lipid shedding, drug release[73] and drug uptake mechanisms[74,75] and valuable information for bubble-mediated medical applications.

5.5.2 Sorting Techniques

Commercial off-the-shelf ultrasound contrast agents typically have a wide size distribution. With the transducers at a rather narrow frequency range, this means that only a small selection of microbubbles participates in the generation of the non-linear echo, *i.e.* only those microbubbles that are resonant to the driving ultrasound frequency. Thus, the scattering efficiency and resulting sensitivity can be improved by narrowing down the size distribution to make the agent more monodisperse. This is particularly useful for techniques now being developed in molecular imaging with ultrasound. As mentioned before, excellent monodispersity can be achieved in flow-focusing devices and the size of the microbubbles can be tuned by tuning the gas and liquid flow rates of the device. One promising approach is to sort the bubbles. Figure 5.11 summarizes various sorting techniques, *e.g.* by filtering[71] or by decantation.[76] A selection of bubbles can also be extracted by means of centrifugation[77] or by using a microfluidic sorting strategy, *e.g.* using pinched flow fractionation.[78] All these approaches lead to a sorting of bubbles by size and as indicated before in Section 5.2, from the non-linear effect of the coating, it could also be advantageous to sort microbubbles based on their acoustical properties, rather than size, as was demonstrated recently.[79]

5.5.3 Acoustical Characterization

From the examples above while discussing damping and energy loss mechanisms it is evident that a high concentration of bubbles at resonance not only delivers a strong (non-linear) echo, but also extracts pulse energy

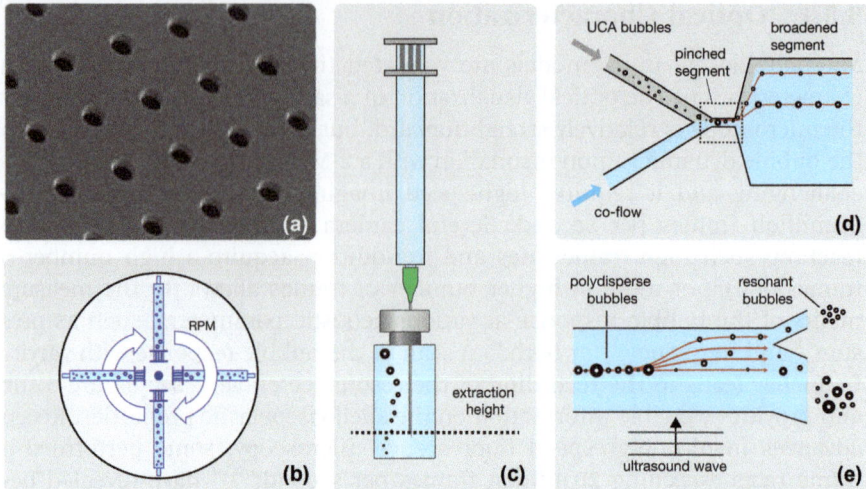

Figure 5.11   Sorting strategies for ultrasound contrast agent enrichment. a) Pore filters can be used to filter bubbles to size.[71,89] Two filters of different size are needed to band-pass filter a bubble suspension. b) Bubbles can be sorted using the difference in gravitational force acting on bubbles of different size. Centrifugation of the bubble suspension can be applied to increase the efficiency of this sorting technique.[77] c) The buoyancy force can also be used through decantation,[76,81] the larger bubbles rise faster in a fluid than the smaller bubbles. d) Pinched flow fractionation[78] is a microfluidic sorting technique which can be utilized to sort microbubbles[90] by pinning them to a wall by a co-flow in the pinched segment. Size-selective sorting is achieved through expansion into the broadened segment by microfluidic amplification. e) Acoustic bubble sorting[79] sorts bubbles to their acoustic property rather than to their size. Bubbles are displaced from the center line in a microfluidic channel, where they are injected, by a traveling acoustic wave through the primary radiation force. The displacement is strongly coupled to the resonance behavior of the microbubbles.

from the driving pulse and the total echo may be attenuated by absorption. Hence it is important to measure the scattering to attenuation ratio[80,81] (STAR) of the microbubble suspension, the suspension being a native commercial agent, freshly produced bubbles in a flow-focusing device, or sorted bubbles extracted from a polydisperse bubble collection. Such a STAR measurement is displayed in Figure 5.12 and can be performed in a small water tank with a central container holding the microbubble suspension under continuous stirring. A transmit pulse is sent through the sample and the pulse is received by a transducer at the opposite end of the water tank and as such it will represent the attenuation. At the same time a transducer placed at a 90° angle to the transmit pulse focused at the sample region of interest measures the scattered signal. To fully characterize the contrast bubble suspension, these measurements must be performed for the full

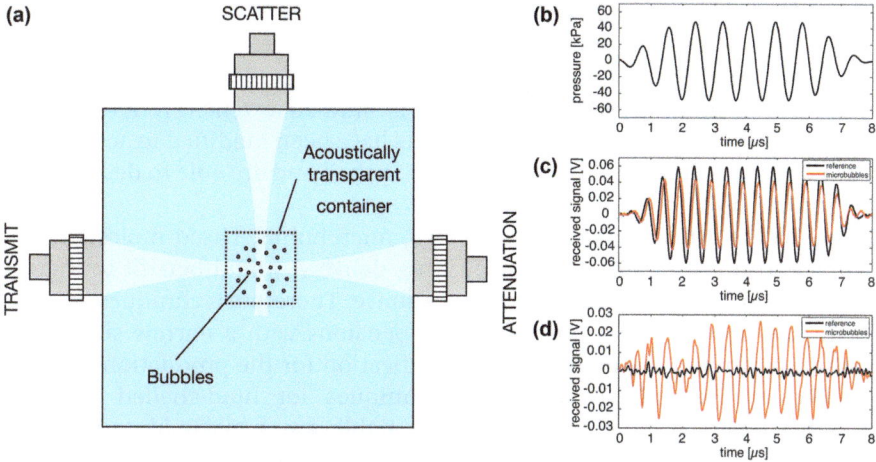

Figure 5.12 Bubble suspensions can be characterized by measuring the scattering and attenuation of acoustic waves by the bubbles. A transmit transducer sends an ultrasound pulse of several acoustic cycles (b) through the bubble sample and it is received by a transducer at the opposite side of the water tank. c) The received signals of a water-filled container and a container filled with a bubble suspension can be divided to find the frequency-dependent, pressure-dependent, and concentration-dependent attenuation of the bubble cloud. At the same time a transducer placed at a 90° angle to the transmit transducer measures the scattered acoustic signal (d). The measured frequency-dependent scattering and attenuation characteristics are divided to find the scattering to attenuation (STAR) ratio as a function of the acoustic parameters (frequency and pressure) and bubble concentration.

parameter space of acoustic pressures and frequencies, in addition to the concentration of the agent.

5.6 Conclusions

Microbubbles are globally used in a clinical setting to perform contrast-enhanced ultrasound imaging. The use of microbubbles for molecular imaging has recently entered the clinical phase where the first targeted bubbles were used for the early detection and localization of prostate cancer.[82] Microbubbles have also been clinically used to successfully induce middle cerebral artery recanalization in combination with rTPA. Ultrasound imaging in combination with microbubbles has therefore shown to have a great potential for molecular imaging, for therapy, and for drug delivery applications.

Microbubbles for *in-vivo* use can be coated with several biocompatible coatings, ranging from hard-shelled polymer agents to soft-shelled, more acoustically active, phospholipid agents. Both coating materials can be used to devise targeted drug-loaded bubbles. Soft-shelled bubbles can transport drugs incorporated in their shell, in liposomes attached to their shell or in an oil layer inside the bubble. Hard-shelled agents are more robust and can carry a variety of drug loads in liquid form. The dynamical behaviour of both hard-shelled and soft-shelled bubbles is studied extensively and is well understood. Rayleigh–Plesset-type models have been modified to account for the non-linear behaviour of the viscoelastic shell and are able to describe the dynamical bubble behavior in detail.

Bubble detection pulse-echo schemes, microbubble-based molecular imaging with ultrasound and targeted drug delivery, would benefit to a great extent from a well-controlled bubble response. Therefore techniques are now being developed to produce bubble suspensions with a narrow size distribution. Examples are membrane emulsification for the production of hard-shelled agents and flow-focusing techniques for lipid-shelled bubbles. Nevertheless, coating the bubbles in a stable way while maintaining the monodispersity over time is difficult and a few hurdles need to be overcome before we see the commercial introduction of monodisperse contrast agents.

Acknowledgments

This work is supported by the Foundation for Fundamental Research on Matter FOM, the Technology Foundation STW, and NanoNextNL, a micro and nanotechnology consortium of the Government of the Netherlands and 130 partners.

References

1. J. R. Lindner, *Nat. Rev. Drug Discov.*, 2004, **3**, 527–533.
2. A. L. Klibanov, *Invest. Radiol.*, 2006, **41**, 354–362.
3. E. Talu, K. Hettiarachchi, S. Zhao, R. L. Powell, A. P. Lee, L. Longo and P. A. Dayton, *Mol. Imaging*, 2007, **6**, 384–392.
4. M. Overvelde, V. Garbin, B. Dollet, N. de Jong, D. Lohse and M. Versluis, *Ultrasound Med. Biol.*, 2011, **37**, 1500–1508.
5. B. Naaijkens, S. J. P. Bogaards, P. A. J. Krijnen, O. Kamp, R. J. P. Musters, T. J. A. Kokhuis, N. De Jong, N. W. M. Niessen, A. Van Dijk and L. J. M. Juffermans, *Eur. Heart J.*, 2013, **34**, 1451.
6. L. Deelman, A. Declèves, J. Rychak and K. Sharma, *Adv. Drug Deliv. Rev.*, 2010, **62**, 1369–1377.
7. A. Carson, C. McTiernan, L. Lavery, M. Grata, X. Leng, J. Wang, X. Chen and F. Villanueva, *Cancer Res.*, 2012, **72**, 6191–6199.
8. R. E. Vandenbroucke, I. Lentacker, J. Demeester, S. C. De Smedt and N. N. Sanders, *J. Contr. Release*, 2008, **126**, 265–273.
9. E. Unger, T. McCreery, R. Sweitzer, V. Caldwell and Y. Wu, *Investig. Radiol.*, 1998, **33**, 886–892.
10. S. Tinkov, C. Coester, S. Serba, N. A. Geis, H. A. Katus, G. Winter and R. Bekeredjian, *J. Contr. Release*, 2010, **148**, 368–372.

11. J. Kang, X. Wu, Z. Wang, H. Ran, C. Xu, J. Wu, Z. Wang and Y. Zhang, *J. Ultrasound Med.*, 2010, **29**, 61–70.
12. B. Geers, I. Lentacker, N. N. Sanders, J. Demeester, S. Meairs and S. C. De Smedt, *J. Contr. Release*, 2011, **152**, 249–256.
13. V. F. K. Bjerknes, *Fields of Force*, Columbio University Press, New York, 1906, pp. 1–136.
14. L. A. Crum, *J. Acoust. Soc. Am.*, 1957, **57**, 1363–1370.
15. P. Marmottant, M. Versluis, N. de Jong, S. Hilgenfeldt and D. Lohse, *Exp. Fluid.*, 2006, **41**, 249–256.
16. L. Rayleigh, *Phil. Mag.*, 1917, **34**, 94–98.
17. O. Lindau and W. Lauterborn, *J. Fluid Mech.*, 2003, **479**, 327–348.
18. T. J. Matula, *Phil. Trans. Roy. Soc. Lond. Math. Phys. Sci.*, 1999, **357**, 225–249.
19. C.-D. Ohl, M. Arora, R. Dijkink, V. Janve and D. Lohse, *Appl. Phys. Lett.*, 2006, **89**, 074102.
20. E. Klaseboer, K. C. Hung, C. Wang, C. W. Wang, B. C. Khoo, P. Boyce, S. Debono and H. Charlier, *J. Fluid Mech.*, 2005, **537**, 387–413.
21. F. Marquet, T. Teichert, S. Wu, Y. Tung, M. Downs, S. Wang, C. Chen, V. Ferrera and E. E. Konofagou, *Plos One*, 2014, **9**, 1–11.
22. N. J. Abbott, L. Ronnback and E. Hansson, *Nat. Rev. Neurosci.*, 2006, 7, 41–53.
23. R. Paolinelli, M. Corada, F. Orsenigo and E. Dejana, *Pharmacol. Res.*, 2011, **63**, 165–171.
24. T. D. Khokhlova, M. S. Canney, V. A. Khokhlova, O. A. Sapozhikov, L. A. Crum and M. R. Bailey, *J. Acoust. Soc. Am.*, 2011, **130**, 3498–3510.
25. C. C. Coussios, C. H. Farny, G. Ter Haar and R. A. Roy, *Int. J. Hyperthermia*, 2007, **23**, 105–120.
26. Z. Xu, T. H. Hall, J. B. Fowlkes and C. A. Cain, *J. Acoust. Soc. Am.*, 2007, **122**, 229–236.
27. O. A. Sapozhnikov, V. A. Khokhlova, M. R. Bailey, J. C. Williams, J. A. McAteer, R. O. Cleveland and L. A. Crum, *J. Acoust. Soc. Am.*, 2002, **112**, 1183–1195.
28. W. Eisenmenger, *Ultrasound Med. Biol.*, 2001, **27**, 683–693.
29. B. Petit, E. Gaud, D. Colevret, M. Arditi, F. Yan, F. Tranquart and E. Allemann, *Ultrasound Med. Biol.*, 2012, **38**, 1222–1233.
30. N. de Jong, F. ten Cate, C. Lancée, J. Roelandt and N. Bom, *Ultrasonics*, 1991, **29**, 324–330.
31. M. S. Plesset, *J. Appl. Mech.*, 1949, **16**, 228–231.
32. K. Vokurka, *Czech. J. Phys.*, 1985, **35**, 28–40.
33. T. Leighton, *The Acoustic Bubble*, Academic Press, London, 1994.
34. A. Prosperetti, *J. Acoust. Soc. Am.*, 1976, **61**, 17–27.
35. M. Minnaert, *Phil. Mag.*, 1933, **16**, 235–248.
36. A. Bouakaz, S. Frigstad, F. J. ten Cate and N. de Jong, *Ultrasound Med. Biol.*, 2002, **28**, 59–68.
37. D. Hope Simpson, C. T. Chin and P. N. Burns, *IEEE Trans. Ultrason. Ferroelectrics Freq. Contr.*, 1999, **46**, 372–382.

38. G. A. Brock-Fisher, M. D. Poland and P. G. Rafter, *US pat.*, 5577505, 1996.
39. P. M. Shankar, P. D. Krishna and V. L. Newhouse, *Ultrasound Med. Biol.*, 1998, **24**, 395–399.
40. P. S. Epstein and M. S. Plesset, *J. Chem. Phys.*, 1950, **18**, 1505–1509.
41. M. Postema, A. Bouakaz, M. Versluis and N. de Jong, *IEEE Trans. Ultrason. Ferroelectrics Freq. Contr.*, 2005, **52**, 1035–1041.
42. GE Healtcare, http://www3.gehealthcare.com.
43. S. B. Feinstein, F. J. Ten Cate, W. Zwehl, K. Ong, G. Maurer, C. Tei, P. M. Shah, S. Meerbaum and E. Corday, *J. Am. Coll. Cardiol.*, 1954, **3**, 14–20.
44. A. Bouakaz, M. Versluis and N. de Jong, *Ultrasound Med. Biol.*, 2005, **31**, 391–399.
45. Bracco Imaging, http://www.braccoimaging.com.
46. Lantheus Medical Imaging, http://www.lantheus.com/.
47. H. Vos, F. Guidi, E. Boni and P. Tortoli, *IEEE Trans. Ultrason. Ferroelectrics Freq. Contr.*, 2007, **54**, 1333–1345.
48. Acusphere, http://www.acusphere.com/.
49. E. Stride and M. Edirisinghe, *Soft Matter*, 2008, **4**, 2350–2359.
50. D. Lensen, E. C. Gelderblom, D. M. Vriezema, P. Marmottant, N. Verdonschot, M. Versluis, N. de Jong and C. M. van Hest, *Soft Matter*, 2011, **7**, 5417–5422.
51. K. Kooiman, M. R. Bohmer, M. Emmer, H. J. Vos, C. Chlon, W. T. Shi, C. S. Hall, S. H. P. M. de Winter, K. Schroen, M. Versluis, N. de Jong and A. van Wamel, *J. Contr. Release*, 2009, **133**, 109–188.
52. M. R. Bohmer, R. Schroeders, J. A. M. Steenbakkers, S. H. P. M. de Winter, P. A. Duineveld, J. Lub, W. P. M. Nijssen, J. A. Pikkemaat and H. R. Stapert, *Colloid. Surface. A*, 2006, **289**, 96–104.
53. A. M. Gañán-Calvo and J. M. Gordillo, *Phys. Rev. Lett.*, 2001, **87**, 274501.
54. S. L. Anna, N. Bontoux and H. A. Stone, *Appl. Phys. Lett.*, 2003, **82**, 364–366.
55. P. Garstecki, H. Stone and G. M. Whitesides, *Phys. Rev. Lett.*, 2005, **94**, 164501.
56. B. Dollet, W. van Hoeve, J.-P. Raven, P. Marmottant and M. Versluis, *Phys. Rev. Lett.*, 2008, **100**, 034504.
57. R. Shih, D. Bardin, T. D. Martz, P. S. Sheeran, P. A. Dayton and A. P. Lee, *Lab. Chip*, 2013, **13**, 4816–4826.
58. K. Hettiarachchi, E. Talu, M. L. Longo, A. Dayton and A. P. Lee, *Lab. Chip*, 2007, **7**, 463–468.
59. E. Talu, K. Hettiarachchi, R. L. Powell, A. P. Lee, P. A. Dayton and M. L. Longo, *Langmuir*, 2008, **24**, 1745–1749.
60. J. Kwan and M. Borden, *Soft Matter*, 2012, **8**, 4756–4766.
61. J. Streeter, R. Gessner, I. Miles and P. Dayton, *Mol. Imaging*, 2010, **9**, 87–95.
62. N. de Jong, R. Cornet and C. Lancée, *Ultrasonics*, 1994, **32**, 447.
63. C. C. Church, *J. Acoust. Soc. Am.*, 1995, **97**, 1510–1521.
64. L. Hoff, P. Sontum and J. Hovem, *JASA*, 2000, **107**, 2272–2280.
65. K. Sarkar, W. T. Shi, D. Chatterjee and F. Forsberg, *J. Acoust. Soc. Am.*, 2005, **118**, 539–550.

66. P. Marmottant, S. Van Der Meer, M. Emmer, M. Versluis, N. De Jong, S. Hilgenfeldt and D. Lohse, *J. Acoust. Soc. Am.*, 2005, **118**, 3499–3505.

67. M. Overvelde, V. Garbin, J. Sijl, B. Dollet, N. de Jong, D. Lohse and M. Versluis, *Ultrasound Med. Biol.*, 2010, **36**, 2080–2092.

68. C. T. Chin, C. Lancee, J. Borsboom, F. Mastik, M. E. Frijlink, N. De Jong, M. Versluis and D. Lohse, *Rev. Sci. Instr.*, 2003, **74**, 5026–5034.

69. E. C. Gelderblom, H. J. Vos, F. Mastik, T. Faez, Y. Luan, T. J. A. Kokhuis, A. F. W. van der Steen, D. Lohse, N. de Jong and M. Versluis, *Rev. Sci. Instrum.*, 2012, **83**, 103706.

70. X. Chen, J. Wang, M. Versluis, N. de Jong and F. S. Villanueva, *Rev. Sci. Instrum.*, 2013, **84**, 063701.

71. M. Emmer, A. van Wamel, D. E. Goertz and N. de Jong, *Ultrasound Med. Biol.*, 2007, **33**, 941–949.

72. S. Van Der Meer, B. Dollet, M. Voormolen, C. T. Chin, A. Bouakaz, N. De Jong, M. Versluis and D. Lohse, *J. Acoust. Soc. Am.*, 2007, **121**, 648–656.

73. Y. Luan, T. Faez, E. Gelderblom, I. Skachkov, B. Geers, I. Lentacker, T. van der Steen, M. Versluis and N. de Jong, *Ultrasound Med. Biol.*, 2012, **38**, 2174–2185.

74. E. C. Gelderblom, Ultra-high-speed fluorescence imaging, PhD thesis, University of Twente, The Netherlands, 2012.

75. Z. Fan, H. Liu, M. Mayer and C. X. Deng, *Proc. Natl Acad. Sci.*, 2012, **109**, 16486–16491.

76. D. E. Goertz, N. de Jong and A. F. W. van der Steen, *Ultrasound Med. Biol.*, 2007, **33**, 1376–1388.

77. J. A. Feshitan, C. C. Chen, J. J. Kwan and M. A. Borden, *J. Coll. Interf. Sci.*, 2009, **329**, 316–324.

78. M. Yamada, M. Nakashima and M. Seki, *Anal. Chem.*, 2004, **76**, 5465–5471.

79. T. Segers and M. Versluis, *Lab. Chip*, 2014, **14**, 1705–1714, http://dx.doi. org/10.1039/C3LC51296G.

80. P. J. A. Frinking and N. de Jong, *Ultrasound Med. Biol.*, 1998, **24**, 523–533.

81. M. Emmer, H. J. Vos, D. E. Goertz, A. van Wamel, M. Versluis and N. de Jong, *Ultrasound Med. Biol.*, 2009, **35**, 102–111.

82. F. Kiessling, S. Fokong, P. Koezera, W. Lederle and T. Lammers, *Molecular and Theranostic Sonography*, 2012, **53**, 345–348.

83. M. Overvelde, Ultrasound contrast agents, dynamics of coated microbubbles, PhD thesis, University of Twente, 2010.

84. J. Sijl, H. J. Vos, T. Rozendal, N. De Jong, D. Lohse and M. Versluis, *J. Acoust. Soc. Am.*, 2011, **130**, 3271.

85. J. Sijl, M. Overvelde, B. Dollet, V. Garbin, N. de Jong, D. Lohse and M. Versluis, *J. Acoust. Soc. Am.*, 2011, **129**, 1729–1739.

86. Nanomi Monosphere Technology, The Netherlands, http://www.nanomi. com/nanomi/index.html.

87. Y. Tan, V. Cristini and A. P. Lee, 2006, **114**, 350–356.

88. Tide Microfluidics, The Netherlands, http://www.tidemicrofluidics.nl.

89. Aquamarijn, The Netherlands, http://www.aquamarijn.nl.

90. M. P. Kok, T. Segers and M. Versluis, *Lab. Chip*, to be submitted, 2014.

CHAPTER 6

Magnetic Particle Actuation in Stationary Microfluidics for Integrated Lab-on-Chip Biosensors

ALEXANDER van REENEN,[a,c] ARTHUR M. de JONG,[a,c]
JAAP M. J. den TOONDER[b,c] AND MENNO W. J. PRINS*[a,c,d]

[a] Department of Applied Physics, Eindhoven University of Technology, 5600 MB Eindhoven, The Netherlands; [b] Department of Mechanical Engineering, Eindhoven University of Technology, 5600 MB Eindhoven, The Netherlands; [c] Institute for Complex Molecular Systems, Eindhoven University of Technology, 5600 MB Eindhoven, The Netherlands; [d] Philips Research, The Netherlands
*Email: m.w.j.prins@tue.nl

6.1 Introduction

In-vitro diagnostics (IVD) play an important role in decision making in a wide range of healthcare settings and in many stages of the patient care pathway.[1,2] Within the IVD market, decentralized diagnostic testing, *i.e.* point-of-care testing (POCT), is a growing segment. POCT helps to provide care-givers with relevant information for making on-the-spot decisions in a single patient interaction, with the possibility to streamline healthcare processes and to enable new models of care, such as remotely monitoring

RSC Nanoscience & Nanotechnology No. 36
Microfluidics for Medical Applications
Edited by Albert van den Berg and Loes Segerink
© The Royal Society of Chemistry 2015
Published by the Royal Society of Chemistry, www.rsc.org

the progress of patients and thereby reducing the number of visits needed to the hospital.[3] Furthermore POCT devices create opportunities to perform testing in less expensive settings such as the doctor's office and the home, helping to improve the accessibility and cost-effectiveness of future healthcare.

The demand for easy to use and cost effective medical technologies inspires scientists to develop innovative lab-on-chip technologies for *in-vitro* diagnostic testing. To fulfill the medical needs, the tests should be rapid, sensitive, quantitative, miniaturizable, and need to integrate all steps from sample-in to result-out. A central challenge in IVD assays is that low concentrations of marker molecules need to be measured within a complex biological fluid – such as blood – containing high concentrations of variable background material. Therefore, in addition to a high sensitivity, assays should have a high molecular selectivity. In case of protein biomarkers, molecular selectivity is generally obtained by making use of antibodies in immunoassays.[4] In case of nucleic-acid testing, purification and biochemical amplification steps are typically applied.[5]

To detect protein biomarkers, several immunoassay sensing technologies have been developed, such as nanoparticle labeling,[6–8] label-free electrical detection,[9] fluorescence detection,[10,11] and oligonucleotide labeling combined with biochemical amplification.[12] While the detection sensitivities of these technologies can be high, the integration of these platforms in cost-effective lab-on-chip devices is complicated because several active fluidic steps are required such as sample-pretreatment steps,[6–12] (bio)-chemical development steps,[6,12] or wash steps using buffer fluids.[6–12] Therefore, it is important to face the challenge of total integration[13] and design solutions that facilitate all assay steps, from sample preparation to final detection.

For several decades magnetic particles[†] have been applied in pipette-based assays, ranging from manual assays for basic research to assays in high-throughput instruments for centralized laboratories.[14] The main advantages of using magnetic particles are that they have a large surface-to-volume ratio, are conveniently bio-functionalized, and can be manipulated by magnetic fields, thereby simplifying extraction and buffer replacement steps. Particles are commercially available with different sizes, magnetic properties and surface coatings. Especially so-called superparamagnetic[15] particles have found major interest in medical applications as these particles exhibit strong magnetic properties only when exposed to magnetic fields. Superparamagnetic particles consist of separate magnetic grains (*e.g.* iron oxides), which are embedded inside a non-magnetic matrix.[16] The non-magnetic matrix can be polymeric or inorganic, and serves as a starting point for bio-functionalization.

[†]In the scientific literature, the terms "magnetic particle" and "magnetic bead" are often interchangeably used. We use "magnetic particle" because it is more general, as "magnetic bead" mostly relates to spherical particles made of composite material.

The availability of magnetic particles and corresponding assay reagents has formed a solid starting point for explorations toward miniaturization, *i.e.* efforts to realize integrated and miniaturized technologies based on magnetic particles.[17,18] At small scales, it is difficult in principle to manipulate fluids due to high flow resistances, dominance of capillary forces, and difficulties to achieve mixing at low Reynolds numbers. Yet, the manipulation of magnetic particles by magnetic fields scales favorably with system miniaturization, because close to field generators the magnetic fields are strong, because magnetic field gradients are large close to structures with high curvature, and in addition only short distances need to be travelled in miniaturized devices. This has led to the concept of stationary microfluidics,[19–21] in which overall fluid manipulation is minimized and the control of assay steps is mainly effectuated by magnetic particles and magnetic forces. Biosensing systems based on stationary-fluidic concepts are very attractive because they do not require continuous fluid actuation to be integrated in the system. Technologies based on continuous fluid actuation generally require large fluid volumes or complex cartridge architectures. Stationary assay concepts also require some kind of fluid actuation, namely to guide the to-be-tested fluid sample into the cartridge. From the perspective of fluid handling, the simplest solution for a magnetically controlled stationary-microfluidic assay is a cartridge in which the initial transport of sample into the cartridge is effectuated by passive capillary forces.

Magnetic particles are highly versatile and have been studied for many process steps that are required in lab-on-chip diagnostic assays. Magnetic particles have been applied (see Figure 6.1) for fluid mixing, selective capture of specific analytes (*i.e.* the biomarkers that need to be detected), analyte concentration, analyte transfer from one fluid to another, analyte labeling, the application of stringency and washing steps, and probing of biophysical properties of the analytes. In this chapter, an overview will be given of the accomplishments of magnetic particles in all these functions within the framework of stationary microfluidics.[22] First we discuss the process of capture of analyte from fluid (Section 6.2) and subsequently the detection step (Section 6.3). While the capture step is general for all particle-based affinity assays, the detection process has three assay variants, as the captured analytes can be (i) directly labeled, (ii) sandwiched between two particles, or (iii) attached to a sensor surface. In Section 6.4 we summarize the status of the *integration* of the different magnetically actuated assay steps with the vision that in the future it will become possible to realize integrated lab-on-chip biosensing assays in which all assay processes are controlled and optimized by magnetic forces. We draw examples from work reported in the peer-reviewed scientific literature including our own research papers. We focus on the application of magnetic actuation in immunoassays and less on nucleic-acid detection assays. Finally, we discuss the current challenges and promising directions for integrated biosensing based on actuated magnetic particles in microfluidic devices.

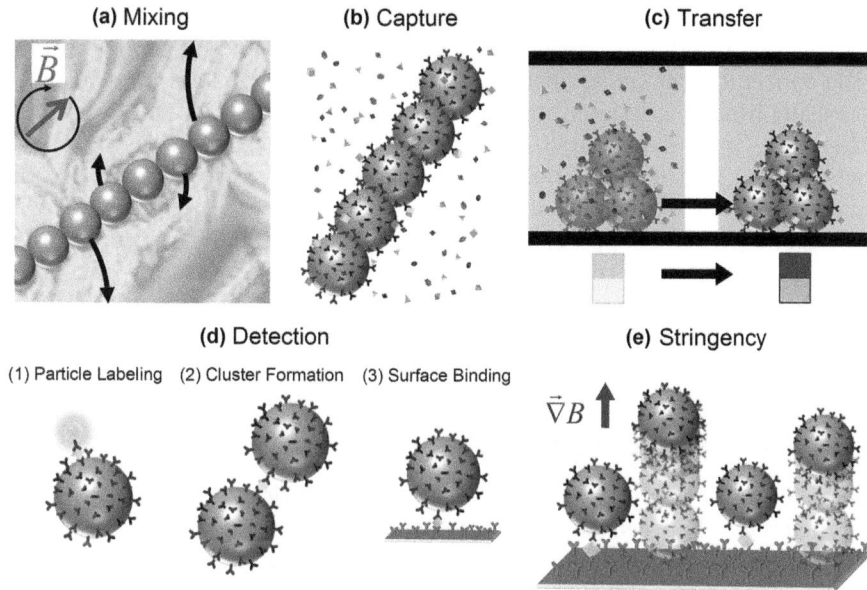

Figure 6.1 Application of magnetic particles in several process steps of a lab-on-chip diagnostic assay. Actuated by applied magnetic fields, magnetic particles have been used (a) to mix fluids, (b) to selectively capture specific analytes, (c) to transfer analytes to another fluid, (d1) to label particles for detection, (d2) to form clusters for detection, (d3) to induce surface binding for detection, and (e) to apply stringency forces in order to improve the signal-to-noise ratio.
(a) Adapted with permission from ref. 41. Copyright 2007, The American Physical Society.

6.2 Capture of Analyte Using Magnetic Particles

The high surface-to-volume ratio and the availability of many bio-functionalization options make magnetic particles well-suited for the capture of analyte from biological samples. The analyte capture can be of specific as well as non-specific nature. Non-specific capture has been mainly developed for the isolation and purification of nucleic acids from lysed samples. In particular magnetic silica particles have been found to be very useful for nucleic acid preparation and detection.[23-27] The capture relies on the physisorption of the nucleic acids to the particles and is followed by fluid exchange steps to achieve isolation and purification. Specific capture requires functionalizations of particles with specific capture molecules, such as antibodies, with a high affinity to the analyte that is to be detected. In both cases, *i.e.* specific and non-specific, the analyte capture rate scales with the total surface area of the suspended particles and therefore with the particle concentration. However, the use of a very high concentration of particles has disadvantages for downstream processes in an integrated multi-step lab-on-chip assay. Increased particle concentrations generally increase non-specific particle–particle and particle–surface interactions, enhance field-induced particle aggregation, cause steric hindrance in particle concentration steps, obstruct chemical reactions on the particles, and sterically hinder reactions between the particles and a biosensing surface. Therefore, it is desirable to decrease particle concentrations while still maintaining high capture rates. To this end, mixing techniques based on magnetic actuation can be applied. In the following paragraphs, we will discuss the effects of applying magnetic actuation for analyte capture, with a focus on the process of *specific* analyte capture. We describe the basic mechanisms underlying particle-based affinity capture of target analytes and review the literature on fluid mixing and analyte capture by magnetic particles in a static fluid.

6.2.1   The Analyte Capture Process

The capture of analytes from a fluid onto magnetic particles is driven by (i) encounters between target analytes and bio-functional molecules on the surfaces of the particles, and (ii) the subsequent biochemical reactions between analytes and the surface-coupled capture molecules. This creates two avenues to accelerate the capture rate, firstly by increasing encounters and secondly by increasing the biochemical reaction rate. In contrast, the specificity of the capture is exclusively generated by the biochemical reaction. For example, in immunoassays antibodies are coupled to the particles for specific capture of analytes from the fluid. The analytes are typically present in very low concentrations within a complex sample containing a high concentration of background material, such as blood or saliva. In such complex matrices, non-specific adhesion of non-targeted molecules to the magnetic particles can seriously hamper the effectiveness of the assay. Therefore, it is essential to have control over the surface properties and to have a detailed understanding of the specific and non-specific surface reactions.

The process of particle-based capture of target analytes is similar to a bimolecular binding process,[28] *i.e.* it consists of an encounter step between the two components, which leads to a transient complex that can subsequently react chemically and become a bound complex. The total process is characterized by the overall rate constant of association, k_a (unit $M^{-1}\,s^{-1}$). For typical protein–protein interactions, association rate constants range between 10^3 and $10^9\,M^{-1}\,s^{-1}$. The association rate constant of specific protein–protein interactions is to a large extent determined by the fact that the two macromolecules can only bind if their outer surfaces are aligned and oriented in a very specific manner.[28] A relative translation by a few Angstroms or a relative rotation of a few degrees is enough to break the specific interactions. In general, the association rate of a protein complex is limited by diffusion, by geometric constraints of the binding sites, and may be further reduced by the final chemical reaction. Often, diffusion or the chemical reaction dominantly limit the reaction rate, but there is no simple test available to determine which process is the most important.

Volume transport   Near-surface alignment   Biochemical binding

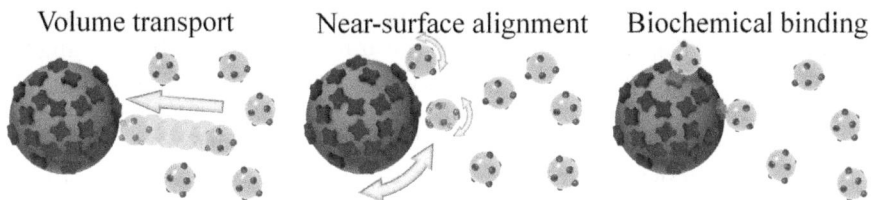

Figure 6.2 Effective affinity capture of target proteins depends on the reactive surface properties of magnetic particles. Schematic representation of the different stages in the capture process of targets (small light-gray spheres) by capture particles (large dark-gray spheres). Targets and capture particles are sketched with multiple binding sites. The stages are as follows: (left) volume transport creates target particle encounters, (middle) near-surface transport creates alignment of binding sites, and (right) bonds are formed by chemical reaction.
Reprinted with permission from ref. 31. Copyright 2013, American Chemical Society.

Nevertheless, there are two indications for a diffusion limitation.[28] First, diffusion-controlled rate constants are usually high ($>10^5$ M^{-1} s^{-1}). Second, diffusion-controlled association involves only local conformational changes between unbound proteins and the bound complex, while reaction-controlled association typically involves gross changes such as loop re-organization or domain movement.[28] Typical antibody-antigen association rate constants are in the range of 10^5–10^7 M^{-1} s^{-1}, which indicates that such reactions are generally diffusion-controlled.[28–30]

While most literature has focused on understanding the bimolecular reaction between two proteins that are free in solution, we here concentrate on the case that one of the proteins is immobilized on a particle surface. In that case, the bimolecular binding reaction is a particle-protein interaction, whose (bio)chemical specificity is determined by the capture proteins that are coupled to the particle. Van Reenen *et al.*[31] performed an experimental study to identify to what extent different stages of the binding process are limiting. In particular, the diffusional encounter step was split up into the process of diffusional transport through the fluid volume and the process of near-surface alignment (see Figure 6.2). Where volume transport generates the first encounters between particles and target proteins, the subsequent near-surface alignment process deals with the alignment rate of the binding sites of the reactants. The volume transport is essentially a translational process, while the alignment is determined by both the translational and the rotational mobility of the reactants. The following reaction equation was used to describe the capture process of a fluorescent target nanoparticle (FT) by a magnetic capture particle (MC):

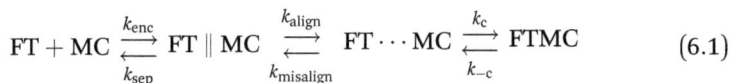

$$\text{FT} + \text{MC} \underset{k_{\text{sep}}}{\overset{k_{\text{enc}}}{\rightleftharpoons}} \text{FT} \parallel \text{MC} \underset{k_{\text{misalign}}}{\overset{k_{\text{align}}}{\rightleftharpoons}} \text{FT} \cdots \text{MC} \underset{k_{-\text{c}}}{\overset{k_{\text{c}}}{\rightleftharpoons}} \text{FTMC} \qquad (6.1)$$

Here the different intermediate reaction products are (i) the encounter complex (FT‖MC), which forms after the initial encounter, (ii) the aligned complex (FT···MC), which forms after alignment of binding sites, and (iii) the bound complex (FTMC), which forms after the chemical reaction has taken place. Between the different states, the forward and backward reaction rate constants are indicated. In experiments, the different processes were studied by varying types of particle actuation, target sizes, types and concentrations of proteins on the particle surface, and ionic strength of the medium. It was found that both volume transport and the alignment of binding sites determine the association rate constants for particle-based target capture.

When free proteins react in solution, the alignment process (*i.e.* rotational diffusion) is an important restriction due to the highly specific alignment constraints,[29,32] but volume transport (*i.e.* translational diffusion) is not a limitation. In case one of the two proteins is attached to a surface, however, volume transport can become a limitation.[33] Depending on the number of binding sites at the surface and the intrinsic chemical reaction rate, reactants can become depleted close to the surface, and depletion can be reduced by actively transporting the fluid over the surface.[34] Therefore, depletion can also play a role in particle-based target capture, which means that reaction rates may be positively influenced by actively applying volume transport processes. Indeed, increased reaction rates have been observed when actively moving particles through the fluid using magnetic fields,[35] and when applying flows to induce fluid perfusion through clusters of magnetic particles.[36] Limitations due to specific alignment constraints can be further overcome by maximizing the number of binding sites on the particle surface and by improving the orientation of the immobilized proteins. It is equally important, however, to optimize the surface properties of the particles to reduce non-specific interactions and to make particle-based assays suitable for operation in complex fluids. In practice, the surface optimization process is more easily performed for particles than for planar surfaces, due to the fact that surface engineering can be applied with a higher throughput to particles in solution than to surfaces in micro-devices.

6.2.2 Analyte Capture Using Magnetic Particles in a Static Fluid

Improving the capture efficiency of target analytes from a fluid means that the interaction between the analyte and the capture agents (*e.g.* antibodies) should be maximized. For example, in surface plasmon resonance biosensing, a surface immobilized with antibodies is used to capture analytes from a fluid that is flowing past the surface.[37] Without a flow, the analyte concentration at the surface can become limited by diffusion, which reduces the binding rate. The application of a fluid flow overcomes the diffusion limitation, maintains a maximum analyte concentration at the surface, and therefore keeps the binding rate at a maximum value. The total binding rate scales linearly with the number of binding sites and therefore also with the available surface.

Immobilizing antibodies on micro- or nanoparticles increases the reactive surface-to-volume ratio during incubation. To improve the binding rate, two options are available: (i) increasing the particle concentration and (ii) enhancing the particle–fluid interaction, *e.g.* by magnetic forces. While the first is easier to apply, care should be taken that the particle concentration does not hinder the further process steps. Analyte capture using non-actuated magnetic particles in a static microfluid was shown by *e.g.* Bruls *et al.*[21] A sample fluid was inserted in a cartridge and the reaction chamber was filled by capillary forces. Thereafter, a dry reagent containing magnetic particles automatically dissolved into the fluid, allowing the magnetic particles to capture analyte from the solution by means of diffusion.

To actively enhance the particle–fluid interaction in a static fluid, several studies have been performed to induce fluid mixing by the application of magnetic fields. Because of the small dimensions in microfluidic devices, viscous forces generally dominate the fluid behavior, resulting in slow and inefficient mixing. Vuppu *et al.*[38,39] were one of the first groups to discover that superparamagnetic particles form rotating chains in a rotating magnetic field. The rotors could be assembled dynamically and the length and speed was found to be varying in time (see Figure 6.3a). To understand the dynamic behavior of the chains in more detail, models have been developed in which particle chains were treated in two[40,41] or in three[42–45] dimensions. To characterize the chain behavior, most studies[40–42] have used the dimensionless Mason number, which is the ratio between the rotational shear forces (*i.e.* hydrodynamic drag) and the magnetic interaction forces (*i.e.* the magnetic forces):

$$Ma = \frac{\eta\omega}{\mu_0\chi_p H^2}, \tag{6.2}$$

with η the dynamic viscosity of the fluid, ω the field rotation frequency, μ_0 the permeability of free space, χ_p the magnetic susceptibility of the particle, and H the magnetic field strength. It has been found[40,41] that for high Mason numbers (*i.e.* low magnetic torques), particle chains split up in small chains that only mix well locally, whereas for low Mason numbers (*i.e.* high magnetic torques) particle chains stay rigid, with little mixing near the center of the particle chain and better mixing towards the ends of the chains (see Figure 6.3b). The best mixing conditions were obtained at intermediate Mason numbers where chains break and reform repeatedly, inducing fluid flow characteristic for chaotic mixing.[46] To prove that the induced mixing is of a chaotic nature, Kang *et al.*[41] computed the Lyapunov exponents[46] at different Mason numbers (see Figure 6.3c), which are a measure for the spatial divergence of two artificial fluid tracers that are initially separated by a very small distance. As shown in Figure 6.3c, for intermediate Mason numbers, the highest Lyapunov exponents are found, indicating the optimal regime for chaotic mixing.

To validate theoretical data, experimental studies on the mixing of fluids have been performed in microliter reaction chambers[43,44,47–49] and in

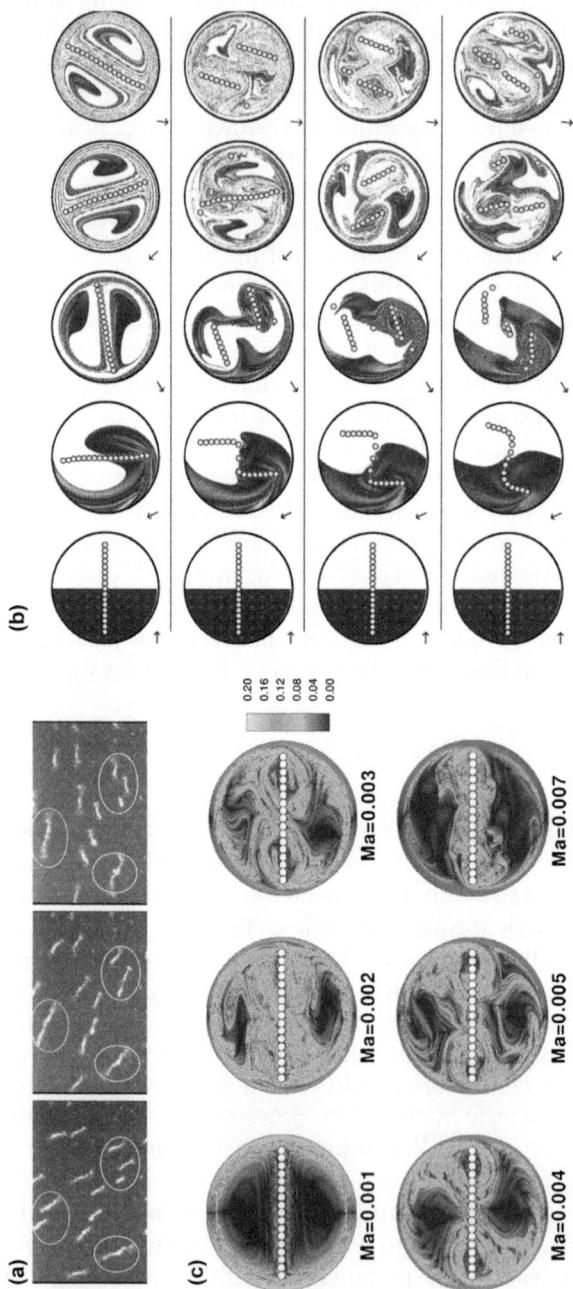

Figure 6.3 Mixing of fluids using chains of magnetic particles. (a) In a rotating field (3.2 Hz), magnetic particles form chains that grow and fragment dynamically. (b) 2-D numerical simulations of a particle chain in a rotating field, showing the progress of mixing at different time points (to the right), for four different Mason numbers: (from top to bottom resp.) Ma = 0.001, 0.002, 0.003, and 0.005. (c) The spatial distributions of the maximum Lyapunov exponent at several Mason numbers. (a) Reprinted with permission from ref. 38. Copyright 2003, American Chemical Society. (b,c) Reprinted with permission from ref. 41. Copyright 2007, The American Physical Society.

droplets.[45] The experiments confirm the computed dynamics of the particles,[43,48] but simulations fail to describe systems in which the particle density is very high. Over time, particle chains grow in length[49] and interact with other chains,[45,47,48] which is not covered in simulations of isolated chains. Still, it is found that optimal fluid mixing is obtained for long chains that exhibit breaking and reformation behavior.

A slightly different application of rotating magnetic particle chains is the capture of biological targets from the fluid. In case of fluid mixing, a multiple fluid compartment is homogenized, and the total asymmetry in the concentration of a particular species is reduced over time. In case of target capture, (bio-)chemical reactions occur at solid-liquid interfaces, resulting in a heterogeneous concentration of the targets within the fluid. When the magnetic particle chains act as the reactive solid-liquid interface, they can be actuated to homogenize the fluid but also to improve the scavenging abilities of the particles. These applications of magnetic particle actuation have been studied less well compared to fluid mixing.

Several papers[48,50,51] have reported on the effect of particle chain rotation on the capture rate (see Figure 6.4). In general, it is found that the capture rate can be improved roughly by one order of magnitude. Gao et al.[48] found that rotating particle chains that exhibit breaking and reformation behavior (resulting in chaotic fluid mixing) enhance the capture rate of target analytes by a factor of 3 compared to rigid rotating particle chains. The breaking and reformation behavior of particle chains was enhanced by alternating the Mason number over time, i.e. by alternating between high- and low-frequency actuation. In this way, global mixing was alternated with local

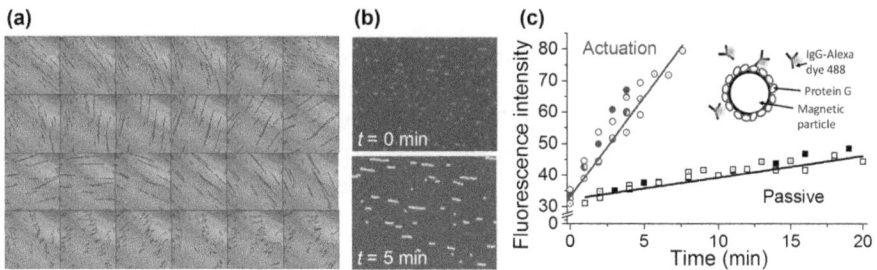

Figure 6.4 Capture of antibodies from a fluid using magnetic actuation of magnetic particle chains. IgG antibody molecules (labeled with Alexa 488 fluorescent dye) are captured by magnetic particles coated with protein G. (a) Microscope images of chain of particles in a magnetic field rotating alternatingly at 1 and 30 Hz in order to break and reform particle chains. (b) Fluorescence images of IgG targets captured by the magnetic particles. (c) Average fluorescence intensity per magnetic particle for actuated and non-actuated particles. In case of actuation, the capture rate is found to be one order of magnitude larger compared to no actuation (passive capture).
Reprinted from ref. 51 with permission from The Royal Society of Chemistry.

mixing. However, the Mason number is not an optimal parameter to describe the system, because it is defined for individual chains, not for large numbers of mutually interacting chains. Where a single isolated chain would remain rigid, a non-isolated chain at the same Mason number would grow further and maintain breaking and reformation behavior.

These studies show that magnetic particle-based target capture can be accelerated by applying time-dependent magnetic fields on particles in a static fluid volume. We note that it would be very useful if future studies reported capture data in terms of the association rates, including the influence of actuation parameters such as field strength, field frequency, and field direction. This would allow for a comparison between the different devices and actuation methods.

6.3   Analyte Detection

After the capture of target analytes, further processing is needed for accurate and specific detection. When the magnetic particles are used as carriers only, then the captured analytes are exposed to further (bio)chemical processes and are typically detected by luminescent labels such as enzymes or fluorescent molecules. Magnetic particles can also serve as a label, to signal molecular binding at a sensing surface, or to signal molecular binding between particles in an agglutination assay. Here we review different methods to perform the detection step and focus on their potential for total lab-on-chip integration of the assay.

To equally evaluate the analytical performance of the different detection methodologies, we analyze the reported dose-response curves in terms of the linearity of the curve, the achieved dynamic range, and the limit-of-detection (LoD) or, better, the limit-of-quantification (LoQ),[52,53] which is the relevant parameter for medical applications. When estimating the LoQ for linear dose-response curves, we will use the definition that the LoQ equals ten times the standard deviation of the blank divided by the slope of the dose-response curve.[53] In general, linear dose-response curves will have similar values for the LoD and the LoQ. Furthermore, we will mention the studied biomarker, the used sample matrix and the incubation times, as all these factors affect the final performance of the detection methodology. For example, an assay based on the streptavidin-biotin affinity reaction in buffer will perform much better than an antigen-antibody assay in plasma.

6.3.1   Magnetic Particles as Carriers

When magnetic particles are used as a carrier or substrate for the detection of target analytes, the particles are first used to capture target analytes, subsequently the captured analytes are labeled, and finally the label is detected.[4] For accurate detection, it is important that only bound analytes are labeled and that only bound labels are detected. This requires several washing or separation steps to be performed, *i.e.* the particles need to be exposed to different fluids. In many studies, microfluidic fluid flows have been used to effectuate fluid replacement.[22] Here we focus on stationary-fluidic concepts, wherein particles are moved between different fluids using magnetic forces.

An example of particle transfer between two stationary fluids is sketched in Figure 6.5. The fluids are separated by a medium that does not mix with the aqueous fluids, for example a non-polar fluid or a gas. When a

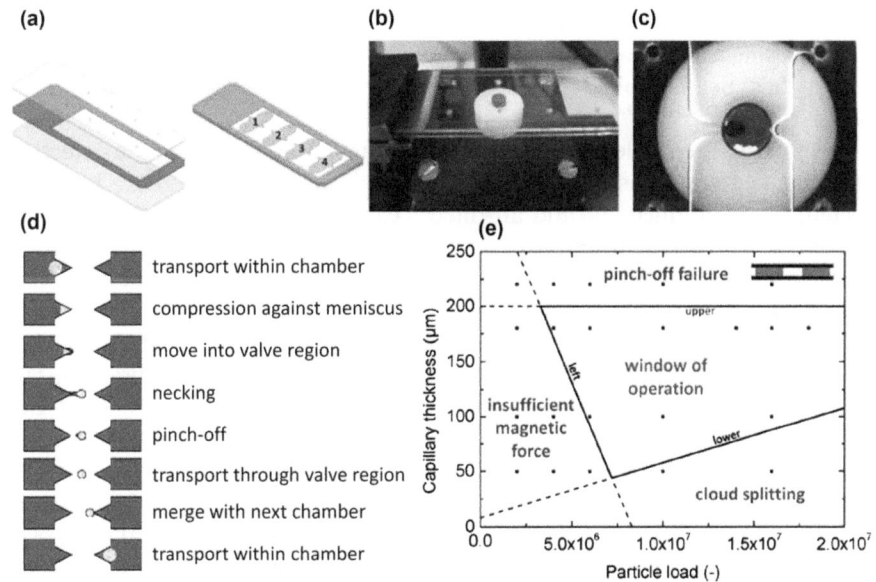

Figure 6.5 Example of Magneto-Capillary Particle Transfer (MCPT). (a) Schematic drawing of a patterned air valve cartridge in exploded view (top) and assembled view (bottom) as developed by Den Dulk *et al.*[24] Shown is double-sided tape that joins the transparent planar top and bottom parts. Aqueous liquids with a typical volume of 15 μL were confined in four chambers by a pattern of hydrophilic and hydrophobic regions. (b) Photo of the MCPT cartridge in the experimental setup, showing the permanent magnet (dark, 4 mm Ø) embedded in a white background, in close proximity to the bottom of the cartridge. (c) Top view microscope image, showing two translucent aqueous chambers and the magnet in the valve region between the chambers. The magnet draws a cloud of magnetic particles (the even darker spot in front of the magnet) from the left chamber to the right chamber. (d) The different phases of the transfer of magnetic particles (white spot) from one liquid (dark) to another through an air-valve. (e) Parameter space of capillary thickness (distance between the planar top and bottom parts) and particle load (for a given geometry). The squares indicate the experiments that were performed. The areas in the diagram describe the behavior of the magneto-capillary valve. In the central region successful MCPT was observed for the given cartridge design. In the three outer regions non-ideal operation was observed.
Reproduced from ref. 24 with permission from The Royal Society of Chemistry.

sufficiently high force is applied on the magnetic particles (see Figure 6.5d), they are pulled out of the fluid, through the interface between the fluid and the separation medium, and are thereafter pulled into the next fluid. The process of transferring magnetic particles between the fluids is controlled by a balance between the magnetic forces on the particles and the forces caused by interfacial tension, so-called capillary forces. We propose to very generally refer to magneto-capillary particle transfer mechanisms as Magneto-Capillary Particle Transfer (MCPT). MCPT concepts have been studied in many different device geometries, like tubes,[54,55] single-plane devices,[19,20,25,56–58] vertical slits,[59,60] arrays of wells,[61] and bi-plane capillary devices.[24,62–65] MCPT has been applied for the enrichment and purification of nucleic acids[19,20,24,55,59,61,63,64,66] as well as proteins.[24,54,62]

Using MCPT, complete lab-on-chip assay integration including the detection step has been demonstrated. Nucleic acids[25,65] and proteins[64] have been detected, using oil-filled[25,65] and air-filled[64] capillary valves. Gottheil *et al.*[64] reported a complete immunoassay with dose-response curves. Magnetic particles were moved over a cartridge through different chambers and incubation in each chamber was performed under actuation by a rotating magnetic field. In the assay, target interleukin-8 in buffer with 10% human serum was incubated for 30 minutes and fluorescently labeled after a washing step. The corresponding dose-response curves had a dynamic range of at least two orders of magnitude and they contained a segment with a linear relationship between signal and concentration, with a LoD/LoQ in the picomolar range.

In the above-mentioned studies the particles were moved through a capillary fluid interface; alternatively a capillary fluid interface can be moved over the particles, *e.g.* by using electrowetting.[57,58,67,68] Ng *et al.*[67,68] used droplets with different fluids (sample fluid, wash buffers, conjugate buffers, *etc.*) to detect Thyroid Stimulating Hormone (TSH) in buffer using an ELISA. It was shown that for droplets surrounded by air, washing efficiencies above 90% could be achieved using the combination of electrowetting and a moving magnet. With an incubation time of 6 minutes, a linear dose-response curve was obtained with a dynamic range of almost two orders of magnitude and a LoD in the picomolar regime.

In the carrier-only assays, the particles are exposed to multiple fluids, as is also done in high-throughput robotic pipetting-based assays. An advantage of carrier-only assays is that one can make use of reagents that are very similar to the ones used in commercial pipetting-based assays. Another advantage is that the detection can occur in the bulk fluid, so the control of cartridge surfaces is not very critical. Intrinsic to magnetic carrier assays is the need for relatively strong magnetic fields to let particles traverse capillary interfaces, with the disadvantage that particles become highly concentrated and non-specific interactions are promoted. In the mentioned papers, redistribution of particles has been effectuated by removing the magnet from the sample chamber, allowing particles to spread by diffusion. The images (see *e.g.* refs. 25,64) show that particle redispersion is incomplete; large clusters break up into small clusters, but not separate particles. We expect that improved disaggregation of particles, *e.g.* by magnetic field actuation[69] and/or by further reducing non-specific interactions, will improve the assays. The carrier-only methods allow miniaturization and integration, however, a series of fluids is always needed, including active control of fluid droplets and/or methods for magneto-capillary particle transfer.

6.3.2 Agglutination Assay with Magnetic Particles

Agglutination assays exploit the fact that aggregates of particles are formed when specific analytes are present in the sample fluid[‡]. The particles are provided with target-binding moieties and the analytes should have at least two epitopes that can react with the particles. The degree of aggregation is a measure for the concentration of analytes within the fluid. In magnetic agglutination assays, the formation of particle clusters is accelerated by bringing particles together under the influence of a magnetic field. Here, an overview will be given of methods that have been studied to perform agglutination assays in a static fluid.

It was shown by Baudry *et al.*[70] that particle aggregates can be formed effectively in a static fluid by the application of static magnetic fields. After target capture, field-induced chains of particles were formed in order to accelerate the formation of target-induced clusters, resulting in a total incubation time of 6 minutes. The particle clustering was detected using turbidimetry, as the scattering properties of particle clusters differ from unclustered particles. A slightly sub-linear dose-response curve was shown for the detection of ovalbumin in buffer, with a dynamic range of less than two orders of magnitude and a LoD in the low picomolar range.

Park *et al.*[71–73] monitored the growth of particle chains in a rotating magnetic field by measuring light transmittance through the sample volume. As particles form aggregates, longer chains are obtained for increased target concentrations, causing larger fluctuations in the transmitted light. For avidin in buffer and an incubation time of 15 minutes, sub-linear dose-response curves were obtained with a dynamic range of three orders of magnitude and a LoD just below the nanomolar.

When the analyte concentration is much smaller than the magnetic particle concentration, only a few particle aggregates are formed, governed by Poisson statistics. Many particles will not form any clusters, some particles will form two-particle clusters, and larger clusters will be rare. Ranzoni *et al.*[74] showed that specific doublet formation in the low-concentration regime can be enhanced by applying a pulsed magnetic field during incubation, *i.e.* to alternatingly bring particles in close contact and let them freely diffuse to form specific bonds (see Figure 6.6a). Furthermore, the optical detection sensitivity of doublets was improved by measuring the optical scattering in a rotating magnetic field.[75] A dose-response curve was shown

---

[‡]In the scientific literature, the terms "particle agglutination", "particle aggregation", and "particle clustering" are interchangeably used.

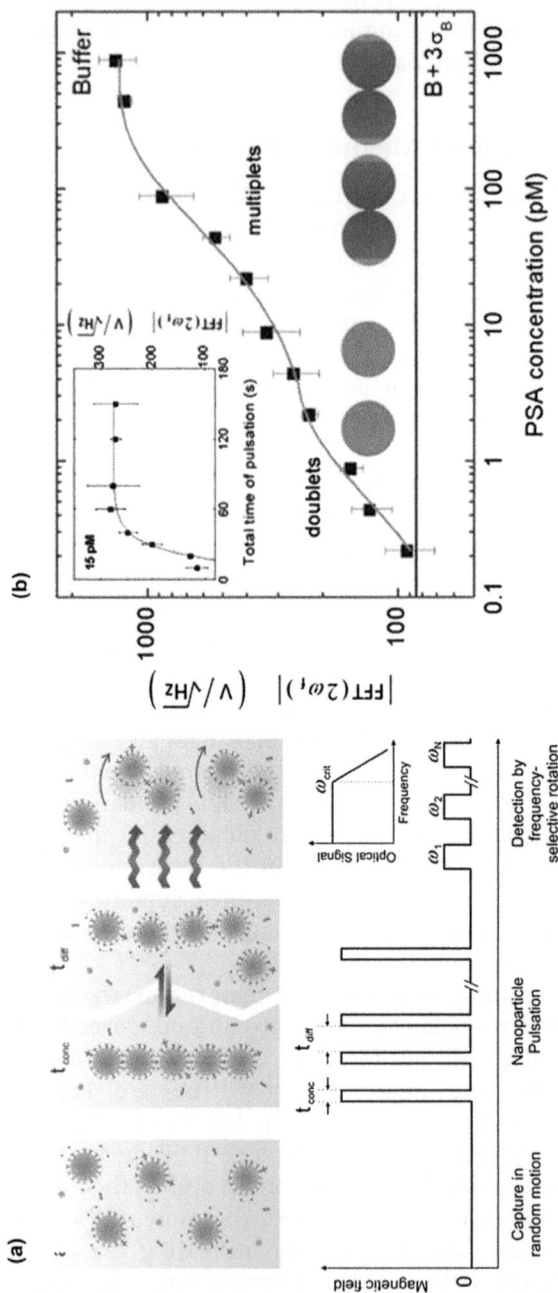

Figure 6.6 One-step magnetic cluster assay using pulsed magnetic fields. (a) Principle of a one-step homogeneous assay technology based on magnetic nanoparticles. After spiking the biological sample with nanoparticles, first targets are captured by diffusive motion. Subsequently cluster formation is accelerated by pulses of the magnetic field to bring particles together (t_{conc}) and subsequently allow relative diffusion to enhance bond formation (t_{diff}). Finally, clusters are detected by applying rotating fields and by measuring the optical scattering signal. The result is a curve of optical scattering signal as a function of frequency (see inset); the plateau reveals the number of clusters in solution, while the critical frequency reveals the cluster size and the viscosity of the sample. (b) Dose-response curve in buffer, where each point has been measured in triplicate. The curve has been fitted based on a model description which accounts for the cluster size. The inset shows the binding signal as a function of the pulsation time. The graphical sketch shows in light-gray the effective areas on the particles for cluster growth, for single particles, and for doublets. The horizontal line shows the level of the blank plus three times the standard deviation of the blank.

(a,b) Reprinted with permission from ref. 74. Copyright 2012, American Chemical Society.

for the detection of PSA directly in undiluted blood plasma with a total in-cubation time of 12 minutes. The curves had an undulating character (see Figure 6.6b), revealing regimes of clusters of different sizes, with a dynamic range of three orders and a LoD around a picomolar.

The two main challenges of particle-based cluster assays are (i) to ensure good contact between the particles in order to increase assay kinetics, and (ii) to minimize non-specific particle clustering in complex biological sam-ples. Magnetic fields help to bring the particles together and thereby en-hance the inter-particle binding kinetics. However, strong magnetic fields may also increase the non-specific binding between the bio-functionalized particles by the strong attractive magnetic dipole–dipole interactions that are induced. Agglutination assays in a non-flowing fluid can, however, be carried out in weak and/or pulsed magnetic fields and thereby avoid strong interactions between the particles.[74] Stationary-fluid magnetic agglutination assays are highly suited for miniaturization and integration, because in principle the assays can be performed in one chamber. Thus far, mag-netically actuated mixing and target capture have not been applied in these types of agglutination assays, but several actuation methods[35,48] seem to be suited. If sample pretreatment steps such as analyte purification or enrich-ment are required, MCPT[24] may in principle be combined with the agglu-tination assay. When further technological improvements are made, we expect that static-fluid magnetic agglutination assays may become a very interesting format for point-of-care applications.

6.3.3 Surface-binding Assay with Magnetic Particles as Labels

In a surface binding assay, magnetic particles are used as labels that bind in a biologically specific manner to a surface and thereby report the presence of a specific molecular species. Most commonly a sandwich format is used, with specific binders being immobilized on the particles and on the surface, which capture the target on two different epitopes. Preferably, all analytes in the fluid become sandwiched between a particle and the surface, which is possible when the concentration of particles exceeds the analyte concen-tration. To ensure efficient capturing and labeling, the magnetic particles need to efficiently (i) capture targets from solution (*cf.* Section 6.2), (ii) be brought to the surface, and (iii) interact with the surface on molecular length scales. The transportation toward the surface can be achieved relatively easily by applying magnetic field gradients towards the sensor surface.[76,77] It is more difficult to control the particle–surface interaction, because particles concentrated at a surface mutually exhibit magnetic particle–particle and steric particle–particle interactions. In addition, rotational exposure of the particles to the surface is important and non-specific interactions between the particles and the surface should be minimized.[78]

Several methods have been developed to optimize the particle–surface interactions. Bruls *et al.*[21] developed an actuation protocol in which in-plane fields, out-of-plane fields, and field-free phases are alternated. In-plane fields bring particles to the surface, out-of-plane fields generate chains, while the field-free phase allows free Brownian motion of the particles in order to optimize their rotational and translation exposure to the surface. In this way, effective specific sandwich formation was shown. The protocol keeps the particles in constant motion relative to the sensor surface, which may also minimize non-specific binding with the surface. Dose-response curves were determined for the detection of cardiac troponin (cTnI) in buffer and in undiluted blood plasma, with total incubation times of less than 5 minutes. The dose-response curves were practically linear with a dynamic range of three orders of magnitude and LoDs around a picomolar.

Gao and Van Reenen *et al.*[69] developed an actuation protocol to induce repulsive magnetic dipole–dipole interactions between particles at a surface. The method consists of aligning particle aggregates with a surface by using field gradients and in-plane oriented magnetic fields, followed by the application of an out-of-plane magnetic field while a field gradient maintains the particles at the surface. In this way, clusters of microparticles were shown to disaggregate (see Figure 6.7). By repeatedly applying these two steps, clusters consisting of tens of particles could be almost completely redispersed over the surface in several tens of seconds. Evaluation of this method in a surface-based assay has, however, not yet been performed.

As described above, the method to assist the binding of particles at the sensor surface is very important for the character of the dose-response

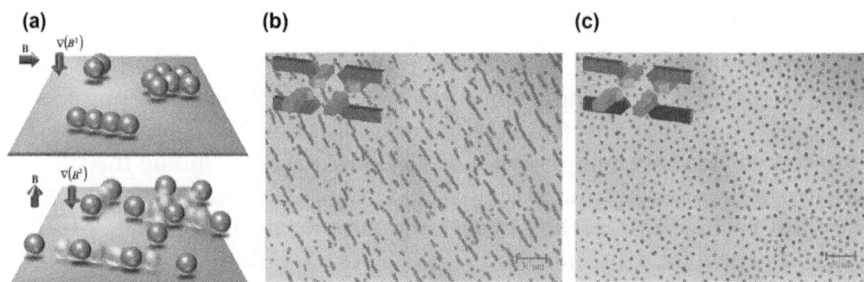

Figure 6.7 Controlling particle behavior at a sensor surface. (a) Schematic representation of disaggregation of magnetic particle clusters. (top) Clustered groups of magnetic particles that are present in solution are drawn to a physical surface by means of a field gradient. An in-plane magnetic field is applied to optimize parallel alignment of the clusters with the surface. (bottom) Application of a magnetic field oriented orthogonal to the surface in combination with a field gradient results in breaking of the particle clusters by magnetic dipole–dipole repulsion. (b) Experimental images of ∅ 2.8 μm magnetic particles aligned to a glass substrate. (c) The application of an out-of-plane magnetic field causes clusters to break up and evenly distribute over the surface.
Reproduced from ref. 69 with permission from The Royal Society of Chemistry.

curve and the resulting detection sensitivity. Now we will discuss the next step, namely the detection of bound particle labels at a sensing surface. A method to detect particle labels should be sensitive but, in addition, one should consider the influence of the different methods on lab-on-chip integration, the cost effectiveness of the resulting disposable cartridge, and the miniaturization potential of the reader instrument. In particular, one should consider the compatibility of the detection methods with the presence of magnets around the microfluidic reaction chamber and near the sensing surface, in order to allow magnetic control of the particle-based assays.

In early reports, magnetic nanoparticle labels were detected by magnetic coils,[79] SQUID,[80] magnetoresistive sensors,[8,76,81–83] and Hall sensors.[84] Although it is possible to combine magnetic sensing with the application of actuating magnetic fields,[76,82] measurements are often perturbed by the applied fields. Furthermore, the use of lithographically made sensing chips adds costs to the disposable cartridge and demands cartridge assembly technologies suited for high numbers of electrical interconnects.

Optical detection methods are not perturbed by the presence of magnetic fields and are compatible with cost-effective mass-manufactured cartridges. Magnetic particles can be optically detected on a transparent surface in several ways, *e.g.* by using bright-field illumination[78,85] or dark-field illumination.[86] A particular challenge is to design the system in such a way that magnet poles can be positioned very close to the sensing surface. Bruls *et al.*[21] described a detection system based on the principle of frustrated total internal reflection (f-TIR) as depicted in Figure 6.8. A light-emitting diode was used to create an evanescent wave at the sensor surface *via* total internal reflection. The presence of magnetic particles at the surface frustrates the evanescent wave, causing a reduction of reflected light. The amount of particles at the sensing surface was recorded as a function of time by monitoring the reflected light intensity. The advantage of using f-TIR is that it is highly surface-sensitive and compatible with close integration of electromagnets.

Surface-binding magnetic particle assays are highly suited for miniaturization and integration, as fluid manipulation is not necessary and the assay can be completely controlled by magnetic fields. A disadvantage of surface-binding assays is that the sensor surface needs to be biofunctionalized, which adds complexity to the assay. Important advantages are that so-called magnetic stringency[21,76,82] can be applied in the assay to discriminate specific bonds from other types of bonds, as will be discussed in detail in the next section. Furthermore, multiplexing can be realized by preparing binding spots with different biochemical compositions. We foresee several avenues to further control and optimize surface-binding assays by magnetic actuation. For example, actuated mixing and capture[35,48] may help to further increase the speed and effectiveness of the capturing process, and magnetic fields may be used to redistribute particles in the assay chamber.[69]

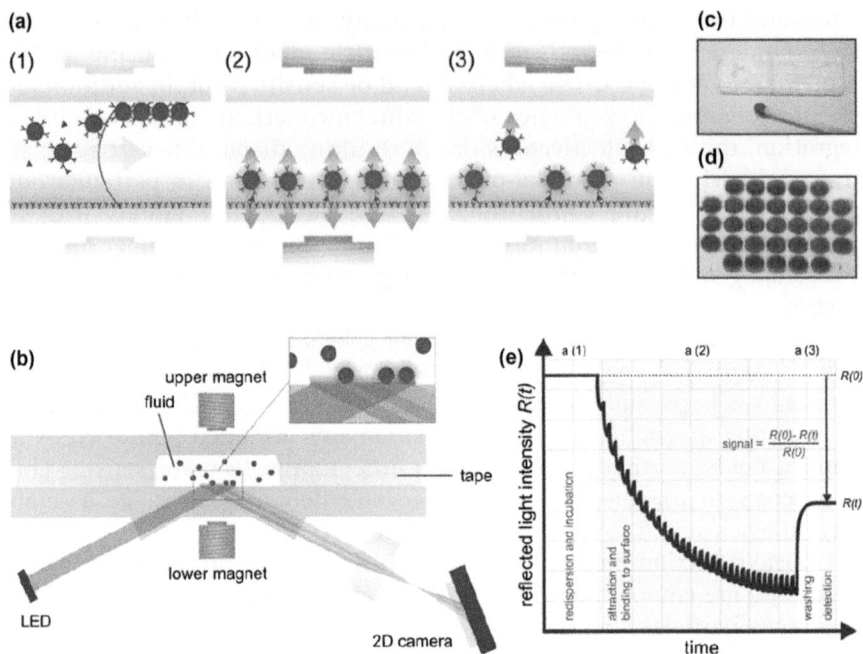

Figure 6.8 Optomagnetic immuno-biosensor based on actuated magnetic particles ($\varnothing$ 500 nm). (a) Schematic representation of the reaction microchamber showing the successive assay processes: (a1) filling of the microchamber, nanoparticle redispersion, and capturing of analyte; (a2) actuation of the particles during the process of binding to the surface; and (a3) removal of free and weakly bound nanoparticles from the sensing surface by magnetic forces. (b) The fluid microchamber placed in the optomagnetic system with electromagnets and detection optics. Light reflects from the sensor surface with an intensity that depends on the concentration of nanoparticles at the sensor surface, by the mechanism of frustrated total internal reflection (f-TIR). (c) Picture of an assembled disposable cartridge (1 cm×4 cm) consisting of two structured plastic parts connected by double-sided adhesive tape. The cartridge contains a sample inlet, a channel, a reaction microchamber (1 μL), and a vent. (d) f-TIR image of magnetic nanoparticles bound to the sensor surface *via* an immunoassay on 31 capture spots of 125 μm diameter each. (e) Schematic real-time curve of the measured optical signal for a single capture spot. The assay phases a1–a3 are indicated. The signal modulation in phase a2 is caused by switched actuation of the magnetic nanoparticles.
Reproduced from ref. 21 with permission from The Royal Society of Chemistry.

6.3.4 Magnetic Stringency

A stringency process aims to improve the specificity of detection, by separating unbound and weakly bound from strongly bound species. In the detection methods discussed above, signals are generated by bonds formed between magnetic particles (agglutination assay) or between particles and a surface (surface binding assay). The bonds should be biologically specific. However, bonds can also have a non-specific nature, *i.e.* the bond is not mediated by an analyte molecule, which results in a false-positive signal. Non-specific bonds can originate from several types of interactions, *e.g.* van der Waals interactions, electrostatic interactions, and hydrophobic interactions.[87] In diagnostic tests, non-specific interactions cause background levels as well as statistical variations of the results, and thus affect the limit of quantification and the precision.

The development of a diagnostic test always involves a series of optimizations in the biochemical and chemical domain in order to improve the specificity of detection, *e.g.* by optimizing the affinity molecules and coupling chemistries, by blocking the surfaces, and by dissolving reagents into the sample, such as pH buffers, salts, surfactants, and blockers for specific macromolecular interferences. The aim of the optimization is to reduce the formation of non-specific bonds and to preserve or improve the specificity of the targeted bonds. In an assay involving magnetic particles, there is an additional degree of freedom in the forces that can be applied to the particles. Magnetic forces can be used to separate bound from unbound particles. Furthermore, when the particle labels are bound to a surface, magnetic field gradients can be applied in order to apply stringency to the bonds and thereby dissociate weak non-specific bonds. Finally, the response of molecular bonds to applied stresses can be recorded, giving even more detailed information about the bonds.

An early report on the application of magnetic stringency to non-covalent bonds was published by Danilowicz *et al.*[88] A permanent magnet was used to apply a constant force to ensembles of bound particles and the dissociation of bonds was recorded as a function of time. Jacob *et al.*[89] used an electromagnet which allowed a larger range of forces to be studied, thereby yielding dissociation rate constants for the biomolecular bonds. Furthermore, they demonstrated that populations of specific and non-specific bonds could be distinguished by the shape of the force-induced dissociation curves. These studies were performed using large magnetic particles as labels (4.5 μm in ref. 88, and 2.8 μm in ref. 89). It is convenient to use large particles for biophysical studies because large forces can be applied to single particles. However, large particles are less suited for integrated biosensing because they diffuse slowly, sediment easily, and they limit the dynamic range of detection due to steric hindrance.

Using particles with diameters of a few hundred nanometers, Bruls *et al.*[21,76,82] have demonstrated the use of magnetic stringency in integrated surface-binding assays, see Figure 6.8a,e. Here, magnetic stringency removes unbound and weakly bound particles from the surface. In fact, it replaces the fluidic wash step in traditional affinity assays. The magnetic stringency obviates the need for fluid manipulation, which simplifies the assay and makes it highly suited for integration in a completely stationary assay concept.

In the future, magnetic stringency may go beyond the application of bound-free separation and the measurement of dissociation properties of molecular bonds. For example, by applying rotating magnetic fields, it has been shown that it is possible to probe the properties of DNA[90,91] or protein complexes[92] that are sandwiched between particles and a surface. Although still very remote from integrated biosensing, the principle of characterizing molecular bonds in a detailed biophysical manner may in the future help to further increase the specificity of biosensing.

6.4   Integration of Magnetic Actuation Processes

Integration is the act of making something into a whole by bringing different parts together. For an engineer, it is the process of (i) defining an overall technological function that needs to be realized, (ii) designing a system architecture and its underlying components, and (iii) quantifying all inter-actions within the system which feeds back to the technological function def-inition. For a given functional aim, one can select different system architectures that each have their own inherent advantages and challenges. As reviewed in the previous paragraphs, the manipulation of magnetic particles by magnetic fields allows one to control in a microfluidic format a number of important assay steps for diagnostic testing. Now we will review how the integration of different assay process steps is proceeding, moving toward integrated assays that perform a series of sophisticated steps, controlled by magnetic forces.

In Table 6.1, we summarize the state-of-the-art in the integration of magnetic particle actuation by magnetic fields for integrated detection assays in stationary-fluidic devices. The top row lists the key assay process steps. The left-most column lists the assay concepts, *i.e.* carrier-only assays, agglutination assays, and surface-binding assays. Within the table matrix, we have classified the type of magnetic actuation used at the intersection between assay-concept and process-step. The gray-scales indicate the type of actuation used: without magnetic fields, with static fields, with dynamic fields, or not applicable. The process-step and assay-concept references serve as examples. The assay-concept references are focused on total assay inte-gration; they report dose-response curves acquired by detection on the microfluidic chip as discussed in the previous paragraphs.

In the field of integrated magnetically actuated assays, individual process steps are being studied as well as the integration of different process steps. The use of magnetic actuation processes for integration purposes is pro-ceeding steadily, as shown in Table 6.1. There are still several white spaces where actuation principles can be applied to further enhance system inte-gration and overall analytical performance. We expect that novel actuation processes will be developed that are based on dynamic rather than static field generation. Scientifically speaking, several magnetic actuation pro-cesses have been qualitatively demonstrated but are not yet well character-ized and modeled. Also, we foresee that magnetic actuation principles will be carefully attuned to specific biomaterials and reagents and, *vice versa*, biomaterials will be designed specifically for use in actuated assays. We foresee that particle-based assays will benefit from the ongoing optimization

Table 6.1 Overview of progress in the integration of magnetic particle actuation for different assay process steps in stationary-fluidic lab-on-chip biosensing. *Top row:* Key assay process steps. *Left-most column:* Assay concepts using magnetic particles. *Matrix:* Classification of the type of magnetic actuation used for the process steps in the different assay concepts. Grayscale indication: ▢ performed without magnetic fields; ▢ performed by applying static magnetic fields; ▢ performed by applying dynamic magnetic fields; blank: not applied in the concept. The mentioned references serve as examples to illustrate the progress in the field; the reference list is not exhaustive. Assay-concept references have integrated the detection step and report dose-response curves.

Assay concepts	References	Overall concept description	Assay Process steps						Detection method
			Magnetic particles are used for mixing	Magnetic particles capture analytes generically[26] or specifically[31,35]	Fluid exchange by magneto-capillary particle transfer (MCPT)	Magnetic particles are redistributed	Molecular sandwich is formed with magnetic particles as label	Labels are magnetically actuated for bound-free separation and/or stringency	
			Vuppu et al.[38] Gao et al.[48]	Berensmeier[26] van Reenen et al.[31,35]	Den Dulk et al.[24]	Gao et al.[69]	Ebersole[94]	Lee et al.[55] Jacob et al.[89]	Wild[4] Boas et al.[96]
Carrier-only assay	Sista et al.[57,58] (2008) Ng et al.[67] (2012)	Magnets are used to manipulate particles, and electrowetting is used to manipulate droplets, surrounded by oil[58] or by air.[67] Particles passively capture targets and enzyme labels.	A magnet is used to move particles into the reagent droplet	idem	idem	Magnet is removed from the droplet to let the particles redisperse by diffusion			Chemilumin-escence
	Gottheil et al.[64] (2013)	Magnets translate magnetic particles through different fluid chambers, separated by air valves. Magnet rotation is applied in each fluid.	The particles are actuated by a rotating permanent magnet	idem	A permanent magnet moves particles through water-air capillary valves	Magnet is removed to let large clusters break up into smaller clusters			Fluorescence in a static field

Table 6.1 (*Continued*)

			Assay Process steps						
Assay concepts	References	Overall concept description	Magnetic particles are used for mixing Vuppu et al.[38] Gao et al.[48]	Magnetic particles capture analytes generically[26] or specifically[31,35] Berensmeier[26] van Reenen[31,35]	Fluid exchange by magneto-capillary particle transfer (MCPT) Den Dulk et al.[24]	Magnetic particles are redistributed Gao et al.[69]	Molecular sandwich is formed with magnetic particles as label Ebersole[94]	Labels are magnetically actuated for bound-free separation and/or stringency Lee et al.[55] Jacob et al.[89]	Detection method Wild[4] Boas et al.[96]
Agglutination assay	Baudry et al.[70] (2006)	Agglutination assay in a static field in absence of flow.	Mixing in a tube	Incubation in a tube by diffusion		Diffusion by switching field off	Particles form chains in a static field		Optical scattering
	Ranzoni et al.[74,75] (2011) (2012)	Agglutination assay in a pulsed rotating magnetic field.	Mixing in a tube	Incubation in a tube by diffusion		Redispersion by diffusion in field-off phase	Particles form chains in a pulsed rotating magnetic field		Optical scattering in a rotating magnetic field
Surface-binding assay	Dittmer et al.[76] (2008) Koets et al.[82] (2009)	Targets are captured passively by magnetic particles and thereafter attracted to a sensor surface by magnetic fields.	Off-chip mixing	Incubation by diffusion		Redispersion by diffusion in field-off phase	Particles are attracted to a surface by a field gradient	Magnetic field gradient	Magnetic field sensing (Giant Magneto-Resistance)
	Bruls et al.[21] (2009)	Targets are captured passively by magnetic particles and thereafter a sandwich is formed under pulsed magnetic fields.	Dried-in particles disperse into the sample fluid; no active mixing	Incubation by diffusion		Particles are redispersed at a surface by a pulsed magnetic field protocol	Particles are attracted to a surface and randomized in a pulsed magnetic field	Magnetic field gradient	Optical scattering (frustrated total internal reflection)

of particles regarding their surface bio-functionalization, surface smooth-ness, and their size and magnetization uniformity.[93]

Current quantitative lab-on-chip biosensing systems mostly consist of a disposable cartridge and a reusable analyzer instrument. Cartridges are single-use objects for reasons of biochemical irreversibility and bio-safety. Therefore, it is important that a system architecture is chosen that limits the complexity of the cartridge. The development of a biosensing cartridge presents challenges in the domains of device technology (*e.g.* fluidics and detection) and biochemistry (*e.g.* reagents and bio-functionalization), and the challenges depend on the architectural choices. Broadly speaking, in the carrier-only concepts, the reagents can be close to the ones developed for pipetting-based assays; however, multiple fluids need to be controlled in the cartridge, which complicates the device technology. The agglutination assays are simpler in terms of device technology, but are demanding on the re-agents because the assays are performed in one step without separation or stringency. The surface-binding assays allow sensitive particle detection and stringency, yet require careful control of the surface bio-functionalization.

The choice of cartridge architecture is also determined by the type of assay. For some assays it is essential to include a purification step, as is for example the case for most nucleic-acid detection assays. A purification step can only be performed within a multi-fluid cartridge concept. Other assays, such as sandwich immunoassays, can be performed in a single step without fluid exchange, which strongly simplifies the cartridge design.

6.5 Conclusions

We have reviewed the use of magnetic particles and magnetic fields to perform key processing steps in integrated stationary microfluidic assays for lab-on-a-chip diagnostic applications. Due to the high surface-to-volume ratio and their adaptable surface (bio-)functionalizations, magnetic particles are effective at achieving rapid and specific capture and labeling of targets. Magnetic particles have been applied to achieve mixing, capture, washing, fluid exchange, labeling, stringency, and detection in several stationary-microfluidic device architectures. Magnetic actuation is an enabler for lab-on-chip integration because it allows a large diversity of sophisticated fluidic and molecular process steps to be controlled by means of externally gener-ated fields, which can strongly simplify the cartridge design. We expect that stationary-fluidic concepts will continue to gain attention in the future, as these concepts maximally exploit the functional properties of the magnetic particles to facilitate lab-on-chip integration.

Overall we see many avenues for further innovation of microfluidic Point-of-Care Testing based on magnetic particles. Magnetic particles are funda-mentally suited for developing miniaturized biosensing systems and allow a range of unique stationary-fluidic system concepts. We expect that inte-grated magnetic actuation-based biosensing systems will have a large impact on society in the future. Such systems will allow quantitative decentralized *in vitro* diagnostic testing in a rapid manner with a user-friendly "sample-in result-out" type of performance, in handheld and desktop-sized instruments. By virtue of these properties, magnetic actuation-based biosensing systems can help to improve patient treatment, patient monitoring, and disease management, with impact on the quality, accessibility, and cost-effectiveness of future healthcare.

Acknowledgements

Part of this work was funded by the Dutch Technology Foundation (STW) under grant number 10458.

References

1. R. W. Forsman, *Clin. Chem.*, 1996, **42**, 813–816.
2. The Lewin Group, The Lewin Group, Inc, 2005.
3. P. von Lode, *Clin. Biochem.*, 2005, **38**, 591–606.
4. D. Wild, *The Immunoassay Handbook*, Elsevier, Amsterdam, 2005.
5. R. Rapley, ed., *The Nucleic Acid Protocols Handbook*, Humana Press, Totowa, New Jersey, 2000.
6. J. M. Nam, C. S. Thaxton and C. A. Mirkin, *Science*, 2003, **301**, 1884–1886.
7. Z. Chen, S. M. Tabakman, A. P. Goodwin, M. G. Kattah, D. Daranciang, X. Wang, G. Zhang, X. Li, Z. Liu, P. J. Utz, K. Jiang, S. Fan and H. Dai, *Nat. Biotechnol.*, 2008, **26**, 1285–1292.
8. S. J. Osterfeld, H. Yu, R. S. Gaster, S. Caramuta, L. Xu, S. J. Han, D. A. Hall, R. J. Wilson, S. Sun, R. L. White, R. W. Davis, N. Pourmand and S. X. Wang, *Proc. Natl Acad. Sci. USA*, 2008, **105**, 20637–20640.
9. G. F. Zheng, F. Patolsky, Y. Cui, W. U. Wang and C. M. Lieber, *Nat. Biotechnol.*, 2005, **23**, 1294–1301.
10. J. Todd, B. Freese, A. Lu, D. Held, J. Morey, R. Livingston and P. Goix, *Clin. Chem.*, 2007, **53**, 1990–1995.
11. R. Fan, O. Vermesh, A. Srivastava, B. K. Yen, L. Qin, H. Ahmad, G. A. Kwong, C. C. Liu, J. Gould, L. Hood and J. R. Heath, *Nat. Biotechnol.*, 2008, **26**, 1373–1378.
12. S. Fredriksson, M. Gullberg, J. Jarvius, C. Olsson, K. Pietras, S. M. Gústafsdóttir, A. Östman and U. Landegren, *Nat. Biotechnol.*, 2002, **20**, 473–477.
13. C. D. Chin, V. Linder and S. K. Sia, *Lab Chip*, 2012, **12**, 2118–2134.
14. J. L. Bock, *Am. J. Clin. Pathol.*, 2000, **113**, 628–646.
15. C. P. Bean and J. D. Livingston, *J. Appl. Phys.*, 1959, **30**, S120.
16. D. Horak, M. Babic, H. Mackova and M. J. Benes, *J. Sep. Sci.*, 2007, **30**, 1751–1772.
17. N. Pamme, *Lab Chip*, 2006, **6**, 24–38.
18. M. A. Gijs, F. Lacharme and U. Lehmann, *Chem. Rev.*, 2010, **110**, 1518–1563.

19. J. Pipper, Y. Zhang, P. Neuzil and T. M. Hsieh, *Angew. Chem. Int. Ed.*, 2008, **47**, 3900–3904.
20. U. Lehmann, C. Vandevyver, V. K. Parashar and M. A. Gijs, *Angew. Chem. Int. Ed. Engl.*, 2006, **45**, 3062–3067.
21. D. M. Bruls, T. H. Evers, J. A. H. Kahlman, P. J. W. van Lankvelt, M. Ovsyanko, E. G. M. Pelssers, J. J. H. B. Schleipen, F. K. de Theije, C. A. Verschuren, T. van der Wijk, J. B. A. van Zon, W. U. Dittmer, A. H. J. Immink, J. H. Nieuwenhuis and M. W. J. Prins, *Lab Chip*, 2009, **9**, 3504–3510.
22. A. van Reenen, A. M. de Jong, J. M. J. den Toonder and M. W. J. Prins, *Lab Chip*, 2014, **14**, 1966–1986.
23. C. E. Smith and C. K. York, *US Pat.*, 6 027 945, 2000.
24. R. C. den Dulk, K. A. Schmidt, G. Sabatte, S. Liebana and M. W. J. Prins, *Lab Chip*, 2013, **13**, 106–118.
25. C. H. Chiou, D. J. Shin, Y. Zhang and T. H. Wang, *Biosens. Bioelectron.*, 2013, **50**, 91–99.
26. S. Berensmeier, *Appl. Microbiol. Biotechnol.*, 2006, **73**, 495–504.
27. R. Boom, C. J. Sol, M. M. Salimans, C. L. Jansen, P. M. Wertheim-van Dillen and J. van der Noordaa, *J. Clin. Microbiol.*, 1990, **28**, 495–503.
28. G. Schreiber, G. Haran and H. X. Zhou, *Chem. Rev.*, 2009, **109**, 839–860.
29. R. R. Gabdoulline and R. C. Wade, *Curr. Opin. Struct. Biol.*, 2002, **12**, 204–213.
30. A. Singhal, C. A. Haynes and C. L. Hansen, *Anal. Chem.*, 2010, **82**, 8671–8679.
31. A. van Reenen, A. M. de Jong and M. W. J. Prins, *J. Phys. Chem. B*, 2013, **117**, 1210–1218.
32. K. S. Schmitz and J. M. Schurr, *J. Phys. Chem.*, 1972, **76**, 534.
33. M. Stenberg and H. Nygren, *J. Immunol. Methods.*, 1988, **113**, 3–15.
34. T. M. Squires, R. J. Messinger and S. R. Manalis, *Nat. Biotechnol.*, 2008, **26**, 417–426.
35. A. van Reenen, Y. Gao, A. M. de Jong, M. A. Hulsen, J. M. J. den Toonder and M. W. J. Prins, *Proc. MicroTAS 2012 Int. Conf.*, 2012, pp. 926–928.
36. Y. Moser, T. Lehnert and M. A. Gijs, *Lab Chip*, 2009, **9**, 3261–3267.
37. S. Sjolander and C. Urbaniczky, *Anal. Chem.*, 1991, **63**, 2338–2345.
38. A. K. Vuppu, A. A. Garcia and M. A. Hayes, *Langmuir*, 2003, **19**, 8646–8653.
39. A. K. Vuppu, A. A. Garcia, M. A. Hayes, K. Booksh, P. E. Phelan, R. Calhoun and S. K. Saha, *J. Appl. Phys.*, 2004, **96**, 6831–6838.
40. R. Calhoun, A. Yadav, P. Phelan, A. Vuppu, A. Garcia and M. Hayes, *Lab Chip*, 2006, **6**, 247–257.
41. T. G. Kang, M. A. Hulsen, P. D. Anderson, J. M. J. den Toonder and H. E. Meijer, *Phys. Rev. E Stat. Nonlin. Soft Matter Phys.*, 2007, **76**, 066303.
42. S. Krishnamurthy, A. Yadav, P. E. Phelan, R. Calhoun, A. K. Vuppu, A. A. Garcia and M. A. Hayes, *Microfluid. Nanofluid.*, 2008, **5**, 33–41.
43. Y. Gao, M. A. Hulsen, T. G. Kang and J. M. J. den Toonder, *Phys. Rev. E*, 2012, **86**, 041503.

44. J. E. Martin, L. Shea-Rohwer and K. J. Solis, *Phys. Rev. E*, 2009, **80**, 016312.
45. T. Roy, A. Sinha, S. Chakraborty, R. Ganguly and I. K. Puri, *Phys. Fluids*, 2009, **21**, 7.
46. J. M. Ottino, *The Kinematics of Mixing: Stretching, Chaos, and Transport*, Cambridge University Press, Cambridge, England, 1989.
47. D. Holzinger, D. Lengemann, F. Gollner, D. Engel and A. Ehresmann, *Appl. Phys. Lett.*, 2012, **100**, 153507.
48. Y. Gao, A. van Reenen, M. A. Hulsen, A. M. de Jong, M. W. J. Prins and J. M. J. den Toonder, *Microfluid. Nanofluid.*, 2014, **16**, 265–274.
49. J. T. Lee, A. Abid, K. H. Cheung, L. Sudheendra and I. M. Kennedy, *Microfluid. Nanofluid.*, 2012, **13**, 461–468.
50. J. T. Lee, L. Sudheendra and I. M. Kennedy, *Anal. Chem.*, 2012, **84**, 8317–8322.
51. A. van Reenen, Y. Gao, A. M. de Jong, M. A. Hulsen, J. M. J. den Toonder and M. W. J. Prins, *Proc. MicroTAS 2013 Int. Conf.*, 2013.
52. D. A. Armbruster and T. Pry, *Clin. Biochem. Rev.*, 2008, **29**(1), S49–52.
53. S. Chandran and R. S. Singh, *Pharmazie*, 2007, **62**, 4–14.
54. M. Shikida, K. Takayanagi, H. Honda, H. Ito and K. Sato, *J. Micromech. Microeng.*, 2006, **16**, 1875–1883.
55. H. Bordelon, N. M. Adams, A. S. Klemm, P. K. Russ, J. V. Williams, H. K. Talbot, D. W. Wright and F. R. Haselton, *ACS Appl. Mater. Interfaces*, 2011, **3**, 2161–2168.
56. Y. Zhang, S. Park, K. Liu, J. Tsuan, S. Yang and T. H. Wang, *Lab Chip*, 2011, **11**, 398–406.
57. R. Sista, Z. Hua, P. Thwar, A. Sudarsan, V. Srinivasan, A. Eckhardt, M. Pollack and V. Pamula, *Lab Chip*, 2008, **8**, 2091–2104.
58. R. S. Sista, A. E. Eckhardt, V. Srinivasan, M. G. Pollack, S. Palanki and V. K. Pamula, *Lab Chip*, 2008, **8**, 2188–2196.
59. K. Sur, S. M. McFall, E. T. Yeh, S. R. Jangam, M. A. Hayden, S. D. Stroupe and D. M. Kelso, *J. Mol. Diagn.*, 2010, **12**, 620–628.
60. B. P. Casavant, D. J. Guckenberger, S. M. Berry, J. T. Tokar, J. M. Lang and D. J. Beebe, *Lab Chip*, 2013, **13**, 391–396.
61. S. M. Berry, E. T. Alarid and D. J. Beebe, *Lab Chip*, 2011, **11**, 1747–1753.
62. R. C. den Dulk, K. A. Schmidt, R. Gill, J. C. B. Jongen and M. W. J. Prins, *Proc. MicroTAS 2010 Int. Conf*, 2010, pp. 665–667.
63. O. Strohmeier, A. Emperle, G. Roth, D. Mark, R. Zengerle and F. von Stetten, *Lab Chip*, 2013, **13**, 146–155.
64. R. Gottheil, N. Baur, H. Becker, G. Link, D. Maier, N. Schneiderhan-Marra and M. Stelzle, *Biomed. Microdevices*, 2014, **16**, 163–172.
65. D. J. Shin, L. Chen and T. H. Wang, *Proc. MicroTAS 2013 Int. Conf.*, 2013, pp. 1350–1352.
66. H. Tsuchiya, M. Okochi, N. Nagao, M. Shikida and H. Honda, *Sensor. Actuat. B Chem.*, 2008, **130**, 583–588.

67. A. H. Ng, K. Choi, R. P. Luoma, J. M. Robinson and A. R. Wheeler, *Anal. Chem.*, 2012, **84**, 8805–8812.
68. K. Choi, A. H. Ng, R. Fobel, D. A. Chang-Yen, L. E. Yarnell, E. L. Pearson, C. M. Oleksak, A. T. Fischer, R. P. Luoma, J. M. Robinson, J. Audet and A. R. Wheeler, *Anal. Chem.*, 2013, **85**, 9638–9646.
69. Y. Gao, A. van Reenen, M. A. Hulsen, A. M. de Jong, M. W. J. Prins and J. M. J. den Toonder, *Lab Chip*, 2013, **13**, 1394–1401.
70. J. Baudry, C. Rouzeau, C. Goubault, C. Robic, L. Cohen-Tannoudji, A. Koenig, E. Bertrand and J. Bibette, *Proc. Natl Acad. Sci. USA*, 2006, **103**, 16076–16078.
71. S. Y. Park, H. Handa and A. Sandhu, *J. Appl. Phys.*, 2009, **105**, 07B526.
72. S. Y. Park, H. Handa and A. Sandhu, *Nano Lett.*, 2010, **10**, 446–451.
73. S. Y. Park, P. J. Ko, H. Handa and A. Sandhu, *J. Appl. Phys.*, 2010, **107**, 09B324.
74. A. Ranzoni, G. Sabatte, L. J. van IJzendoorn and M. W. J. Prins, *ACS Nano*, 2012, **6**, 3134–3141.
75. A. Ranzoni, J. J. H. B. Schleipen, L. J. van IJzendoorn and M. W. J. Prins, *Nano Lett.*, 2011, **11**, 2017–2022.
76. W. U. Dittmer, P. de Kievit, M. W. J. Prins, J. L. Vissers, M. E. Mersch and M. F. Martens, *J. Immunol. Methods*, 2008, **338**, 40–46.
77. V. N. Morozov and T. Y. Morozova, *Anal. Chim. Acta*, 2006, **564**, 40–52.
78. R. De Palma, G. Reekmans, W. Laureyn, G. Borghs and G. Maes, *Anal. Chem.*, 2007, **79**, 7540–7548.
79. C. B. Kriz, K. Radevik and D. Kriz, *Anal. Chem.*, 1996, **68**, 1966–1970.
80. Y. R. Chemla, H. L. Grossman, Y. Poon, R. McDermott, R. Stevens, M. D. Alper and J. Clarke, *Proc. Natl Acad. Sci. USA*, 2000, **97**, 14268–14272.
81. D. R. Baselt, G. U. Lee, M. Natesan, S. W. Metzger, P. E. Sheehan and R. J. Colton, *Biosens. Bioelectron.*, 1998, **13**, 731–739.
82. M. Koets, T. van der Wijk, J. T. W. M. van Eemeren, A. van Amerongen and M. W. J. Prins, *Biosens. Bioelectron.*, 2009, **24**, 1893–1898.
83. R. S. Gaster, D. A. Hall and S. X. Wang, *Lab Chip*, 2011, **11**, 950–956.
84. T. Aytur, J. Foley, M. Anwar, B. Boser, E. Harris and P. R. Beatty, *J. Immunol. Methods*, 2006, **314**, 21–29.
85. F. Colle, D. Vercruysse, S. Peeters, C. Liu, T. Stakenborg, L. Lagae and J. Del-Favero, *Lab Chip*, 2013, **13**, 4257–4262.
86. V. N. Morozov, S. Groves, M. J. Turell and C. Bailey, *J. Am. Chem. Soc.*, 2007, **129**, 12628–12629.
87. D. Leckband and J. Israelachvili, *Q. Rev. Biophys.*, 2001, **34**, 105–267.
88. C. Danilowicz, D. Greenfield and M. Prentiss, *Anal. Chem.*, 2005, 77, 3023–3028.
89. A. Jacob, L. J. van IJzendoorn, A. M. de Jong and M. W. J. Prins, *Anal. Chem.*, 2012, **84**, 9287–9294.
90. I. D. Vilfan, J. Lipfert, D. A. Koster, S. G. Lemay and N. H. Dekker, *Magnetic Tweezers for Single-Molecule Experiments: Springer Handbook of Single-Molecule Biophysics*, Springer, New York, 2009.

91. I. De Vlaminck and C. Dekker, *Annu. Rev. Biophys*, 2012, **41**, 453–472.
92. A. van Reenen, F. Gutierrez-Mejia, L. J. van IJzendoorn and M. W. J. Prins, *Biophys. J.*, 2013, **104**, 1073–1080.
93. A. van Reenen, Y. Gao, A. H. Bos, A. M. de Jong, M. A. Hulsen, J. M. J. den Toonder and M. W. J. Prins, *Appl. Phys. Lett.*, 2013, **103**, 043704.
94. R. C. Ebersole, *US Pat.*, 4 219 335, 1978.
95. G. U. Lee, S. Metzger, M. Natesan, C. Yanavich and Y. F. Dufrene, *Anal. Biochem.*, 2000, **287**, 261–271.
96. D. A. Boas, C. Pitris and N. Ramanujam, *Handbook of Biomedical Optics*, CRC Press, Boca Raton, FL, 2011.

CHAPTER 7

Microfluidics for Assisted Reproductive Technologies

DAVID LAI, JOYCE HAN-CHING CHIU, GARY D. SMITH AND
SHUICHI TAKAYAMA*

Biointerfaces Institute, University of Michigan, A183 Bldg. 10 NCRC,
2800 Plymouth Rd, Ann Arbor, MI 48109, USA
*Email: takayama@umich.edu

7.1 Introduction

Infertility affects almost 10% of couples in the United States and the causes
are roughly equal between male- and female-derived infertility.[1] As reproduction is a major quality of life issue, there is a large motivation for the
advancement of assisted reproductive technology. Although a relatively new
field, microfluidics has already demonstrated the benefits of automation,
consistency, sensitivity, and precision to the field of assisted reproductive
technologies (ARTs). Efforts in this field utilize the high-throughput nature
of microfluidics as well as the potential to perform highly sensitive non-
destructive tests for gamete/embryo quality to overcome some of the
toughest and most laborious challenges faced by assisted reproductive
technologies.

One of the most common benefits of microfluidics is its cost efficiency due
to the low amounts of materials used; however, in the field of ART, this
benefit is of secondary importance. The advantage of highly accurate high-
throughput gamete sorting and the identification of the most suitable male
and female gamete to use for fertilization justify the use of microfluidics
even if it were more expensive to operate.

RSC Nanoscience & Nanotechnology No. 36
Microfluidics for Medical Applications
Edited by Albert van den Berg and Loes Segerink
© The Royal Society of Chemistry 2015
Published by the Royal Society of Chemistry, www.rsc.org

Male gamete quality assessment chips typically operate at a relatively high-throughput range due to the large quantities of available material, while female gamete quality assessment chips are the complete opposite in terms of sorting speed as the low amounts of available sample justify the use of sophisticated biophotonic non-destructive tests of gamete quality. Fertilization-on-chip devices either mimic existing ART techniques or incorporate selectivity barriers within the microfluidic device based on sperm motility to mimic the motility selection of the uterus and fallopian tube.

Cryopreservation has been key to increasing the efficiency of fertility treatments in ART. Cryopreservation devices focus on reducing osmotic stress, particularly for vitrification where concentrations of cryoprotectant agents (CPA) are particularly high. Microfluidic cryopreservation devices also reliably standardize the vitrification protocol with precision and accuracy to the fluid transfer of the CPA exchange, addressing the large unknown that is the variable fluid behavior in current manual pipetting even with the best trained and skilled personnel.

Microfluidic embryo cultures have also employed biomimicry of the fallopian tubes in multiple strategies to assist embryos into developing with superior developmental competence. Such strategies include mimicry of the mechanical stimulation provided by fallopian tube flow as well as the recirculation of secreted factors that chemically stimulate embryo growth. As developmental competence increases with technological advancement of each stage whether using conventional clinical lab techniques or lab on a chip (LOC) techniques, it is increasingly more sensible and feasible to perform single embryo transfers (SETs). The chances of multiple births, typically an unfavorable outcome, exponentially increase with each embryo transferred beyond one. It is therefore paramount in this stage to select the best embryo to transfer so as to maximize the chance of a successful SET. To address this issue, multiple non-destructive tests have been developed to determine embryo quality after embryo culture and these include advanced spectroscopy techniques that directly examine the embryo, indirect metabolite analysis of spent culture media as well as real-time embryo development monitoring combined with predictive algorithms for selecting the best embryo.

7.2   Gamete Manipulations

Gamete manipulations using lab-on-a-chip technologies focus on non-destructive quality analysis of sperm and oocytes. Andrology strategies take advantage of the high-throughput characteristics of microfluidics due to the high number of gametes available and are particularly useful for individuals with low numbers of viable sperm. On the contrary, gynecology strategies focus on in-depth analysis due to the limited number of gametes and the high impact of oocyte quality on subsequent embryo development. Despite the widespread acceptance that morphology is a poor indication of future developmental success,[2] it is still the primary measure of quality. Female lab-on-a-chip devices employ the precision and accuracy provided by microfluidic devices to develop more sophisticated measures of oocyte quality assessment.

7.2.1 Male Gamete Sorting

The determination of viability for sperm is rather simple as the sole concern is DNA integrity as advanced ARTs such as intracytoplasmic sperm injection (ICSI) are capable of overcoming numerous infertility issues that complicate natural conception. ICSI is such a powerful technique that it is possible to use dying or even dead sperm to achieve fertilization. However, it is well known that the use of dead or dying sperm for ICSI results in adverse outcomes including low fertilization, poor embryo development, high abortion rates, and prevalent childhood diseases.[3] Due to the large number of sperm available, it is rare that live and motile sperm cannot be found even if in low number for the majority of male infertility cases. As ICSI only requires a single viable sperm and there is a significant adverse result of using dead or dying sperm, the first conventional sorting criterion is that the sperm be motile: evidence that it is alive. The second criterion is that it has normal morphology in the absence of defects such as multiple heads, multiple tails, large or small heads, or abnormal mid-sections. Such defects in morphology indicate defected genotype on which it is based and are discarded during conventional sperm sorting.

For lab-on-a-chip applications, mobility provides a simple mechanism for separation as non-motile sperm and debris are incapable of crossing streamlines in laminar flow regime. The enclosed system of microfluidic devices as well as the high level of fluid control allow sorting of non-motile sperm and debris with high accuracy and at high-throughput speeds. Conveniently for microfluidic applications, the second conventional criterion for sorting, abnormal morphology, also often affects motility. When designing the microfluidic device with high safety factors (where only highly motile sperm are selected), the device would also eliminate abnormal albeit motile sperms simultaneously from the sample.

One passively driven microfluidic device to sort sperm by motility uses gravity to create two streams flowing in a laminar regime[4] (Figure 7.1). It relies on the motile sperm's ability to cross enough streamlines by traveling in the transverse direction to flow. The passive flow of the microfluidic device delivers the sperm to the collection chamber separate from the waste collection chamber. Viable sperm sorted into the collection chamber can then be collected for ICSI. The continuously separating nature of the microfluidic device allows for reasonable volumes of sample to be sorted for easy identification of viable sperm even in cases of severe male infertility.

In another example, a microfluidic device with a long hydrostatic channel connected to a detection cuvette allowed for the screening of male subfertility with the staining of sperm with calcein-AM[5] (Figure 7.2A and B). The long hydrostatic channel served as a selective barrier for motile sperm: allowing only the sperm of healthy motility to traverse the long distance to

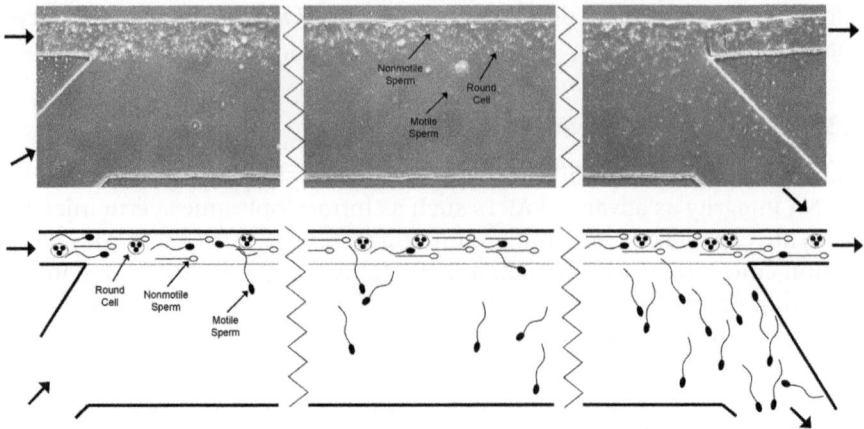

Figure 7.1   Image captures and illustrations of the movement of sperm sample during sorting. Non-motile and dead sperm will be transported along the laminar streamline from the top left inlet reservoir to the top right channel while motile sperm can move across the streamlines into the collection outlet at the bottom right corner.

reach and populate the detection cuvette (Figure 7.2C–F). The fluorescence profile is monitored by microfluorometer over a 50-minute period (Figure 7.2G and H). As the detection cuvette is populated by more sperm, the fluorescence from the combined calcein-AM increases the fluorescence intensity proportionally to the number of sperm in the detection cuvette over time. This fluorescence profile of each sample can then be screened relative to a reference sample to determine their subfertility level. The device, being small and disposable, has shown promise as an accurate in-home screening technique for male subfertility.

It was a microfluidic study that definitively demonstrated that sperm behave predictably in certain well-defined flow conditions. The study determined that the sperm would tend to swim either towards or against flow depending on the intensity of the flow velocity.[6] This device demonstrated the predictability of this behavior and applied it to separate, align, and orient sperm of mouse, bull, and human species.

One recent advancement of note in male infertility microfluidic devices employs an integrated charge-coupled device (CCD) to the microfluidic device to provide a wide field of view to sort and track sperm populations[7] (Figure 7.2I) The sophisticated automated sperm monitoring algorithm is capable of determining sperm average velocity, straight line velocity, straightness of swim path, and average acceleration. The attention given by the software to each individual sperm provides the capability to identify the most motile sperm, which by conventional wisdom is the healthiest, for potential applications in ICSI. Such detailed quantification of sperm quality eliminates the arbitrary elements of conventional sperm selection when there are numerous viable sperm available.

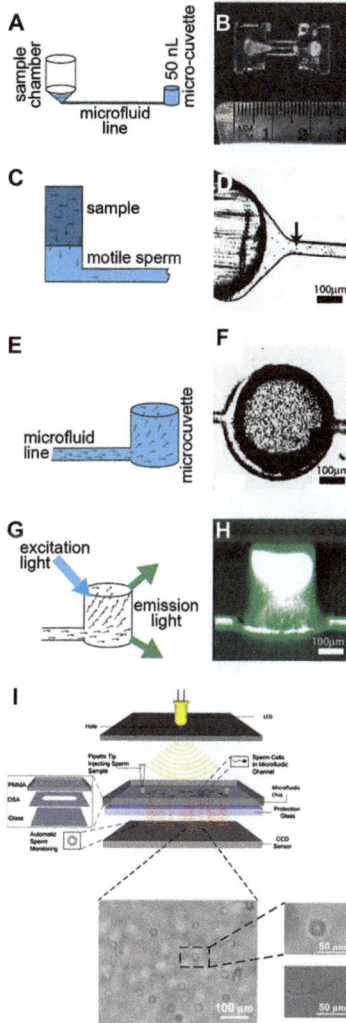

Figure 7.2 (A–H) Sperm assessment with microfluorometer. (I) Schematic representation of the multilayer microfluidic chip for simultaneous sperm motility sorting and CCD monitoring.

Even more recent advances in sperm quality assessment by microfluidic
devices include a device that quantifies both sperm concentration and
motile percentage[8] (Figure 7.3a–i). The device uses a microfluidic component for the separation of motile sperm of similar mechanisms to its
predecessors and is designed to be centrifuged to sediment the sperm
sample for quantification. Concurrently, a sophisticated method for quantifying human sperm quality was developed using a microfluidic resistive
pulse technique (RPT).[9] The change in current over time from the resistive
properties of the sperm is measured as it travels along a long narrow

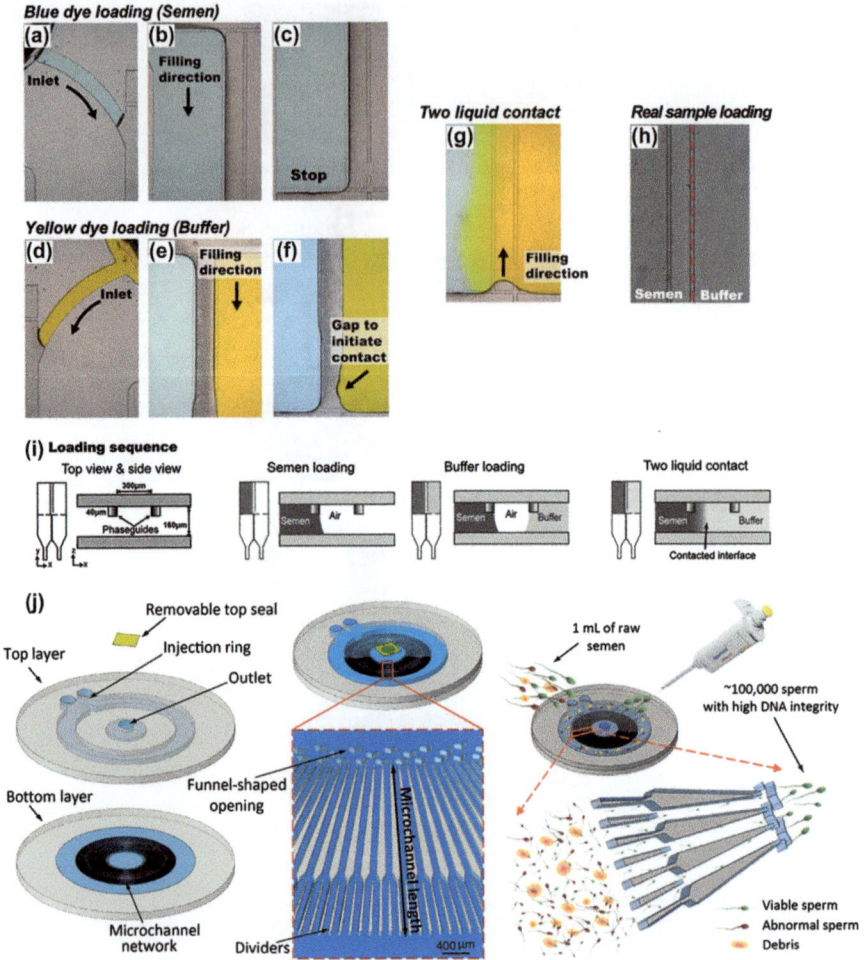

Figure 7.3 (a–i) Motility sorting based on phase-guide structure. The motile sperm will cross the interface barrier to the buffer zone. (j) Sperm motility selection based on micro-channels filled with viscoelastic medium are used to mimic reproductive tracts.

aperture where the voltage is applied. The signal profile can be easily analyzed for the sperm's swimming behavior in terms of beat frequency and swim velocity as well as the sperm's volume. Such proposed techniques offer economic advances to comparable sperm analysis provided by conventional computer assisted sperm analysis (CASA).

The latest microfluidic assay for determining sperm DNA integrity demonstrates strong evidence that sperm motility is a strong indication of high chromatin and DNA integrity[10] (Figure 7.3j). The device is capable of using raw semen and uses a viscoelastic medium and long hydrostatic channels that mimic the motility selection process of the female reproductive tract.

A B C

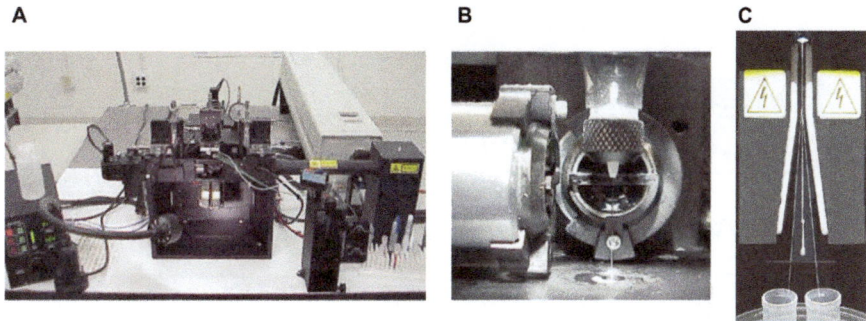

Figure 7.4 (A) Overview of sperm sorter setup. (B) Fluorescence detector integrated with hydrodynamics system. (C) Droplets containing spermatozoa are sorted to positive or negative charge depending on its sex chromosome.

The length of the device can be adjusted according to the subfertility of the sample for an optimal level of motility-based sorting.

Although not strictly lab-on-a-chip, fluorescence activated cell sorting (FACS) uses a cytometer in combination with DNA-binding dyes and fluorescence detection. It has been used for detecting minute differences in fluorescence signal originating from a 3.8% difference in total DNA content between X and Y chromosome bearing bovine spermatozoa[11] (Figure 7.4). When the difference in DNA content is detected, an applied electric field sorts the droplet containing the spermatozoa into X and Y chromosome bearing bins. The system can reach up to 25 000 individual spermatozoa per second and is useful in agricultural industries. However it is known that high shear stress and high voltage cause extensive DNA fragmentation.[12,13] This is a concern that should not be ignored as both high shear stress and voltage exist in the cytometer and sorting region of FACS.

7.2.2 Female Gamete Quality Assessment

Oocyte quality is a large factor that affects later developmental success during and after fertilization. Despite the importance of oocyte selection, especially for SETs, the determination of oocyte quality is typically based on highly subjective morphological properties. However, a more sophisticated analysis of oocyte quality must not compromise the quality of the oocyte as it needs to be retrieved for further use. As such, there are significant challenges in developing the appropriate tools to probe the oocyte for a quantitative analysis of its quality.

One such microfluidic device that studied this problem developed an optical solution to study the refractive index and optical absorption of the oocyte to determine its maturity as well as a microindentor to measure the hardness and reduced Young modulus of the oocyte, a tool that can be used for further study of the oocyte[14] (Figure 7.5E and F). The optical probe was capable of detecting a significant difference in absorption spectra between

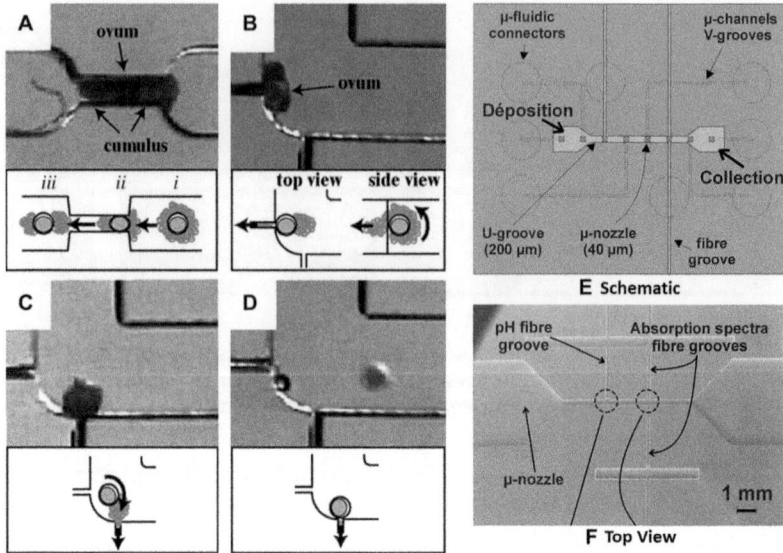

Figure 7.5 (A–D) Stepwise denuding of zygote. (E and F) Schematic and top view of optical sensor.

mature and immature oocytes. This development provides a solution to alleviate some of the subjectivity that exists in oocyte selection for ART.

During common ART procedures such as *in vitro* maturation (IVM), *in vitro* fertilization (IVF), and *in vitro* embryo culture (IVC), the oocyte or zygote undergoes a process where its cumulus cells are removed, sometimes called denuding. The most common form of denuding requires a technician to repeatedly flush the cumulus oocyte complex (COC) into and out of a narrow pipette tip. Such a procedure is known to cause high mechanical stress.[15] As such, a novel micromanipulation device was designed using the precise control provided by microfluidics. The device is capable of reorienting the cumulus into a ring around a zygote. The cumulus ring surrounding the zygote is then attached to a removal port, which, combined with the suction from an adjacent port, rotates the COC to remove the ring, leaving the denuded zygote with less mechanical stress[16] (Figure 7.5A–D). The ability to precisely control the zygote position by device geometry and fluid control enables predictable and reliable cumulus removal.

Recent advances in microelectromechanical systems (MEMS) for ART developed a set of magnetically driven microtools (MMTs) that upon actuation enucleated oocytes.[17] Enucleation is a process where the nucleus is removed from the cell body, a key component in the cloning process. Such a technique is normally performed by highly trained technicians but typically still suffers from low success rates and repeatability. The use of this device overcomes limitations of conventional enucleation by providing precision and reproducibility. The group also demonstrated that the combination of MMT with ultrasonic vibration doubled the output force of the MMT to improve the positioning accuracy by 100-fold[18] (Figure 7.6).

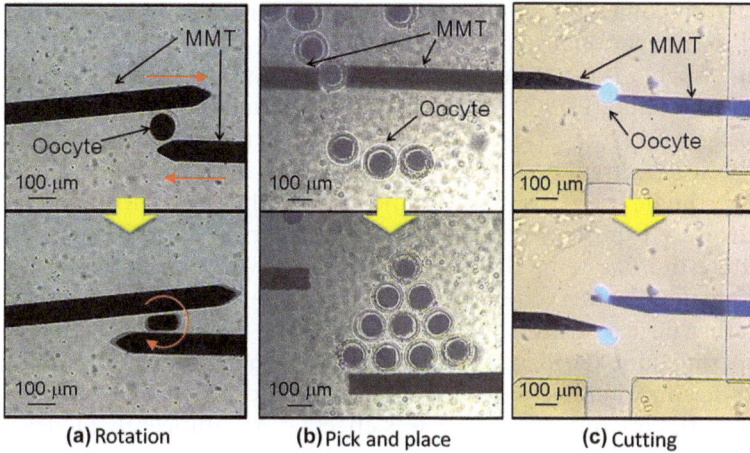

Figure 7.6 Magnetically driven microtools with ultrasonic vibration for oocytes manipulation in a microfluidic environment.

Optofluidic devices are a specialized research field that combines microfluidic and optics designs. One such optofluidic device was developed using optical ablation for microsurgery of oocytes.[19] While microfluidic channels have been shown to be effective at cell positioning as well as long-term and real-time monitoring, optical ablation is an effective non-invasive method to destroy intracellular organelles. Such microsurgery techniques are a valuable tool that can provide insight about the functions of the organelle and developmental repercussions for damaged oocytes.

7.3 *In Vitro* Fertilization

There are two prevalent methods of fertilization in ART that are modified in lab-on-a-chip devices. The first method is the use of fertilization medium and conventional insemination with sperm and the default method of IVF. However, if it is deemed that conventional insemination is likely to be unsuccessful, ICSI will be used in ART. The conventional insemination IVF technique is relatively simple where medium containing washed sperm is mixed and incubated with medium containing oocytes. A sperm will make contact with oocyte where it penetrates the zona pellucida as it would naturally for fertilization. The second method is ICSI, a powerful technique that overcomes numerous male and female infertility challenges. ICSI overcomes many stages of fertilization where complications may occur by assisting the sperm in penetrating the egg through direct injection of the sperm into the egg cytoplasm.

There have been a significant amount of microfluidic fertilization devices that enhance the conventional insemination and IVF technique. Depending on the design, it can boast higher fertilization efficiency with lower total number and concentrations of sperm needed. The device can also increase primary outcomes by lowering instances of polyspermic penetration compared to conventional fertilization medium.

One such IVF microfluidic device was designed using polydimethylsiloxane (PDMS)/borosilicate material and was driven passively by hydrostatic pressure[20] (Figure 7.7A). The device employed a now commonly used weir-type trap design for oocytes while sperm are allowed to flow through the device. This device demonstrated that by limiting the time the oocytes were exposed to the sperm, as the sperm had a limited chance to penetrate the oocyte as they flowed by, the device design allowed for significantly lower rates of polyspermy.

Another device of similarly inspired weir-type design focused its study on low-sperm-count fertilization. The device demonstrated that although the rate of fertilization decreases in a microfluidic device for high-sperm-count samples, the microfluidic device was effective at significantly increasing the fertilization rate of low-sperm-count samples[21] (Figure 7.7B). This result is primarily due to the geometry and scale of the microfluidic device that increases the effective concentration of sperm in the area surrounding the oocyte, thus greatly increasing the chances of fertilization.

More sophisticated devices later emerged that also used weir-type traps to individually compartmentalize oocytes within the microfluidic device[22] (Figure 7.7C). The device applied the microfluidic enhancement of the conventional insemination IVF technique and the individual compartments were capable of providing single embryo development tracking over 96 hours. Such a device demonstrates great potential for a microfluidic device to provide a fully integrated and automated system for ART research and clinical applications.

The microfluidic enhancement to conventional insemination IVF is clear, but the advancement in microfluidic system in ICSI has just begun. Recently, there has been effort in producing a fully integrated and automated microfluidic system for ICSI.[23] The device uses sperm separation by motility to select viable sperm for ICSI and is further combined with a chemical and

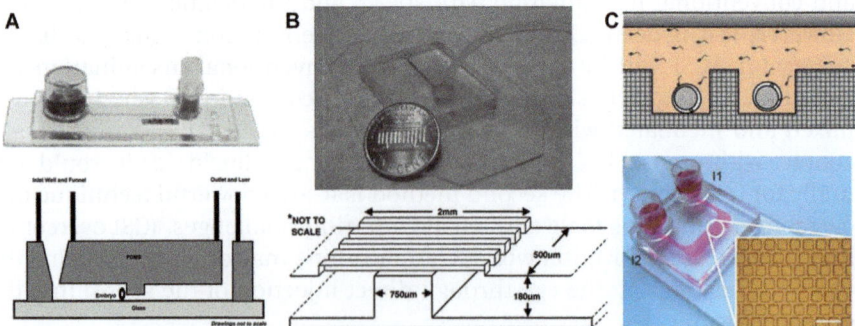

Figure 7.7 (A) Weir-type trap design for fertilization. (B) Small-scale design of microfluidic device promotes the effectiveness of fertilization. (C) Compartmentalized fertilization chambers for multiple embryos.

mechanical cumulus denuding compartment. A series of embedded electrodes are used to orient the oocyte and laser trapping is used to immobilize the selected sperm. The actual ICSI procedure is performed on-chip by a piezo-actuator driven needle. The device then releases the fertilized zygote to individualized compartments for single embryo long-term culture and monitoring. This fully integrated system is one of the first to fully integrate both male and female gamete processing and selection for usage, ICSI, and long-term single embryo culture and monitoring.

7.4 Cryopreservation

Cryopreservation is now an essential part of ART and substantially increases the efficiency of techniques/processes such as SETs, donor-recipient cycle synchronizations, and oncofertility preservation with both embryo[24] and oocyte[25] cryopreservation. There are two prevalent strategies of cryopreservation: 1) slow-rate freezing or equilibrium freezing; and 2) ultra-rapid cooling or vitrification.

The microfluidic designs used to explore microfluidic cryopreservation, like the chronological development of conventional cryopreservation, started with slow-rate freezing. The unprecedented control of fluid osmolality was capable of minimizing osmotic stress using diffusion and laminar flow[26] (Figure 7.8). By lowering the osmotic stress, the device improved post-thaw survivability by up to 25% on average over conventional cryopreservation techniques.

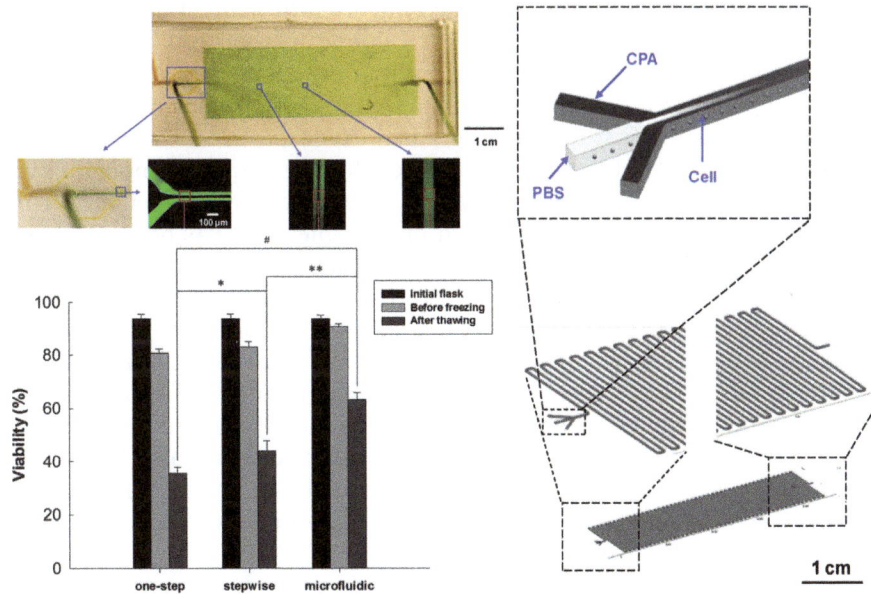

Figure 7.8 Increased viability is achieved by reducing osmotic shock during cryopreservation process through gradual exposure of CPA concentration to the cells.

With recent technological advancements to vitrification, recent studies on mature human oocyte cryopreservation demonstrated that vitrification has enhanced outcome in terms of higher survival rates, cleavage, and pregnancy rates compared to its slow-rate freezing alternative.[27] The osmotic fluid control provided by microfluidics gives an unprecedented level of precision and osmolality control not available with manual pipetting. This is particularly important in vitrification where higher concentrations of CPAs are used. In an effort to lower the osmotic stress the cells experience, a protocol for fluid exchange was developed that moves the cell into solutions of sequentially increasing concentrations of CPAs so that it lessens the osmotic change per step and divides the osmotic change necessary for vitrification over several steps over time. Some groups have demonstrated that the increase in the number of steps, and therefore the less osmotic stress introduced per step, increases cryosurvival[28] and development.[29] However, the significantly increased number of pipetting steps in a fixed amount of time became increasingly difficult to perform manually. Microfluidics provides the capability of continuous perfusion, essentially an infinite amount of steps, and is capable of providing real-time and continuous optical microscopy and lab-on-a-chip automation for reproducibility by eliminating operator variability. Facing these current limitations, the motivations for using microfluidics for CPA exposure of oocytes and embryos for vitrification is clear and many groups have begun microfluidic designs for study.

A computerized microfluidic control of media exposure for human cleavage embryos was designed to permit gradual cryoprotectant exposure to limit the large osmolality changes currently used in manual pipetting vitrification. It demonstrated that the microfluidic method produced similar embryo survival, morula formation, cavitation, and blastocyst formation rates compared to its conventional manual pipetting method alternative.[30] This device was the first to explore gradual CPA exposure for vitrification.

Osmotic stress can be accurately predicted using computer modelling by Kedem–Katchalsky equations and known permeability parameters.[31] This was first explored using a device with only permeable CPAs demonstrating that the Kedem–Katchalsky model, which has historically been used to model step-wise increases in CPA concentration, can also be reliably used to model osmotic stress behavior of cells exposed to continuously changing CPA concentrations by microfluidic perfusion[32] (Figure 7.9). The microfluidic device design, however, had limited access to cells for loading and removal, which limited its ability to explore cryosurvival and developmental response to the different CPA exposure regime. The device also lacked impermeable solutes that are highly recommended and present for practical vitrification solutions due to their low toxicity and its thermodynamic contributions towards vitrification.[33]

Most recently, another microfluidic device using both permeable and impermeable solutes was designed to further confirm the remarkable accuracy of the Kedem–Katchalsky equations to model osmotic stress of oocytes and zygotes for vitrification. By using a combination of both permeable

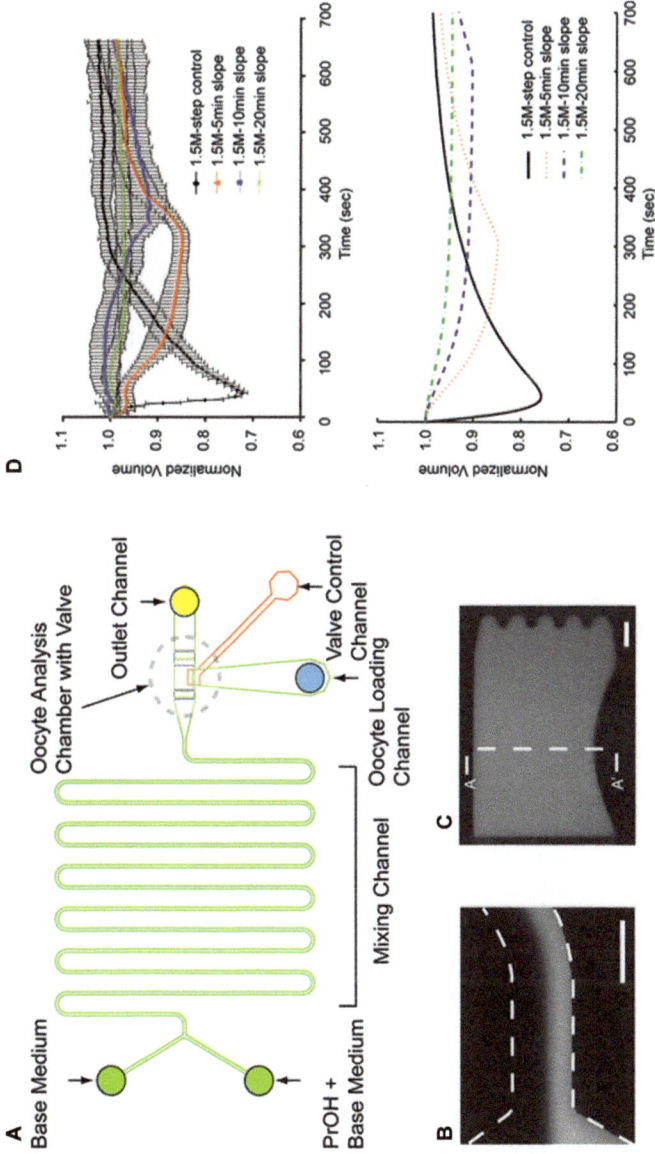

Figure 7.9 (A)schematic of microfluidic network. (B) Separation of two streams at the inlet channel. (C) Mixing of streams in the analysis chamber. (D) Volumetric response showed decreasing volume excursion with increasing linear change in CPA.

and impermeable solutes, the device was capable of controlling the shrinkage rate of the cell independently of its minimum cell volume. The study demonstrated the importance of shrinkage rate and its effect on sub-lethal damage.[34] The minimization of shrinkage rate had no effect on cryosurvival of murine zygotes as both manual and microfluidic CPA exposure had 100% cryosurvival. The shrinkage rate however did have an effect on sub-lethal damage. The lowering of shrinkage rate, and therefore sub-lethal damage, significantly increased the developmental competence of cryopreserved murine zygotes. This study is the first to experimentally and mathematically separate the effects of shrinkage rate from minimum shrinkage volume and is the first to demonstrate a sub-lethal improvement in cryopreserved embryos.

7.5   Embryo Culture

Microfluidic techniques provide embryo culture devices with mechanical fluid stimulation, automated culture media renewal, real-time and long-term imaging, as well as autocrine factor retention. Currently, only up to half of human embryos fertilized by IVF develop into the blastocyst stage in *in vitro* culture with approximately another 50% that attain successful implantation. The motivation of microfluidic design typically aims to simulate the natural *in vivo* environment. As in cryopreservation, automation also eliminates much of the user variability. Such control of physical and chemical parameters is needed for the careful dissection of experimental parameters necessary to further improve embryo culture outcomes[35] (Figure 7.10A).

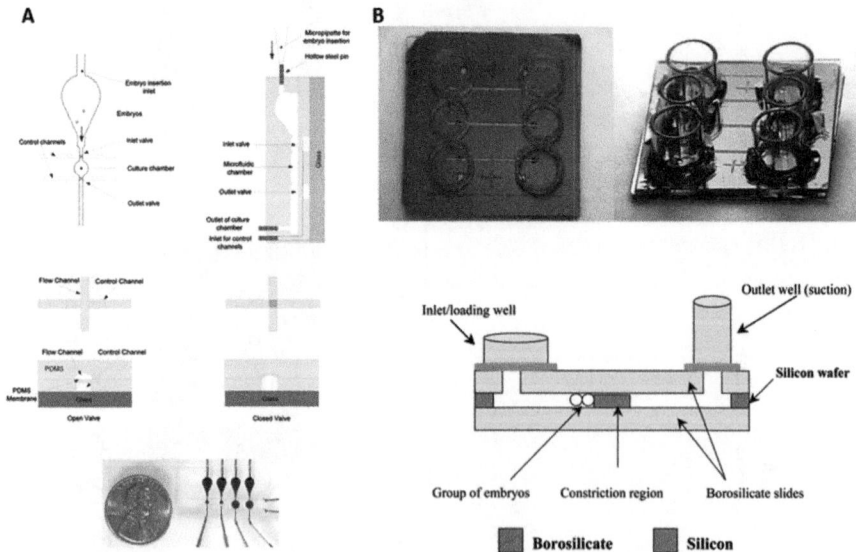

Figure 7.10   (A) Flow channels design for sub-microliter volume culture. (B) Silicon/borosilicate device for micro-channel culture.

Early studies of microfluidic embryo culture devices studied the effect of micro-channel culture as well as the material selection for microfluidic device fabrication. The foundation for microfluidic *in vivo* embryo culture was laid with the demonstration of advanced blastocyst development, higher survival rates, and fewer degenerated embryos in both silicon/borosilicate and PDMS/borosilicate devices[36] (Figure 7.10B). This study employed static (no flow) micro-channel culture allowing for a direct comparison with static microdrop systems. The improved developmental outcome is presumably due to the lower volume and the embryo's enhanced ability to regulate its microenvironment with paracrine and/or autocrine factors.

Although the addition of flow conditions is more similar to *in vivo* environments of embryo culture, it was discovered that simple continuous media perfusion within a micro-channel and weir-type trap during embryo development was detrimental to embryo development across a range of flow rates.[37] Multiple reasons for the poor outcome were suggested, but the primary reason may be the loss of paracrine and autocrine factors and the inability of embryo self-regulation of its microenvironment. Subsequently developed was a dynamic microfunnel embryo culture system that provides mechanical stimulation from fluid flow while retaining biochemical stimulations by allowing the retention of autocrine factors from the embryo's attempt to regulate its microenvironment[38] (Figure 7.11A). The murine embryos subjected to microfunnel pulsatile culture had a significantly higher number of blastomeres per blastocyst. The more advanced embryo development was also shown to result in enhanced implantation and pregnancy rate than static culture controls.

The embryo culture devices so far have been preimplantation embryos. Microfluidic devices can also be used for zebrafish embryo development into zebrafish larvae. The use of the devices showed no increase in gross

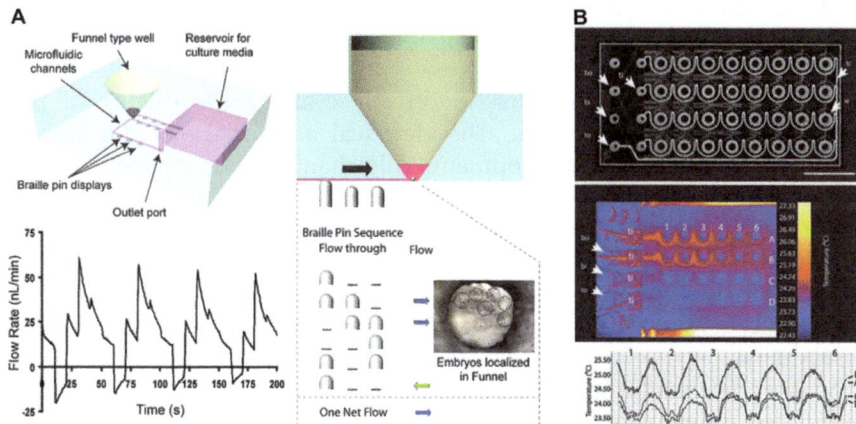

Figure 7.11 (A) Microfunnel with Braille pin piezoelectric actuators providing dynamic controls to enhance embryo culture. (B) Zebrafish embryo culture to larvae in a continuous buffer flow microchip.

malformations of zebrafish larvae[39] (Figure 7.11B). The cost effectiveness of the zebrafish model combined with low reagent volumes used during the microfluidic device embryo culture could be used for new analytic assays for the pharmaceutical industry.

More recently, more sophisticated microfluidic automation demonstrates fully integrated devices to culture and analyze developing zebrafish embryos.[40] Microfluidic automation demonstrated the fluid control for automatic embryo loading, immobilization, medium perfusion, and release of viable embryos. It also had embedded control systems to control chemical and mechanical fluid stimulations, as well as temperature regulation. The ability to simultaneously perform automatic image acquisition and real-time monitoring provided the functionality to analyze subtle indications of developmental quality.

7.6 Embryo Analysis

As the practice of SETs becomes more prevalent in ART, the importance of embryo selection for the best embryo becomes paramount to improve implantation rates. Current research direction is toward non-destructive analysis of embryo quality. While non-destructive direct analysis on the embryo is uncommon, some recent solutions have been developed that use advanced photonics such as light induced dielectrophoresis (DEP) or coherent anti-stokes Raman spectroscopy (CARS). Neither technique currently has lab-on-a-chip applications, but has a great potential for on-chip integration. Current microfluidic devices for non-destructive embryo analysis employ indirect analysis of embryo quality such as spent medium metabolite analysis. Current limitations of such devices, however, result from the devices being designed as targeted assays that measure specific metabolites of interest, therefore the microfluidic design becomes exponentially complex with each additional metabolite of interest. Such limitations highlight the strength of advanced spectroscopy capable of analyzing a broad spectrum of metabolites.

DEP devices have been used to directly evaluate the quality of pre-implantation murine embryos. The electrical admittance of the embryo changes according to its development. This change, when controlled using the appropriate optoelectronic tweezer (OET) medium and in the presence of an OET induced DEP, can attract early stage development embryos while repelling late stage development embryos.[41] Such a non-invasive sorting mechanism uses a reproducible and quantitative measurement to determine embryonic developmental stages without the subjectivity of morphology assessment.

Another recent advancement in spectroscopy that has proved powerful for direct measurements of cell metabolism is CARS. The technique is capable of non-invasively monitoring lipid concentration and distribution within the cell.[42] This label-free method of monitoring lipid content over time allows for the assessment of lipid metabolism of embryos.

In an effort to avoid exposure of embryos to intense UV light needed for metabolite assays, a reasonable strategy is to physically separate the embryo development with spent medium aspirate analysis into two separate microfluidic devices. One group demonstrated such potential with a fully automated analysis for performing the necessary aliquotting, mixing, data acquisition, and data analysis operations to quantify glucose, pyruvate, and lactate levels[43] (Figure 7.12).

Another strategy to avoid harmful UV light exposure to developing embryos is to use longer emission and excitation wavelengths of UV light and to spatially segregate optical interrogation from embryo culture. The device was capable of fluid priming, fluid mixing, chemical reaction, fluid washing and injections to analyze glucose metabolism[44] (Figure 7.13). With integrated

Figure 7.12   Multilayer design for a fully automated system.

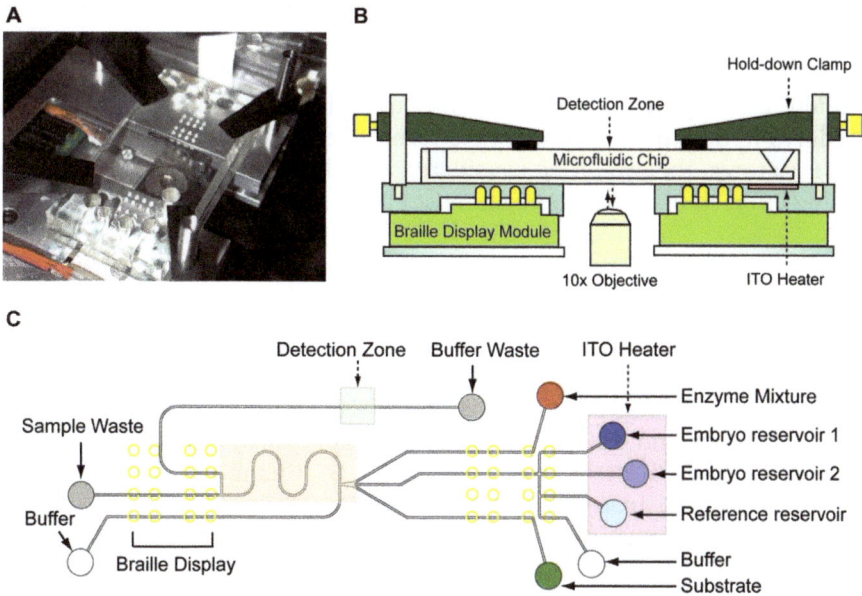

Figure 7.13   Braille-display.

embryo culture, this device is capable of performing real-time and continuous culture aspirate measurements for over 6 hours.

Although not yet integrated into microfluidic devices, many powerful broad spectrum analysis techniques have been used to analyze embryo metabolism. Such techniques not only simplify the analysis of multiple metabolites of interest, but also provide the possibility to study metabolites that one would not initially expect to be interesting. They also benefit from high sensitivity in addition to versatility and include nuclear magnetic resonance (NMR), mass spectroscopy (MS), near-infrared spectroscopy (NIR), and regular Raman spectroscopy.[45]

7.7   Conclusion

One of the most pressing challenges for ART as the trend for SET continues to grow is a method to non-invasively evaluate embryo and gamete quality. This challenge is also compounded by the now accepted realization that morphology, the most conventional method, is a poor indication of quality. Microfluidics has already shown promise to provide alternative methods of quality measurement with strategies tailored to the gamete or embryo analyzed using automation and sensitivity and sometimes combined with sophisticated biophotonic analysis. It is worthwhile to note that the use of microfluidic devices will not eliminate the need for an embryologist but will enhance outcome and reduce variability. The operation of complex microfluidic processes such as a series of mixing, dilutions, and sorting as well as sophisticated biophotonic techniques require a large amount of supporting equipment. Research effort in embedded microfluidic designs aims to decrease the amount of supporting equipment needed. Future device design should consider usability and adoptability for embryologists. Embryologists would also require additional training to include a change in current procedures to incorporate microfluidic techniques. As microfluidic devices for ART matures, this would be considered as enhanced job training as an ongoing effort to adopt the newest, most effective techniques.[46]

References

1. V. M. Brugh, 3rd and L. I. Lipshultz, *Med. Clin.*, 2004, **88**, 367–385.
2. A. Borini, C. Lagalla, M. Cattoli, E. Sereni, R. Sciajno, C. Flamigni and G. Coticchio, *Reproductive BioMedicine Online*, 2005, **10**, 653–668.
3. S. E. Lewis and R. J. Aitken, *Cell Tissue Res.*, 2005, **322**, 33–41.
4. B. S. Cho, T. G. Schuster, X. Zhu, D. Chang, G. D. Smith and S. Takayama, *Anal. Chem.*, 2003, **75**, 1671–1675.
5. M. C. McCormack, S. McCallum and B. Behr, *J. Urol.*, 2006, **175**, 2223–2227.
6. D.-b. Seo, Y. Agca, Z. C. Feng and J. K. Critser, *Microfluidics and Nanofluidics*, 2007, **3**, 561–570.

7. X. Zhang, I. Khimji, U. A. Gurkan, H. Safaee, P. N. Catalano, H. O. Keles, E. Kayaalp and U. Demirci, *Lab on a Chip*, 2011, **11**, 2535–2540.
8. C. Y. Chen, T. C. Chiang, C. M. Lin, S. S. Lin, D. S. Jong, V. F. Tsai, J. T. Hsieh and A. M. Wo, *Analyst*, 2013, **138**, 4967–4974.
9. Y. A. Chen, K. C. Chen, V. F. Tsai, Z. W. Huang, J. T. Hsieh and A. M. Wo, *Clin. Chem.*, 2013, **59**, 493–501.
10. R. Nosrati, M. Vollmer, L. Eamer, M. C. San Gabriel, K. Zeidan, A. Zini and D. Sinton, *Lab on a Chip*, 2014, **14**, 1142–1150.
11. G. E. Seidel and D. Garner, *Reproduction*, 2002, **124**, 733–743.
12. D. H. Triyoso and T. A. Good, *J. Physiol.*, 1999, **515**, 355–365.
13. M. Stacey, J. Stickley, P. Fox, V. Statler, K. Schoenbach, S. J. Beebe and S. Buescher, *Mutat. Res. Genet. Toxicol. Environ. Mutagen.*, 2003, **542**, 65–75.
14. R. Zeggari, B. Wacogne, C. Pieralli, C. Roux and T. Gharbi, *Sensor. Actuator. B Chem.*, 2007, **125**, 664–671.
15. Y. Agca, J. Liu, J. J. Rutledge, E. S. Critser and J. K. Critser, *Mol. Reprod. Dev.*, 2000, **55**, 212–219.
16. H. C. Zeringue, D. J. Beebe and M. B. Wheeler, *Biomed. Microdevices*, 2001, **3**, 219–224.
17. N. Inomata, T. Mizunuma, Y. Yamanishi and F. Arai, *J. Microelectromech. Syst.*, 2011, **20**, 383–388.
18. M. Hagiwara, T. Kawahara, Y. Yamanishi, T. Masuda, L. Feng and F. Arai, *Lab on a Chip*, 2011, **11**, 2049–2054.
19. C. Chandsawangbhuwana, L. Z. Shi, Q. Zhu, M. C. Alliegro and M. W. Berns, *J. Biomed. Optic*, 2012, **17**, 015001.
20. S. G. Clark, K. Haubert, D. J. Beebe, C. E. Ferguson and M. B. Wheeler, *Lab on a Chip*, 2005, **5**, 1229–1232.
21. R. S. Suh, X. Zhu, N. Phadke, D. A. Ohl, S. Takayama and G. D. Smith, *Hum. Reprod.*, 2006, **21**, 477–483.
22. C. Han, Q. Zhang, R. Ma, L. Xie, T. Qiu, L. Wang, K. Mitchelson, J. Wang, G. Huang, J. Qiao and J. Cheng, *Lab on a Chip*, 2010, **10**, 2848–2854.
23. A. Giglio, S. H. Cheong, Q. V. Neri, Z. Rosenwaks and G. D. Palermo, *Fertil. Steril.*, 2013, **100**, S479.
24. A. Trounson and L. Mohr, *Nature*, 1983, **305**, 707–709.
25. M. Kuwayama, G. Vajta, O. Kato and S. P. Leibo, *Reproductive BioMedicine Online*, 2005, **11**, 300–308.
26. Y. S. Song, S. Moon, L. Hulli, S. K. Hasan, E. Kayaalp and U. Demirci, *Lab on a Chip*, 2009, **9**, 1874–1881.
27. G. D. Smith, P. C. Serafini, J. Fioravanti, I. Yadid, M. Coslovsky, P. Hassun, J. R. Alegretti and E. L. Motta, *Fertil. Steril.*, 2010, **94**, 2088–2095.
28. M. Kuwayama, S. Hamano and T. Nagai, *Reproduction*, 1992, **96**, 187–193.
29. T. Otoi, K. Yamamoto, N. Koyama, S. Tachikawa and T. Suzuki, *Cryobiology*, 1998, **37**, 77–85.
30. L. Meng, X. Huezo, B. A. Stone, K. Baek, G. Ringler and R. P. Marrs, *Fertil. Steril.*, 2011, **96**, S207.

31. O. Kedem and A. Katchalsky, *Biochim. Biophys. Acta*, 1958, **27**, 229–246.
32. Y. S. Heo, H. J. Lee, B. A. Hassell, D. Irimia, T. L. Toth, H. Elmoazzen and M. Toner, *Lab on a Chip*, 2011, **11**, 3530–3537.
33. L. L. Kuleshova, D. R. MacFarlane, A. O. Trounson and J. M. Shaw, *Cryobiology*, 1999, **38**, 119–130.
34. D. Lai, J. Ding, G. D. Smith and S. Takayama, *Fertil. Steril.*, 2013, **100**, S107.
35. J. Melin, A. Lee, K. Foygel, D. E. Leong, S. R. Quake and M. W. Yao, *Dev. Dynam.*, 2009, **238**, 950–955.
36. S. Raty, E. M. Walters, J. Davis, H. Zeringue, D. J. Beebe, S. L. Rodriguez-Zas and M. B. Wheeler, *Lab on a Chip*, 2004, **4**, 186–190.
37. D. L. Hickman, D. J. Beebe, S. L. Rodriguez-Zas and M. B. Wheeler, *Comp. Med.*, 2002, **52**, 122–126.
38. Y. S. Heo, L. M. Cabrera, C. L. Bormann, C. T. Shah, S. Takayama and G. D. Smith, *Hum. Reprod.*, 2010, **25**, 613–622.
39. E. M. Wielhouwer, S. Ali, A. Al-Afandi, M. T. Blom, M. B. Riekerink, C. Poelma, J. Westerweel, J. Oonk, E. X. Vrouwe, W. Buesink, H. G. vanMil, J. Chicken, R. van't Oever and M. K. Richardson, *Lab on a Chip*, 2011, **11**, 1815–1824.
40. K. I. Wang, Z. Salcic, J. Yeh, J. Akagi, F. Zhu, C. J. Hall, K. E. Crosier, P. S. Crosier and D. Wlodkowic, *Biosens. Bioelectron.*, 2013, **48**, 188–196.
41. J. K. Valley, P. Swinton, W. J. Boscardin, T. F. Lue, P. F. Rinaudo, M. C. Wu and M. M. Garcia, *PloS One*, 2010, **5**, e10160.
42. A. Enejder, C. Brackmann and F. Svedberg, *IEEE J. Sel. Top. Quant. Electron.*, 2010, **16**, 506–515.
43. J. P. Urbanski, M. T. Johnson, D. D. Craig, D. L. Potter, D. K. Gardner and T. Thorsen, *Anal. Chem.*, 2008, **80**, 6500–6507.
44. Y. S. Heo, L. M. Cabrera, C. L. Bormann, G. D. Smith and S. Takayama, *Lab on a Chip*, 2012, **12**, 2240–2246.
45. D. Lai, G. D. Smith and S. Takayama, *J. Biophotonics*, 2012, **5**, 650–660.
46. J. E. Swain, D. Lai, S. Takayama and G. D. Smith, *Lab on a Chip*, 2013, **13**, 1213–1224.

CHAPTER 8

Microfluidic Diagnostics for Low-resource Settings: Improving Global Health without a Power Cord

JOSHUA R. BUSER,[†] CARLY A. HOLSTEIN[†] AND PAUL YAGER*

University of Washington, Dept. of Bioengineering, Box 355061, 3720 15th Ave. NE, Seattle WA 98195
*Email: yagerp@uw.edu

8.1 Introduction: Need for Diagnostics in Low-resource Settings

8.1.1 Importance of Diagnostic Testing

The ability to diagnose a patient quickly and accurately is of paramount importance in the management of most diseases, as the appropriate treatment cannot be administered until the cause has been identified. In the developed world, accurate diagnosis can often be achieved, especially in hospitals and large clinics where sophisticated equipment and trained laboratory staff are available.[1] In these settings, the available infrastructure supports advanced diagnostic procedures such as the enzyme-linked immunosorbent assay (ELISA) and nucleic acid amplification assays such as

---

[†]These authors contributed equally to this chapter.

RSC Nanoscience & Nanotechnology No. 36
Microfluidics for Medical Applications
Edited by Albert van den Berg and Loes Segerink
© The Royal Society of Chemistry 2015
Published by the Royal Society of Chemistry, www.rsc.org

polymerase chain reaction (PCR). Additionally, bacterial culture remains widely used for microbe identification, epidemiology, and drug resistance testing.[2] While diagnostics account for only about 1.6% of total Medicare expenditures in the US, an estimated 60–70% of medical decisions are based on their results,[3] highlighting the role of diagnostics as an extremely valuable and cost-effective tool in medicine.

8.1.2   Limitations in Low-resource Settings

In low-resource settings, such as those found in many developing countries, clinics are equipped with only minimal infrastructure, often lacking consistent electricity and refrigeration, and do not have access to highly trained personnel.[2] The lack of these resources and, often, the lack of rapid and effective transportation also make it difficult to establish a cold chain, which is required for transport of many diagnostics that use aqueous reagents.[4] Additionally, laboratories that are fortunate enough to possess diagnostic equipment often cannot support the required maintenance schedules, leaving much advanced equipment in an unmaintained and unusable condition.[5] For these reasons, performing expensive, sophisticated laboratory testing is usually not possible in resource-poor areas of the developing world, often leaving clinics without the means for accurate diagnosis. This is especially unfortunate, since treatments are often available, if only the diagnosis could be made.[2,6]

Overall, there is an urgent need for affordable diagnostics that can be used at the point of care (POC) of the patient. Diagnostic techniques utilized in high-income regions are often resource-intensive, requiring climate-controlled usage and storage, calibration, reliable electrical power, and trained personnel. The POC, from a global health standpoint, is often lacking these amenities.[5] The fact that diagnostic tests for prevalent diseases are not available in formats compatible with low-resource settings not only contributes to lowering the quality of life in these areas, but also results in compounded health problems, such as incorrect treatments and increased drug resistance.[2] To guide the development of diagnostic devices for low-resource settings, the World Health Organization (WHO) compiled the **ASSURED** criteria, stating that devices should be **A**ffordable, **S**ensitive, **S**pecific, **U**ser-friendly, **R**apid and robust, **E**quipment-free, and **D**eliverable to end users.[7] The ability to manufacture high-performance diagnostics inexpensively would lower one of the critical barriers to their adoption in clinics across the globe, thereby increasing access to accurate diagnostic information. This would, in turn, improve disease management, individual patient outcomes, and potentially public health outcomes.[6,8,9]

8.1.3   Scope of Chapter

The goal of this chapter is to provide the reader with a concise and holistic assessment of the need for and use of microfluidic diagnostics in low-resource settings. Since comprehensive review articles already exist on the topic of point-of-care diagnostics,[8,10] our goal is not to review all microfluidic technologies in detail, but instead to offer an overview of the field to highlight both the successes of and opportunities for microfluidic diagnostics in global health. This overview intentionally incorporates our own perspective on these issues, including an emphasis on paper-based microfluidics, which we view as an important and rapidly growing component of the microfluidics field with significant potential to revolutionize diagnostic testing in low-resource settings. Most importantly, we aim to provide a useful context with which to think about the development of microfluidic diagnostics. To do so, we first categorize and describe the different types of diagnostic testing that are needed in low-resource settings (Section 8.2), many of which are often overlooked by test developers. We then provide an overview of the major types of microfluidic diagnostics that are being pursued for global health applications (Section 8.3), which we broadly categorize into channel-based microfluidics and paper-based microfluidics. Finally, we break down the diagnostic tests into many of the components that must be considered, and ideally integrated, when developing a diagnostic test for use in low-resource settings (Section 8.4). In this section, we highlight microfluidic technologies that have addressed these aspects particularly well, in addition to pointing out areas that are rife with potential for microfluidic innovation.

8.2 Types of Diagnostic Testing Needed in Low-resource Settings

There is a good deal of diagnostic testing underway today to improve and maintain the health of individuals in low-resource settings in venues ranging from large hospitals in urban settings to small clinics in villages and rural homes by itinerant healthcare workers. However, there is a substantial outstanding need for more and better testing, at a manageable cost. The most imperative types of testing comprise the diagnosis of disease, including the identification of drug-resistance, and monitoring of therapy, which are discussed in Sections 8.2.1 and 8.2.2, respectively. Additionally, tests that can determine the quality of drugs used for treatment and identify counterfeits would help improve health in the developing world, as described in Section 8.2.3. Finally, tests that can detect environmental contaminants related to health issues are also needed in resource-poor settings, as discussed in Section 8.2.4.

8.2.1 Diagnosing Disease

The ability to identify the cause(s) of a given patient's disease is the most pressing diagnostic need everywhere. Diseases caused by pathogens are top contributors to the global burden of disease, due to the high prevalence of infectious diseases in parts of the developing world.[11,12] For example, the 2010 Global Burden of Disease study led by the Institute for Health Metrics and Evaluation ranks lower respiratory infections, diarrhea, HIV/AIDS, and malaria in the top ten causes of global disease burden, accounting for an estimated 337 million DALYs (disability-adjusted life years) in total.[11,12] All four of these infectious diseases can have multiple pathogenic causes, including an array of viruses and bacteria for lower respiratory infections and diarrhea[11,12] and several different strains of the specific pathogens for malaria (five different species of the *Plasmodium* parasite[13]) and HIV/AIDS (two different types of HIV, each with multiple sub-types.[14,15]) Diagnostic tests that can detect and identify specific disease-causing pathogens from biospecimens are therefore of extreme importance for managing a patient's disease and getting him/her on the correct path to treatment. While infectious diseases afflict many parts of both the developed and developing worlds, no region is more severely impacted than sub-Saharan Africa, where infectious diseases are still the leading causes of disease burden,[11,12] as illustrated in Figure 8.1. The resource-poor areas of sub-Saharan Africa therefore present a unique intersection of the need for infectious disease diagnostics and tight design constraints (see ASSURED criteria in Section 8.1.2) that is well suited for low-cost microfluidic technologies.

Point-of-care testing would be invaluable for the diagnosis, treatment, and epidemiological strategy for a variety of infectious agents, but tuberculosis (TB) represents a prime example of a disease that is widely globally prevalent and in need of better diagnostic testing. According to the WHO, roughly one-third of the world's population is infected with TB. While most of these cases are dormant, the WHO estimates that there were 1.3 million deaths from TB and 8.6 million new cases of TB in 2012 alone,[16] indicating the severity of this disease. A recent report from UNITAID on the TB diagnostic technology landscape outlines the current status of TB testing along with unmet needs.[17] The illustration in Figure 8.2, reproduced from that report, highlights laboratory capacity for TB diagnostics in resource-limited settings, along with the types of testing performed at each level of the health system. This figure highlights a huge problem and area of need: there are currently no suitable TB diagnostics available at the peripheral level of health systems, where 60% of the patients are seen.[17]

Like many other infectious diseases, diagnosing TB early in the disease progression and providing effective treatment is critical to patient outcomes and for controlling spread of the disease. Sputum smear microscopy and culture are the conventional diagnostic techniques, with drug susceptibility testing performed on cultures, but results often take weeks.[16] Failing to test for drug resistance can lead to inappropriate treatments, prolonged disease, increased mortality, and increased drug resistance (discussed in detail next). More timely and specific detection of TB can be achieved using nucleic acid amplification assays, but much of the technology required to perform these assays has not traditionally been functional in low-resource settings, where diagnostics must be portable and operable without advanced training.

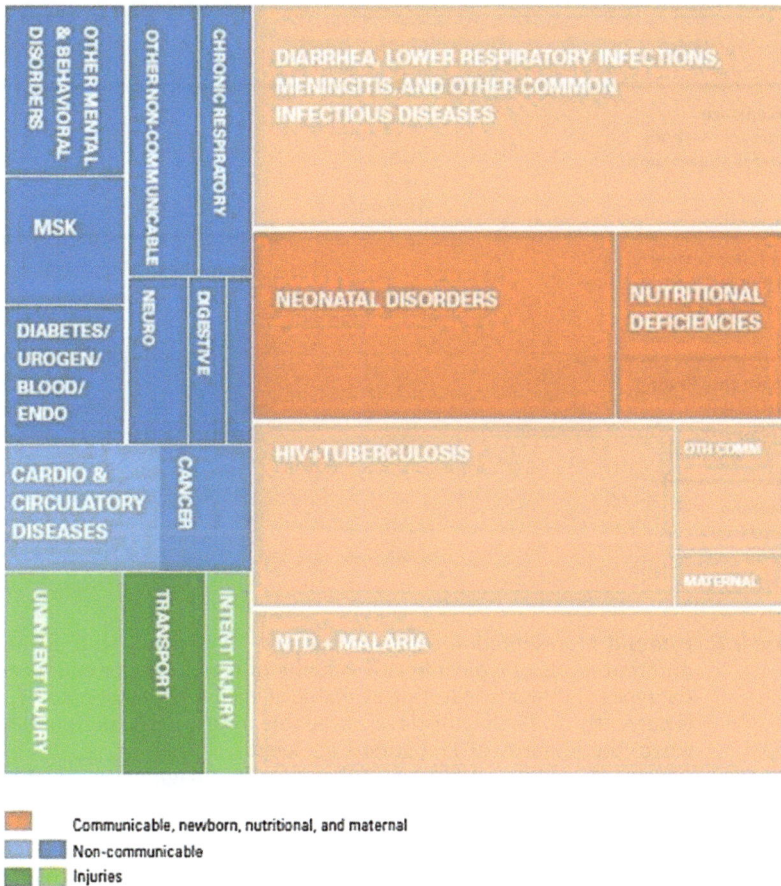

Figure 8.1 of the figure content:

			DIARRHEA, LOWER RESPIRATORY INFECTIONS, MENINGITIS, AND OTHER COMMON INFECTIOUS DISEASES	
OTHER MENTAL & BEHAVIORAL DISORDERS	OTHER NON-COMMUNICABLE	CHRONIC RESPIRATORY		
MSK				
DIABETES/ UROGEN/ BLOOD/ ENDO	NEURO	DIGESTIVE	NEONATAL DISORDERS	NUTRITIONAL DEFICIENCIES
CARDIO & CIRCULATORY DISEASES	CANCER		HIV+TUBERCULOSIS	OTH COMM
				MATERNAL
UNINTENT INJURY	TRANSPORT	INTENT INJURY	NTD + MALARIA	

Communicable, newborn, nutritional, and maternal
Non-communicable
Injuries

Figure 8.1 Illustration of the relative contributions of various diseases and causes to disease burden (quantified in DALYs) in sub-Saharan Africa in 2010. Infectious diseases, including HIV, TB, and malaria, are the major contributors to disease burden in this region.
Reprinted from the 2010 Global Burden of Disease report,[11] with permission. (Institute for Health Metrics and Evaluation, *The Global Burden of Disease: Generating Evidence, Guiding Policy*, University of Washington, Seattle, WA, IHME, 2013.)

In addition to pathogen detection, the ability to determine whether or not a given pathogen is resistant to drugs used for treatment is critical to both managing a patient's disease effectively and monitoring the prevalence of such drug-resistance for public health purposes. Drug resistance is a major consideration when choosing the treatment for a patient's disease, since it is critical to know if the infecting pathogen will be resistant to the drug of choice. Diagnostics that cost less than the difference between potentially ineffective, cheap drugs and efficacious, more expensive drugs are critical for implementing effective patient management.[2] Further, limiting the use of effective second-line drugs helps to keep the pathogens from developing

| **Types of testing** | **Health system levels** | **Fraction of patients seen at given level** |

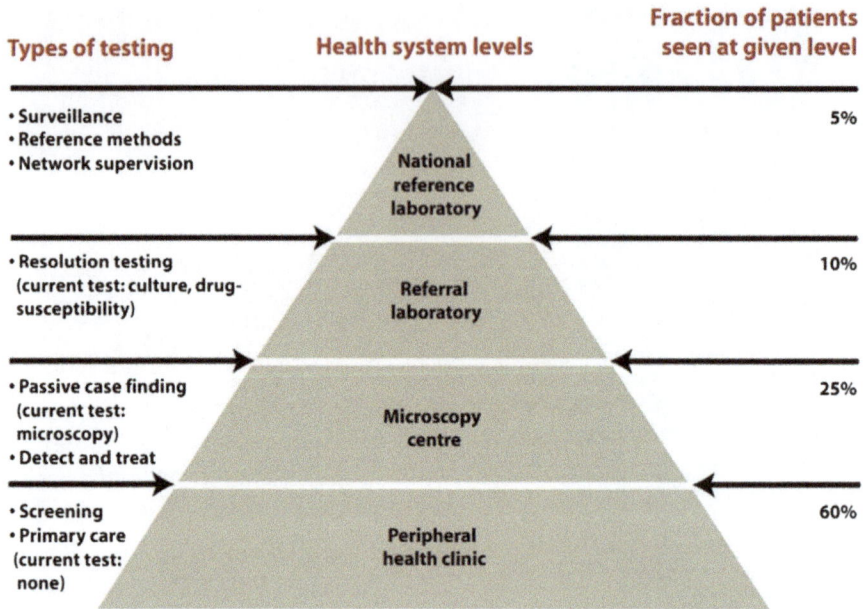

• Surveillance
• Reference methods
• Network supervision
                    National reference laboratory   5%

• Resolution testing
(current test: culture, drug-susceptibility)
                    Referral laboratory   10%

• Passive case finding
(current test: microscopy)
• Detect and treat
                    Microscopy centre   25%

• Screening
• Primary care
(current test: none)
                    Peripheral health clinic   60%

Figure 8.2 Pyramid representation of the fraction of TB patients seen at the four different levels of typical health systems in resource-poor countries and the types of diagnostic testing available at each level. Most notably, there is currently no TB diagnostic test available for use at the peripheral level, where the majority of TB patients are seen.
Reprinted from UNITAID–Tuberculosis diagnostics technology landscape report,[17] with permission.

resistance to these drugs, preserving them as treatment options for a longer period of time.[2] Finally, tests for drug-resistant pathogens could also be used to monitor a region for drug-resistance, which can be an important public health strategy.[2] For these reasons, monitoring drug resistance is a critical effort in the control of prevalent infectious diseases such as malaria, TB, HIV, and bacterial infections.[2]

Finally, as infectious diseases are controlled in the developing world, lifetimes increase. Consequently, chronic diseases such as diabetes and cancer increase, generating a need for low-cost diagnostic tests for these non-infectious (also called non-communicable) diseases as well. According to the 2010 Global Burden of Disease report, mortality due to chronic diseases has increased 30% since 1990 for people worldwide, reflecting a global shift in disease burden from infectious to non-infectious diseases.[11,12] As shown in Figure 8.3, non-communicable diseases actually account for the majority of disease burden (DALYs) in most countries across the globe, including many developing countries.[11] The one major exception is the sub-Saharan region of Africa, in which infectious disease remains the major contributor to disease burden,[11,12] as discussed above.

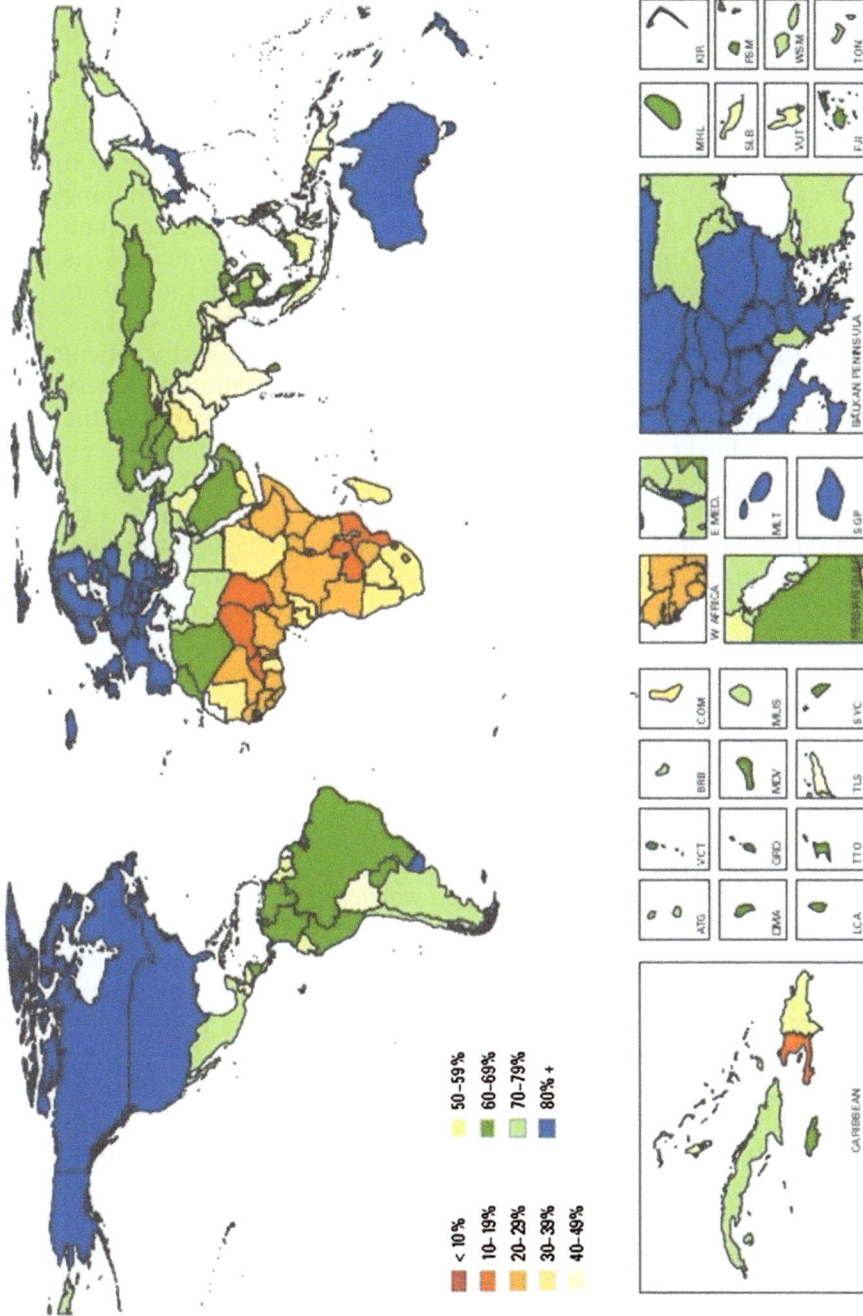

Figure 8.3 Percentage of disease burden due to non-communicable diseases in 2010, by country, quantified in terms of DALYs. In most countries, non-communicable disease is the major cause of disease burden. Reprinted from the 2010 Global Burden of Disease report,[11] with permission. (Institute for Health Metrics and Evaluation, *The Global Burden of Disease: Generating Evidence, Guiding Policy*, University of Washington, Seattle, WA, IHME, 2013.)

Diabetes represents one prominent and growing chronic disease world-wide, experiencing a 69% increase in global disease burden over the past 20 years.[11] Monitoring blood glucose levels through point-of-care test strips is a common practice for the *management* of diabetes, and will be discussed further in Section 8.2.2. The *diagnosis* of diabetes represents a distinct and important area of diabetes testing in need of better point-of-care options, especially for use in the developing world.[18,19] In particular, diabetes diag-nostics that detect glycated hemoglobin (HbA1c) have grown in popularity in the developed world, but are still widely unavailable in the developing world.[19] Glycated hemoglobin is considered to be a more convenient and robust biomarker for diabetes than blood glucose, since blood glucose levels fluctuate day-to-day, while HbA1c is more stable, representing the average blood glucose content of the patient over the previous 8 to 12 weeks.[19] Because of the robust nature of HbA1c as a diabetes biomarker, the patient does not need to fast prior to the test, making the diagnostic process much more convenient.[19] These advantages of HbA1c are magnified in the developing world, where access to healthcare is often already limited and inconvenient, making HbA1c tests an important area of need in global health diagnostics. Additionally, gestational diabetes – or diabetes that occurs during pregnancy–is another area in need of better diagnostic testing, due to the severity of complications that can arise for both mother and baby and to the lack of appropriate screening methods for low-resource settings.[20] One approach, under development by Weigl and colleagues at PATH, is to screen for glycated albumin using a lateral flow test.[20] This type of low-cost, microfluidic diagnostic offers much greater convenience and ease-of-use than traditional glucose challenge tests, which require fasting, venous blood draws, and laboratory analysis.[20]

8.2.2 Monitoring Disease

In contrast to *diagnosing* a new disease, *monitoring* a patient's existing disease is critical for the management of many long-term illnesses. In these circumstances, the need for tests to be inexpensive is even more crucial, since the test must be repeated many times during the duration of the patient's condition. Monitoring can be used to help control the patient's health and determine when medication is needed, as is the case for diabetes. Alternatively, monitoring can be used to evaluate a patient's response to treatment, which is typically done for HIV patients on anti-retroviral therapy, and could be applied to compliance with the long-term course for treatment of tuberculosis.

For diabetes management, monitoring blood glucose levels is crucial for enabling patients to manage their disease, as they can know when to adjust their sugar intake or administer insulin to maintain a safe blood glucose level. Although chronic diseases like diabetes were once referred to as "diseases of affluence",[7] diabetes incidence has increased in the developing world over the last few decades, with 80% of diabetics now living in low- and middle-income countries.[21] The 2010 Global Burden of Disease study highlights this change, showing that diabetes rose from the #21 global disease burden rank in 1990 to the #14 position in 2010, experiencing a 69% increase in DALYs during this time.[11,12] Glucose monitoring devices have fundamentally changed the lives of diabetics in the developed world, allowing them to reduce or eliminate devastating consequences of un-controlled blood glucose. Consequently, it is no surprise that these devices lead the way in terms of commercial volume for point-of-care tests, given the prevalence of the disease and the need for multiple daily measurements. According to Gubala *et al.*, nearly 10^{10} glucose test strips are manufactured per year, with some production lines producing 10^6 test strips per hour.[8] While glucose monitoring devices and test strips are fairly low-cost and translatable to the developing world, these low-resource settings carry their own unique challenges that remain unaddressed. In particular, the lack of doctors, health centers, and healthcare workers in remote areas of the de-veloping world make it difficult for patients to connect with healthcare providers to review their blood glucose data and to obtain testing supplies in some cases.[22] For these reasons, combined with the prevalence of diabetes in the developing the world, the WHO seeks to promote improved methods for the diagnosis, surveillance, and control of diabetes in low- and middle-income countries.[18,22] Incorporating mobile health technology to connect patients to doctors would be one important step in helping patients manage their disease (more on data integration into health systems in Section 8.4.6), and enabling continuous glucose monitoring for patients in these regions would be an even greater feat.[22]

Disease monitoring is also an important process for patients infected with HIV, as proper monitoring of HIV treatment is key to optimal disease management and prognosis.[23–26] A key component of treatment monitoring is regular testing of HIV viral load,[23–27] which represents the level of circu-lating virus in the blood. Since HIV is highly mutagenic, it commonly evades treatment regimens by mutating into drug-resistant forms, which can then resume viral replication.[28] By monitoring for an increase in a patient's viral load over time, clinicians can determine when treatment is failing and therefore when to switch the patient to an alternative drug regimen.[23–27]

While HIV viral load testing is commonplace in the developed world, its current use is severely limited in low-resource settings.[23–26] This disparity is particularly problematic because the majority of global HIV prevalence is in low- and middle-income countries.[29] More specifically, the WHO estimates that of the 34 million people living with HIV in 2010, 30 million were in low-resource countries.[30] The unfortunate lack of viral load testing in these countries stems from the inappropriate nature of current testing methods for low-resource settings. The most common viral load test, by far, is nucleic acid testing by PCR to measure the number of copies of HIV RNA present in a blood sample. Since PCR requires sophisticated equipment, highly trained personnel, and specialized rooms and workflow to prevent contamination, it can only be performed in the highest-level laboratories and hospitals in developing-world countries.[25,26] One proposed solution has been to collect samples as dried blood spots and send them to a reference laboratory for testing,[25,31] but this approach is considered infeasible in many low-resource settings.[25] Furthermore, even if such reference laboratory testing can be accomplished logistically, the turnaround time (on the order of weeks) is prohibitively long, causing patients to be lost to follow up.[24,25] Current alternatives to PCR-based viral load testing, such as the reverse transcriptase activity assay and the heat-denatured p24 antigen assay, represent improvements in point-of-care viral load testing, but are still not considered to be perfect solutions due to poor usability and reproducibility.[26,32,33]

There is therefore tremendous need for a POC test for HIV viral load that can be used in low-resource settings.[23–26,29] In fact, in its most recent report on the global AIDS epidemic, UNAIDS touts the development of POC diagnostics as one of the five pillars of its "Treatment 2.0" plan to simplify and increase access to HIV treatment.[29] While a previous unavailability of drugs for antiretroviral therapy (ART) in the developing world may have stagnated the search for low-cost HIV viral load tests, these drugs are becoming increasingly available.[23,24,26,29,30] From 2004 to 2009, access to antiretroviral (ARV) drugs increased 13-fold in low- and middle-income countries,[29] and currently 6.6 million HIV-infected individuals in these countries are receiving ART.[30] While there are still nearly 10 million more people in low- and middle-income countries in need of ART,[29,30] the ever-increasing number of patients receiving therapy necessitates POC tests for monitoring treatment adherence and efficacy.[25,26] If anything, the limited supply of ARV drugs in low-resource settings should prompt better treatment monitoring and viral load testing, since efficient use of these drugs is all the more critical.[23,24]

While CD4 + T-cell counting is also performed for monitoring of ART, viral load testing alone would represent a significant improvement in HIV disease management. The combination of CD4 + T-cell and viral load monitoring would be ideal,[27,34,35] but it is known that CD4 + T-cell monitoring alone is not sufficient to predict treatment failure,[25,34,36] and many experts support the use of viral load testing alone, if a choice must be made between the two methods.[23,26,36] Overall, the development of a POC viral load test represents a logical first step toward low-cost ART monitoring.

Finally, in addition to treatment monitoring, HIV viral load testing is also used for the diagnosis of HIV in infants. Since maternally derived antibodies persist in babies until approximately 18 months of age, traditional diagnosis based on the detection of patient-derived anti-HIV antibodies is not suitable for young infants.[24,25] Conversely, diagnosis by viral load testing is much easier in infants than in adults, since HIV-infected infants have extremely high viral loads ($>10^5$ RNA copies mL^{-1}) compared to the low levels that can circulate in adults with latent infection (10^2–10^3 RNA copies mL^{-1}).[25,37] A low-cost diagnostic device for viral load testing would therefore be extremely useful for infant diagnosis and prompt disease management for these young children.[24,25] Moreover, a qualitative test for determining a yes-or-no diagnosis of HIV for infants could represent an ideal entry point for POC viral load testing, since the detection limit required is much higher than that needed for quantitative treatment monitoring.[24,25]

8.2.3 Counterfeit Drug Testing

Another important area in need of better testing capabilities is the identification of counterfeit medicines. While counterfeit drugs are not unique to the developing world, they are much more prevalent in these regions than in the developed world due to the lack of regulatory oversight in developing countries.[38,39] The absence of required active drug compounds, the introduction of erroneous active drug compounds, and the use of binder and filler materials are all common problems found in counterfeit medicines.[39] Having testing devices that meet the ASSURED criteria (see Section 8.1.2) and can detect the presence or absence of important compounds for a given drug would therefore offer an important tool in the fight against counterfeit drugs, which currently contribute to morbidity, mortality, and the waste of precious medical and financial resources.[38,39]

The Lieberman group at the University of Notre Dame has made significant strides in counterfeit drug testing, using paper-microfluidic devices. In a recent publication from this group, Weaver *et al.* describe their novel paper analytical device (PAD) platform for rapid field testing of drugs used for disease treatment.[39] In these PADs, the drug is swiped across the bottom of a device, and several distinct lanes perform chemical analyses of the drug that generate colorimetric signals, resulting in a "color bar code" that describes the quality of the drug[39] (Figure 8.4). In particular, the authors demonstrate PADs for the evaluation of β-lactam-based antibiotics and anti-TB drugs that can identify wanted and unwanted compounds with mostly high sensitivities and specificities, all within a matter of minutes.[39]

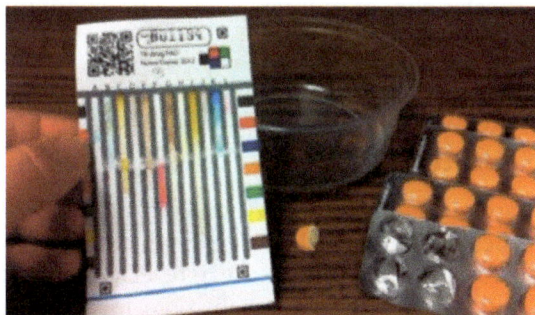

Figure 8.4   Example of a paper analytical device (PAD) developed by the Lieberman group for counterfeit pharmaceutical testing, in this case for TB drugs.[39] Reprinted with permission from *Anal. Chem.*, **85**(13), 6453–6460. Copyright 2013, American Chemical Society.

8.2.4 Environmental Testing

Detecting a substance of interest in an environmental sample as opposed to a patient sample provides unique challenges in handling highly variable, complex sample matrices.[40] The outbreak of cholera in Haiti following the already devastating 2010 earthquake highlights the power of water-borne contaminants' affect on large sections of a population. Between October 2010 and December 2012, the number of cholera cases in Haiti exceeded 630 000, killing over 7900 people.[41] In addition, nearly 30 000 cholera cases were recorded after spreading to the Dominican Republic and Cuba, killing over 400 more people.[41] Better methodologies for monitoring cholera contamination are needed for low-resource settings like Haiti to help implement prevention and care efforts.[42] One promising opportunity is the monitoring for *V. cholerae* O1 and vibriophages, which have been shown to indicate a cholera outbreak one month in advance, providing an opportunity for the implementation of prevention measures.[43]

Additionally, the biochemical, industrial, pharmaceutical, and medical industries have seen much growth in the past few decades. As a result, the amount of waste product from these industries has increased dramatically, elevating the need for monitoring of environmental contaminants, preferably without the need for complex and expensive peripheral equipment.[40] Microfluidic systems may be able to address some of these needs, but will need to handle sample preparation from a wide variety of sample matrices while maintaining compatibility with the detection of a variety of contaminants, including microorganisms, hydrocarbons, herbicides, and toxic metals,[40] as well as emerging contaminants such as Splenda, siloxanes (widely used in consumer products), and synthetic musks (widely used fragrance additives).[44] The fact that many of these pollutants are only present at extremely low concentrations further complicates the problem, introducing the need for sample pre-concentration in many cases. The unique capabilities of microfluidic devices may enable low-cost monitoring of these and other pollutants, allowing studies to monitor both acute and chronic effects of how we as a species choose to interact with our environment.

8.3 Overview of Microfluidic Diagnostics for Use at the Point of Care

Currently, there are many point-of-care (POC) diagnostic tests on the market and even more emerging through the research pipeline as the need for these tests becomes increasingly recognized. Current POC tests vary in type, cost, sophistication, and efficacy, leaving much room for improvement in most cases. The two major types of current POC diagnostics on the market are microfluidics-based tests, reviewed in Section 8.3.1, and lateral flow tests, reviewed in Section 8.3.2.2. The emerging platform of paper-based microfluidic tests is reviewed in Section 8.3.2.3.

8.3.1   Channel-based Microfluidics

Microfluidics refers to the use of systems that manipulate fluid through channels with at least one dimension less than 1 mm. Microfluidics date back to at least the 1970s,[45] with an uptick in interest in the 1990s and the push toward "micro total analysis systems" (MicroTAS).[46] These systems utilized defined channels fabricated in materials such as glass, PDMS, and plastic laminates, combined with external pumps to deliver fluid through the channels.[47–49] These microfluidic systems feature several attributes that are ideal for rapid diagnostic testing, including process automation, fast diffusion times, and the need for only small (microliter) sample volumes.[1,50] For these reasons, several microfluidics-based POC tests have been developed. We refer to these channel-based microfluidic systems as "traditional microfluidics", as opposed to the paper-based systems discussed in Section 8.3.2. For detail on the principles of channel-based microfluidics, we refer the reader to the other chapters of this book.

8.3.1.1   Current Microfluidics-based Platforms

The GeneXpert from Cepheid is perhaps the biggest success story in global health diagnostics, as it has greatly aided the diagnosis of tuberculosis (TB), including drug-resistant forms.[51] The GeneXpert MTB/RIF test, endorsed by the WHO in 2010, utilizes microfluidics to perform nucleic acid testing of *Mycobacterium tuberculosis*, obtaining a result within two hours.[52] The GeneXpert system consists of a fairly sophisticated instrument that performs most of the testing functionality and disposable, one-time-use cartridges on which the testing reagents are stored.[53] While the GeneXpert has greatly increased the speed of TB diagnosis, reducing mean testing-to-treatment times from 56 to 5 days in one study,[51] the instrument is too expensive and requires too much infrastructure (*e.g.* electricity) to be used in very remote settings. Even at special negotiated prices for low-resource countries, the instrument still costs $17k, plus $10 for every one-time-use cartridge.[54] For these reasons, the GeneXpert is recognized as a solution primarily for centralized laboratories within developing countries.[9,55] As of the third financial quarter of 2012, 4660 modules and 1 482 550 cartridges had been sold, with half of the cartridge sales in South Africa. Unfortunately, this technology will not reach the 60% of patients at the peripheral level (see Figure 8.2), and according to Pantoja *et al.*, will require further funding increases and/or price reductions to be financially viable in low-income countries.[56]

The iSTAT from Abbot Diagnostics is a true POC device, serving as a portable analyzer of a panel of common blood analytes used for patient monitoring.[9,57] This handheld device is based on a microfluidic format that couples the fluid flow to electrochemical detection systems for the measurement of blood chemistries and electrolytes.[9,57] Despite its utility, the iSTAT is still too expensive for many low-resource settings and is rendered useless in settings where maintenance is not possible when it breaks.

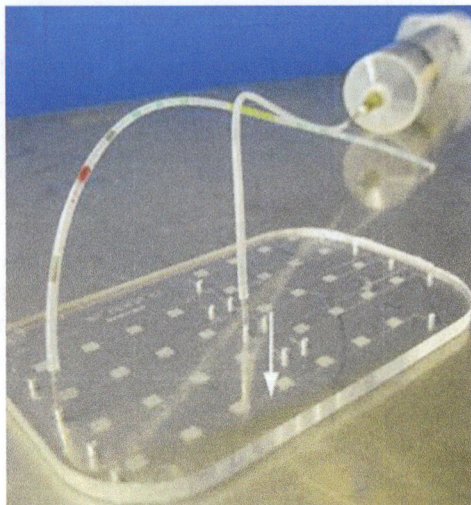

Figure 8.5   The "mChip" by Sia and colleagues, a low-cost, microfluidic replacement
for ELISA testing.[58]
Reprinted by permission from Macmillan Publishers Ltd, *Nature
Medicine*, **17**, 1015–1019, copyright 2011.

Finally, the Sia group demonstrated the mChip (Figure 8.5), a great example
of a low-cost microfluidic diagnostic system.[58] The mChip essentially performs the enzyme-linked immunosorbent assay (ELISA), but in a simple
microfluidic format that includes an automated reader and comes at
one-tenth of the cost of a traditional laboratory ELISA.[58] In particular, the
Sia group used injection-molded chips from polystyrene and cyclic olefin co-
polymer, allowing for significantly reduced cost compared to microfluidic
chips created from polydimethylsiloxane (PDMS) *via* photolithography.[58]
During field-testing in Rwanda, the mChip was able to simultaneously detect
HIV and syphilis from 1 µL of whole blood with sensitivities and specificities
comparable to laboratory ELISA.[58] To eliminate the need for pumps, the first
version of the mChip utilizes a simple plastic syringe to drive fluid through the
chip.[58] This creative approach is extremely low-cost, which is important for
resource-poor settings, but the manual manipulation of reagents required by
this method is not ideal, especially for untrained users. A second iteration of
the mChip, recently described by the Sia group, addressed this issue by incorporating the fluidic handling into the handheld reader, creating a fully
integrated portable device.[59] While much more user-friendly, this integrated
mChip device is significantly more expensive (estimated <$1000), demon-
strating the critical trade-off that exists for channel-based diagnostics.

8.3.2   Paper-based Microfluidics

Based on the continued need for low-cost, easy-to-use POC diagnostics, a
new type of test has emerged: the paper-based diagnostic test. The use of

paper in biological testing is actually a decades-old technique, most notably used in lateral flow test strips, which are discussed in detail in Section 8.3.2.2. The newly emerging paper-based diagnostic tests, however, are based on the concept of paper microfluidics–or Microfluidics 2.0[60]–which combines the simplicity of lateral flow tests with the sophistication of microfluidic-based tests to achieve an intermediate format that is highly suitable for POC use. The principles of paper microfluidics and current paper-based diagnostic platforms are discussed in Sections 8.3.2.1 and 8.3.2.3, respectively.

8.3.2.1 Principles of Paper Microfluidics

The hallmark of all paper-based diagnostic systems is the use of "paper" as the primary assay substrate. While paper in the literal sense refers to the everyday writing material made from cellulose pulp, the paper microfluidics community broadly defines "paper" as any porous material that takes the form of a thin sheet.[61] This broad definition of paper will be used herein.

Paper is an ideal substrate for POC bioassays for many reasons. First, paper materials are generally inexpensive, allowing paper-based devices to be widely affordable and accessible.[61,62] Second, paper is highly compatible with biological and chemical reagents and amenable to the storage of such reagents in dry form.[61,62] These features enable long shelf lives for paper-based diagnostics, as have already been demonstrated by lateral flow tests.[63] Third, the porous structure of paper provides both a high internal surface area and short diffusion distances, allowing for high adsorptive capacity and fast reaction times, respectively.[61,62,64] Finally, and perhaps most importantly, paper affords the passive wicking of fluids by capillary action, eliminating the need for external pumps or power sources and thereby significantly reducing the cost of paper-based systems relative to their microfluidic counterparts.[61,62,65]

8.3.2.2 The Ubiquitous Paper-based Diagnostic: The Lateral Flow Test

Lateral flow tests (LFTs) are the most ubiquitous commercially available point-of-care diagnostic. First gaining popularity in the 1980s for pregnancy testing,[66–68] LFTs are now used for testing of many analytes, including disease markers in humans and animals, environmental and agricultural contaminants, drugs of abuse, and biowarfare agents.[66,67] The widespread use of LFTs has propelled this technology into a $2.3 billion market, with pregnancy tests still comprising the largest market share.[66] In fact, 10 million lateral flow tests are manufactured each year for pregnancy testing alone.[8]

A lateral flow test is based on a simple strip of porous membrane, typically nitrocellulose, which both allows fluid flow through the strip and serves as

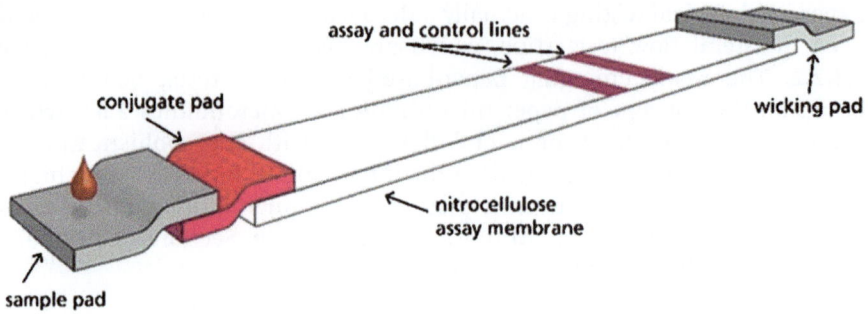

Figure 8.6   Illustration of a typical lateral flow test (LFT). Courtesy of Gina Fridley.

the substrate on which the detection reaction takes place[63,67] (Figure 8.6). This porous strip is in contact with an absorbent pad on the distal end, which promotes wicking, and a series of sample and conjugate pads on the proximal end, where the assay begins. The conjugate pad contains dried detection antibody for the target analyte that has been conjugated to a visible label, typically gold nanoparticles. Two lines are pre-printed on the porous strip: 1) a test line containing antibody that specifically captures the target analyte of interest and 2) a control line containing antibody that captures excess gold-labeled detection antibody. To start the assay, the user simply applies the patient sample (usually blood, urine, or saliva) to the sample pad, which rehydrates the gold-labeled detection antibody and wicks through the porous membrane, allowing this antibody to bind the target analyte, if present. This antigen-antibody complex (if present) is then bound by the capture antibody at the test line, generating a pink color in this region due to the accumulation of gold nanoparticles. The control line also turns pink due to the binding of excess gold-labeled antibody, indicating that the test has functioned properly and the reagents have reached the end of the strip.

Due to their low cost and ease of use, lateral flow assays have found utility for diagnostic testing in both low-resource settings, such as clinics in the developing world and the battlefield, and POC settings in high-resource countries, such as doctor's clinics, emergency rooms, ambulances, and the home.[2,66,69,70] In the developing world, lateral flow tests for infectious diseases such as malaria, influenza, and dengue fever have been important, yet imperfect, tools for the diagnosis of disease.[71-73] In the developed world, lateral flow assays have most commonly been used for at-home ovulation screening and pregnancy testing,[66] but are gaining increasing reach. For example, OraQuick,[74] the first commercially available home HIV test, is now available in United States pharmacies without a prescription, and a preliminary study has found that usage of these at-home tests has resulted in decreased HIV transmission in a high-risk population.[75]

Lateral flow tests have therefore played a significant role in establishing POC testing as a valid means of medical diagnosis, and the widespread success of lateral flow assays is undoubtedly related to their simplicity.

Despite this success, LFTs often suffer from low sensitivity, as compared to gold standard laboratory-based tests,[66] driving the need for more accurate POC tests.

8.3.2.3   Current Paper-based Platforms

While paper chromatography, blotting assays, and lateral flow tests have existed for decades, the recent renaissance of novel paper-based diagnostic platforms began in the late 2000s with the work of George Whitesides' group.[65,76] Recognizing the benefits of paper-based systems and their suitability for POC use, several different paper-based platforms and subsequent iterations have since been developed. These platforms fall under two main categories: microfluidic paper analytical devices (µPADs) and two-dimensional paper networks (2DPNs), as reviewed below. All platforms utilize some form of patterning to create flow channels in a porous medium that are hydrophilic relative to a hydrophobic barrier.[62] In general, paper-based microfluidic platforms are poised to bridge the gap between sophisticated channel-based systems and simple but less accurate lateral flow tests. While both channel-based microfluidic diagnostics and lateral flow tests can be highly effective for certain applications, we view paper-based diagnostics as the current platform that most widely fulfills the ASSURED criteria, as enumerated in Table 8.1.

8.3.2.3.1 Microfluidic Paper Analytical Devices (µPADs). The original paper-based device of the paper resurgence was the microfluidic paper analytical device (µPAD), developed in the laboratory of George Whitesides.[65,76] Basic two-dimensional µPADs feature a flow path consisting of a single inlet channel that diverges into multiple analyte channels, as shown in Figure 8.7A. These flow paths were initially created using photolithography to pattern hydrophobic channel barriers onto a cellulose substrate, a technique that was carried over from PDMS microfluidics.[76] Since then, most two-dimensional µPADs have been fabricated using wax printing to deposit the hydrophobic barriers that define the flow channels within the cellulose substrate.[77] Three-dimensional µPADs have also been developed (Figure 8.7B), created by the stacking of multiple layers of porous media (typically cellulose) and adhesive material.[78–80] The flow channels within each layer are designed to interact strategically with those of the other layers to achieve three-dimensional flow paths that fit compactly into a small material footprint.

Many examples of diagnostic tests have been developed using µPADs. These tests include the colorimetric detection of total glucose and protein in urine (*e.g.* urinalysis),[76,78] the fluorescence-based detection of β-galactosidase using a fluidically activated battery,[81] and the colorimetric detection of particulate metal in aerosols as a monitoring tool for occupational exposure.[82] Perhaps the most successful application of µPADs to date has been the liver enzyme test to monitor the health of patients who are

Table 8.1 Comparison of the ability of three microfluidic diagnostic platforms to fulfill the ASSURED criteria. This assessment is based on each platform as a whole, while individual devices may vary in their characteristics. Overall, emerging paper-based microfluidic technologies seem best-suited to serve as diagnostics in low-resource settings.

	Channel-based microfluidics	Traditional lateral flow tests	Emerging paper-microfluidic platforms
	E.g. Cepheid GeneXpert (www.cepheid.com)	*E.g.* Dengue NS1 Ag LFT (Courtesy of Yager Lab)	*E.g.* 2DPN (Courtesy of Tinny Liang)
Affordable		✓	✓
Sensitive	✓		✓
Specific	✓		✓
User-friendly	✓	✓	✓
Rapid & robust		✓	✓
Equipment-free		✓	✓
Deliverable to end user		✓	✓

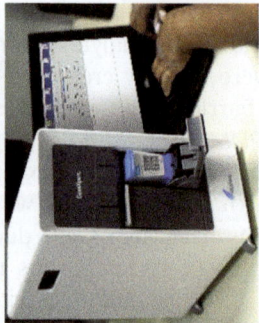

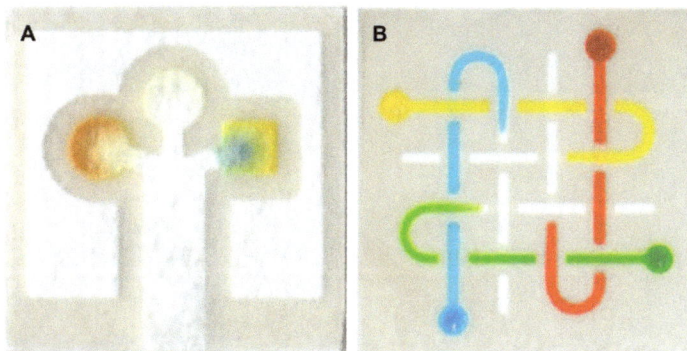

Figure 8.7 Examples of (A) two-dimensional[65] and (B) three-dimensional[78] μPADs. (A) Adapted with permission from *Anal. Chem.*, **82**(1), 3–10. Copyright 2009, American Chemical Society. (B) Adapted with permission from *PNAS*, **105**(50), 19606–19611. Copyright 2008, National Academy of Sciences, USA.

taking combinatorial medications for HIV and/or tuberculosis,[83] which is currently in field-testing.[84] Additionally, μPAD technology is continually being developed, and there is currently a large focus on incorporating printable electronic elements on μPADs.[61,85,86] Overall, the μPAD represents a useful paper-based platform, especially when multi-analyte detection is desired. However, demonstrations of μPADs to-date have been mostly limited to detection steps that employ simple, single-step chemical or enzymatic reactions.

8.3.2.3.2 Two-dimensional Paper Networks (2DPNs). Our own group in the Yager, Lutz, and Fu laboratories pioneered the development of the two-dimensional paper network (2DPN). This paper-based platform utilizes the shape and spatial arrangement of the paper substrate to control fluid flow and achieve automated, multi-step processing. These devices are made from porous nitrocellulose, in which flow paths are physically cut from the parent sheet, typically using a CO_2 laser cutter. This method represents a simple way to create flow paths within the porous substrate and eliminates the need for patterning of additional materials to create hydrophobic barriers.

To date, our group has demonstrated many capabilities of 2DPNs. In particular, Fu *et al.* demonstrated that the timing of fluid delivery could be tuned based on the geometry of the network.[87,88] Importantly, the use of multiple inlet legs and a single outlet leg allows for the delivery of multiple reagents over a given detection region (Figure 8.8).[88,89] This multiple input leg format has been used, in combination with a folding-activated card platform, to demonstrate the chemical amplification of gold nanoparticle-based signal,[90] the improvement in sensitivity of a lateral flow test for pregnancy,[91] and the achievement of a malaria assay with sensitivity comparable to bench-top ELISA.[92] Additional tools have since been

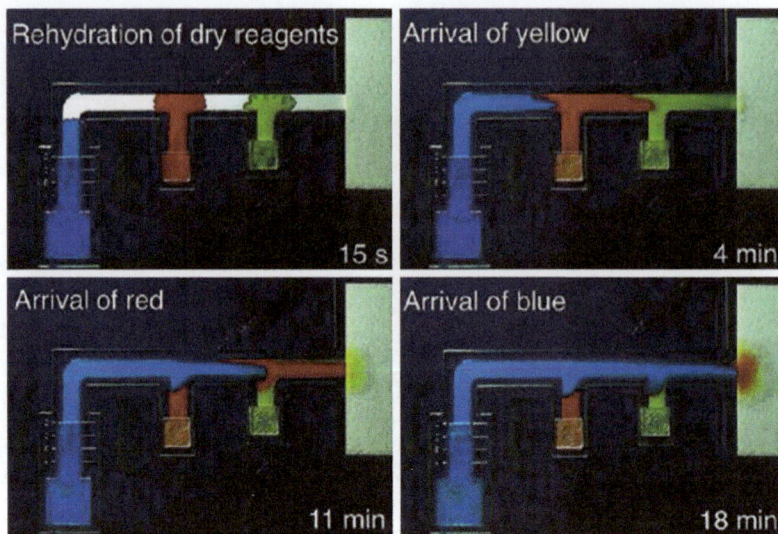

Figure 8.8   Demonstration of the multi-step reagent delivery afforded by 2DPNs.[92] Reprinted with permission *Anal. Chem.*, 2012, **84**(1), 4574–4579. Copyright 2012, American Chemical Society.

demonstrated by our group for use in 2DPN devices, such as mixing and dilution,[93] flow visualization,[94] and the use of sugar barriers to achieve timing delays without the expansion of the 2DPN footprint.[88] The Richards-Kortum group has also made significant contributions to the 2DPN platform, demonstrating a 2DPN folding card that performs isothermal amplification and subsequent detection of HIV DNA.[95] Finally, Apilux *et al.* recently demonstrated a nitrocellulose-based device for an automated ELISA.[96] Although these investigators employed inkjet printing of hydrophobic barriers, instead of laser cutting, the device utilizes the geometry of the paper network to control flow and achieve multi-step processing, making it a prime example of 2DPN technology. Overall, the 2DPN format is ideal for bioassays employing biological detection reagents and offers the advantage of multi-step processing, which can be used to achieve sample pre-processing, signal enhancement, and even simple rinsing to improve signal-to-noise ratios.

8.3.2.3.3   Other Paper-based Devices.   Since paper microfluidics is still a new and growing field, many opportunities for innovation still exist, both in terms of expanding the μPAD and 2DPN platforms and in regard to developing entirely new platforms. One prime example is the recent work of the Derda group demonstrating paper as a low-cost scaffold for cell culture[97] (Figure 8.9), proving that porous media can do more than provide capillary pressure to drive fluid flow. In this demonstration by Funes-Huacca *et al.*, hydrophobic inks are patterned onto sheets of paper,

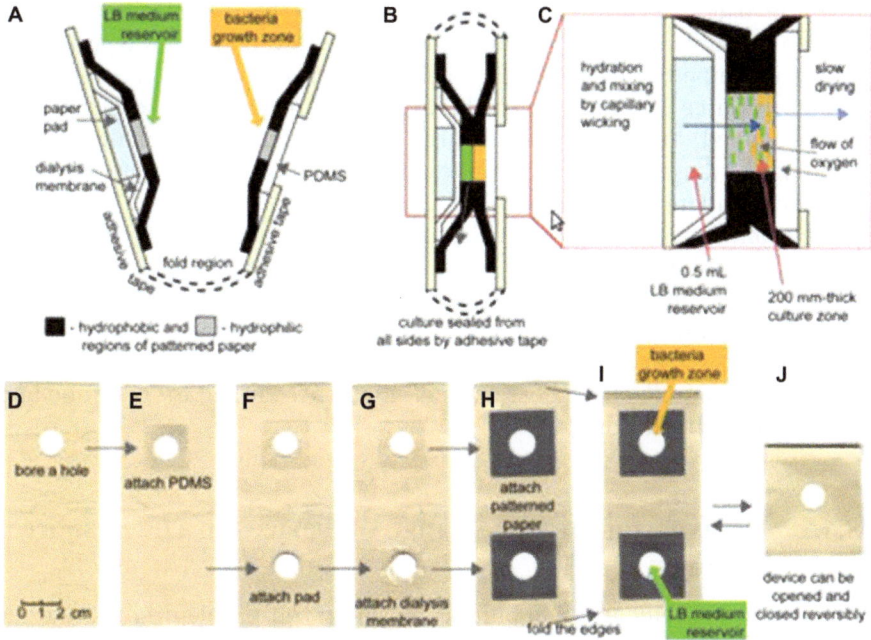

Figure 8.9 Paper-based cell culture, demonstrated by Funes-Huacca *et al.* [97] Reprinted with permission from the Royal Society of Chemistry.

sandwiched between packing tape on one side and a PDMS membrane on the other. This setup provides a bacterial culture grown on the paper substrate with necessary gas exchange, but prevents evaporation of the cell culture media. The authors went a step further, showing that the devices could be fabricated using simplified techniques amenable to the high-school classrooms in which they were optimized. Based on retail pricing, Funes-Huacca *et al.* estimate that a simplified production facility could be set up for $1–2k. In addition, on-site production was demonstrated at the 1st annual Diagnostic Workshop in Nairobi, Kenya, highlighting the feasibility of paper-based cell culture in low-resource settings when culture-based diagnostic testing is required.

8.4 Enabling All Aspects of Diagnostic Testing in Low-resource Settings: Examples of and Opportunities for Microfluidics (Channel-based and Paper-based)

No matter the number of clever techniques developed by microfluidics-focused academic groups and startup companies regarding fluid handling, signal detection, and fabrication, a successful diagnostic assay requires that the user transform a sometimes complex biological signal into a human-readable output.[98] Complicating this effort, many of the peripheral amenities (pipettes, clean water, reliable electrical power, *etc.*) common in higher-resource laboratories (including those in which these techniques are developed) are not available at the point of care. Additionally, even if appropriate diagnostics and supplies are available, adoption and scale-up can be hindered by economic, regulatory, policy-related, and user-perception issues, along with cultural barriers.[55] The manufacturing and disposal of a diagnostic device must also be considered and designed for. Following is a non-comprehensive overview of some of the factors specially affecting point-of-care diagnostics in low-resource settings that should be considered by technology developers. Importantly, technologies that address and integrate all of these aspects will be best-suited for adoption and use in global health diagnostics.

8.4.1 Transportation and Storage of Devices in Low-resource Settings

Cold chain capacity and reliability is a well-known problem in vaccine delivery.[4] If the required reagents of a diagnostic assay require refrigeration, a reliable cold chain must be in place for a given region to have access to the diagnostic. The problems regarding temperature control do not stop when the supplies are delivered; communities are often not equipped or educated regarding best practices regarding storage of perishable reagents.[99]

In the absence of a reliable cold chain, alternative approaches are necessary for keeping assay reagents functional. Storing reagents in anhydrous form on the device is one option, and can have the additional benefit of reducing the number of steps the assay user needs to perform.[100] The dried form of the reagent weighs less, and takes up less volume. Stevens *et al.* show that anhydrous gold-antibody conjugates stored on a device in sugar matrices can be rehydrated after 60 days at elevated temperature while retaining 80–96% of their activity. The assay developed by Fu *et al.*[92] demonstrates that dried reagents can also be incorporated into a two-dimensional paper network (2DPN) containing glass fiber conjugate pads (within which the reagents are stored in a sugar matrix), a nitrocellulose section comprising the "legs" and detection region of the assay, and a cellulose wicking pad (Figure 8.10c).

Storing reagents on separate pads from the nitrocellulose media used for the backbone of the assay requires additional parts and manufacturing steps. Fridley *et al.*[101] have demonstrated a methodology that eliminates some of these complexities while enabling precise control over reagent temporal and spatial release by preserving the reagents directly on the nitrocellulose membrane (Figure 8.10). Novel functions enabled by this format include the creation of uniform reagent pulses, mixing of rehydrated reagents, and tuning of temporal reagent delivery. A gold enhancement

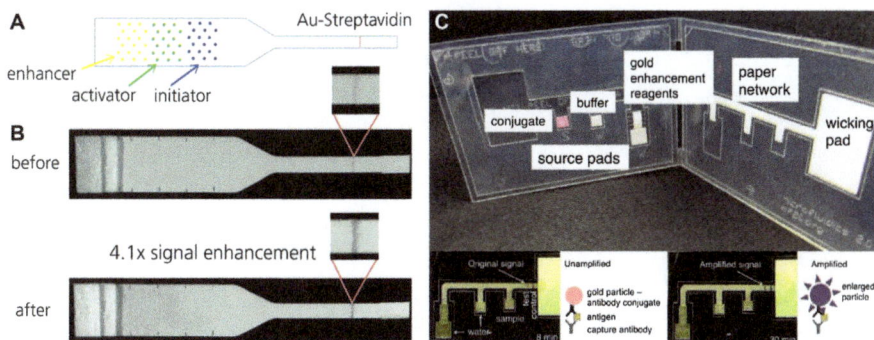

Figure 8.10 Dry reagent storage and rehydration. In both of these examples, reagents are stored in dried form and stabilized in a sugar matrix. On the left, reagents are stored directly on nitrocellulose membranes,[101] whereas on the right, discrete glass fiber pads are used.[92]
(A–B) Reprinted with permission from the Royal Society of Chemistry. (C) Reprinted with permission from *Anal. Chem.*, 2012, **84**(1), 4574–4579. Copyright 2012, American Chemical Society."

assay is demonstrated where gold enhancement reagents are sequentially combined to produce a 4.1× signal enhancement.

8.4.2 Specimen Collection

Specimen-collection strategies utilized in well-equipped clinics will not necessarily translate directly to lower-resource settings, and device performance is directly impacted by how samples are obtained. Cira *et al.* describe a device that only requires a single pipetting step to load the sample into multiple chambers[102] (Figure 8.11). The pressure differential used to drive the fluid flow through the device is provided by gas-adsorbing properties of poly(dimethylsiloxane) (PDMS): the device contains multiple wells molded in PDMS, and is vacuum-degassed after assembly. The user removes the device from its vacuum-sealed packaging, and pipettes a drop of sample onto the inlet port. The degassed PDMS then provides a net-negative pressure, pulling sample into each of the wells. These wells are pretreated with antibiotic and used to test for minimum inhibitory concentration of the antibiotic. Each self-loaded well holds roughly 1 mL. The device was shown by the authors to mimic the protocol developed by the US Clinical Laboratory and Standards Institute more closely than its less automated counterparts.

Begolo *et al.* describe an additional automated sample collection device, which has the added capability of long-term stabilization of samples through desiccating in the presence of a stabilization matrix.[103] This microfluidic device is designed using the SlipChip-based methodology, which has repeatedly been demonstrated by the Ismagilov group to translate simple user steps into complex microfluidic channel switching.[104,105] Briefly,

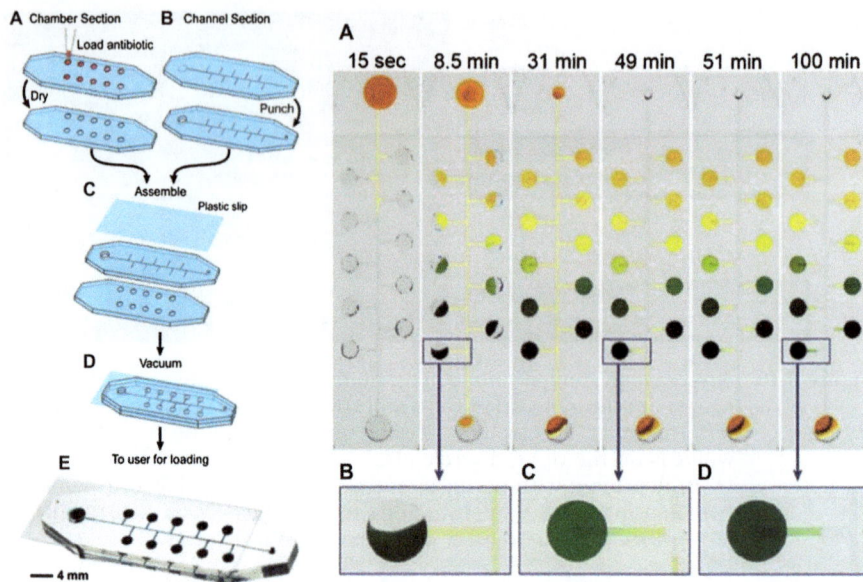

Figure 8.11   Sample collection demonstrated by Cira *et al.*[102] After assembly, the device is vacuum-degassed and placed in a sealed package. The sample is drawn into the chambers by the negative pressure provided by the degassing.
Reprinted with permission from the Royal Society of Chemistry.

this style of device is comprised of multilayer channels, and a one-dimensional "slip" action results in channel connections to be broken and established, rerouting any number of microfluidic channels in the process. The sample collection and preservation device described by Begolo *et al.* fills automatically, due to a pressure developed by the closure of the device lid after sample addition. After the automated filling of sequential fluid chambers, a pre-loaded desiccant dries the sample within the chambers. This methodology is shown by the authors to result in highly reproducible sample volumes, and the preserved samples are shown to be recoverable with minimal degradation without refrigeration, demonstrating compatibility with environments beyond the reach of an established cold chain.

8.4.3   Sample Preparation

Blood, plasma, saliva, serum, sputum, stool, and urine are sample types from which clinically relevant analytes could be quantified. Before an analyte can be quantified, often the sample will need to be concentrated, filtered, or purified from substances (including proteins[106]) that may interfere with the assay.[8] Integrating these functions into the diagnostic device (rather than assuming the user has the time, knowledge, and amenities to prepare the sample before introducing it to the diagnostic) is preferred. For example,

Marshall *et al.* demonstrate an integrated device capable of lysing cells through heating and purifying nucleic acids using isotachophoresis.[107] Cell lysis and DNA extraction kits (*e.g.* Qiagen) are often used in laboratories to prepare samples, but require centrifugation, heating steps, and quite a bit of waiting, making other methods of lysis and DNA isolation necessary for POC nucleic acid amplification tests. The Klapperich group at Boston University demonstrated an on-chip sample preparation platform for lysis and DNA extraction of both Gram-negative and Gram-positive bacteria.[108] This platform utilizes a hybrid chemical/mechanical method of cell lysis, where a blood/bacteria suspension is first treated with a chaotropic buffer containing guanidinium thiocyanate (GuSCN), proteinase K, and detergents. GuSCN, used as part of the lysis buffer, is additionally necessary for the solid phase extraction, causing the DNA to preferentially bind to the silica particles. After extraction and rinsing, the bacterial nucleic acids are purified and ready for amplification.

As a more universal alternative to chemical techniques, which often require purification steps to keep from interfering with the DNA amplification, mechanical techniques inspired by bench-top bead beaters have been developed to lyse cells for nucleic acid amplification.[109] The OmniLyse cell lysis device currently being sold by ClaremontBio has been shown by Vandeventer *et al.* to have similar lysis capabilities to bench-top bead beaters.[110] Mechanical lysis methods like bead beating have the advantage of not using chemicals (for example, guanidinium thiocyanate salts) that need to be removed from the sample before subsequent assay steps, and also need to be safely handled and disposed of after completion of the assay; it is important to remember the disposal method of choice in many lower resource settings is incineration.[2] Benchtop bead beaters (Figure 8.12A) require electricity, and are quite heavy, making chemical lysis methods on the surface more

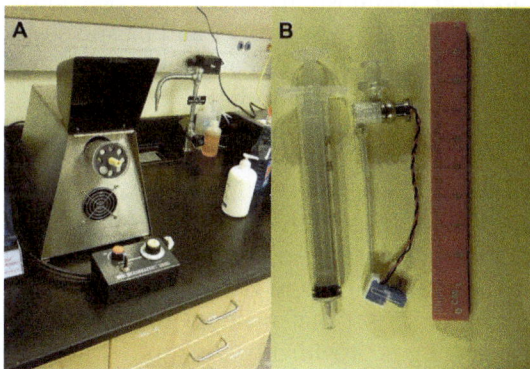

Figure 8.12 (A) Bench-top bead beater compared to (B) OmniLyse mechanical cell disruptor. These two devices, while being quite different in size, weight, and power requirements, were shown by Vandeventer *et al.*[110] to have similar efficacy in lysing *Bacillus subtilus* spores and *Mycobacterium bovis* BCG cells.

desirable than mechanical means. The OmniLyse device shows that bead beater-like lysis efficiency can be accomplished with a very small device, smaller than the three AA batteries used to power it (Figure 8.12B). The device consists of a small motor, plastic lysis chamber containing beads and a rotor, and a syringe used to pass the sample through the lysis chamber multiple times. A battery pack with an on/off switch powers the device, with one battery pack providing sufficient energy to run many sequential samples. Though the manufacturer's claim that "the kits can then be simply thrown away without any harm to the environment"[111] is questionable, the disposal of alkaline batteries and other associated waste products may be simpler than agents such as chaotropic salts.

8.4.4   Running the Assay

A great number of assays have been developed that utilize microfluidic technologies, covered elsewhere in this book and in review articles. Rather than attempt to do all of these technologies justice within the confines of this book chapter, the reader can refer to two comprehensive review articles which do a great job covering microfluidic point-of-care diagnostics.[8,10] This section will focus on some unique challenges of running diagnostic assays in lower-resource settings, namely the lack of reliable electrical power, reagents, and supplies. Unlike in better-equipped laboratories, the technician in a low-resource setting cannot be expected to have a supply of pipette tips, sterilized water, and a 120 V receptacle.

Material compatibility with assay chemistry is perhaps no more complex than in nucleic acid amplification assays. Nevertheless, the potential value of a self-powered microfluidic device with the sensitivity and specificity of nucleic acid amplification in both low-resource settings and in the point-of-care in higher-resource settings explains the high level of interest in this technology. Rohrman *et al.* demonstrated isothermal Recombinase Polymerase Amplification (RPA) of HIV DNA in a paper and plastic device[95] (Figure 8.13).

Many assays, like the isothermal RPA demonstrated by Rohrman *et al.*, require precision temperature control. PATH has demonstrated several platforms for this purpose. The non-instrumented nucleic acid amplification (NINA) platform[112,113] is designed to heat PCR tubes containing reagents for isothermal nucleic acid amplification, including isothermal loop mediated isothermal amplification (LAMP) to the required 63 °C. This device uses a reusable thermos-shaped double-walled container as an insulating shell. Tubes are placed into an aluminum block, where they are surrounded by phase change material (PCM) that melts at the target temperature of the assay. Energy for heating is provided by the exothermic reaction of calcium oxide (CaO) and water. The authors show that this setup can retain 63 °C for an hour under the conditions tested (see Figure 8.14).

A collaborative effort between PATH and our group at the University of Washington has produced a similar device[114] designed to heat reagents in a paper microfluidic format, which is optimized for a flat heated zone

A

base layer (with paper covering)

acetate layer

wax

sample wick strip

master mix pad

magnesium acetate pad

1 cm

C 1. add master mix

2. add magnesium acetate

3. dip wick in sample

4. fold wick

5. fold device in half

B

1. place enzyme pellet on exposed adhesive

2. place master mix pad and sample wick strip

3. remove remaining paper to expose adhesive

4. place acetate layer over exposed adhesive

5. place magnesium acetate pad in window

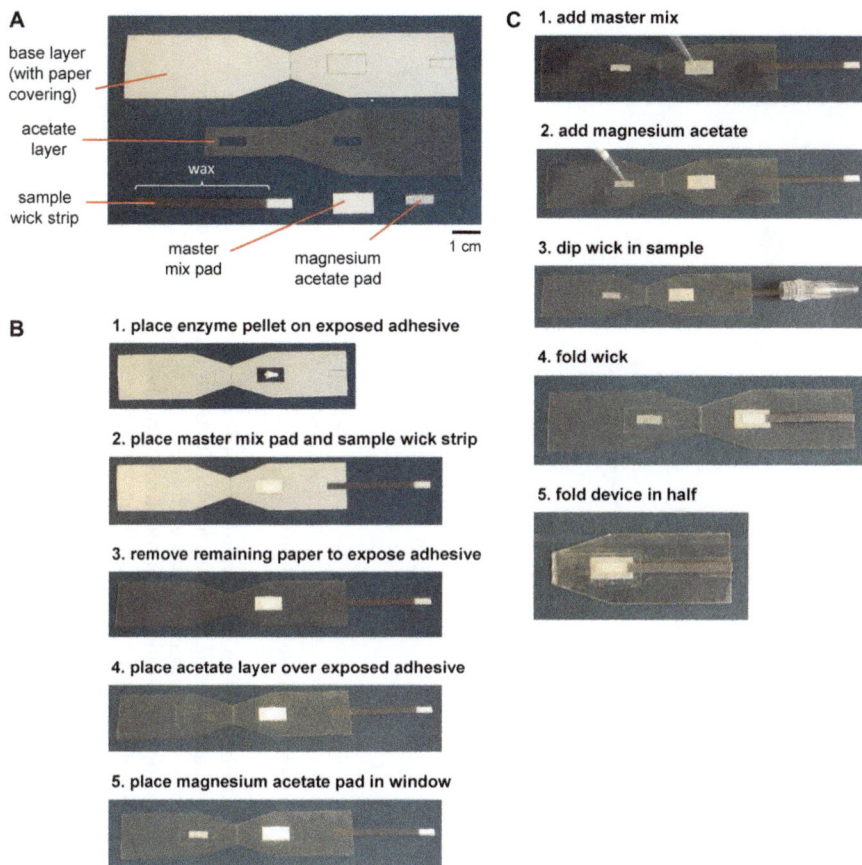

Figure 8.13 Isothermal nucleic acid amplification in porous media, demonstrated by Rohrman *et al.*[95] Here, the steps of the simple device construction and assay preparation are illustrated.
Reprinted with permission from the Royal Society of Chemistry.

appropriate for heating reagents in a 2DPN. This prototype device was designed to be integrated into a fully disposable device, dubbed Multiplexable Autonomous Disposables for Nucleic Acid Amplification Testing and is currently being funded by the Defense Advanced Research Projects Agency (DARPA). The exothermic process chosen for this device is the reaction of a magnesium iron alloy and saline, the same fuel used in meal ready-to-eat (MRE) technology employed to provide hot meals away from the amenities of a kitchen. Shown in Figure 8.14, the fuel chamber is supplied saline by a blister pack located above. The temperature rise of the fuel is buffered by PCM, providing a tightly controlled temperature to the sample region, which contains the assay reagents.

Another chemical heater capable of heating without external power was developed by the Bau group at the University of Pennsylvania.[115] Here, the

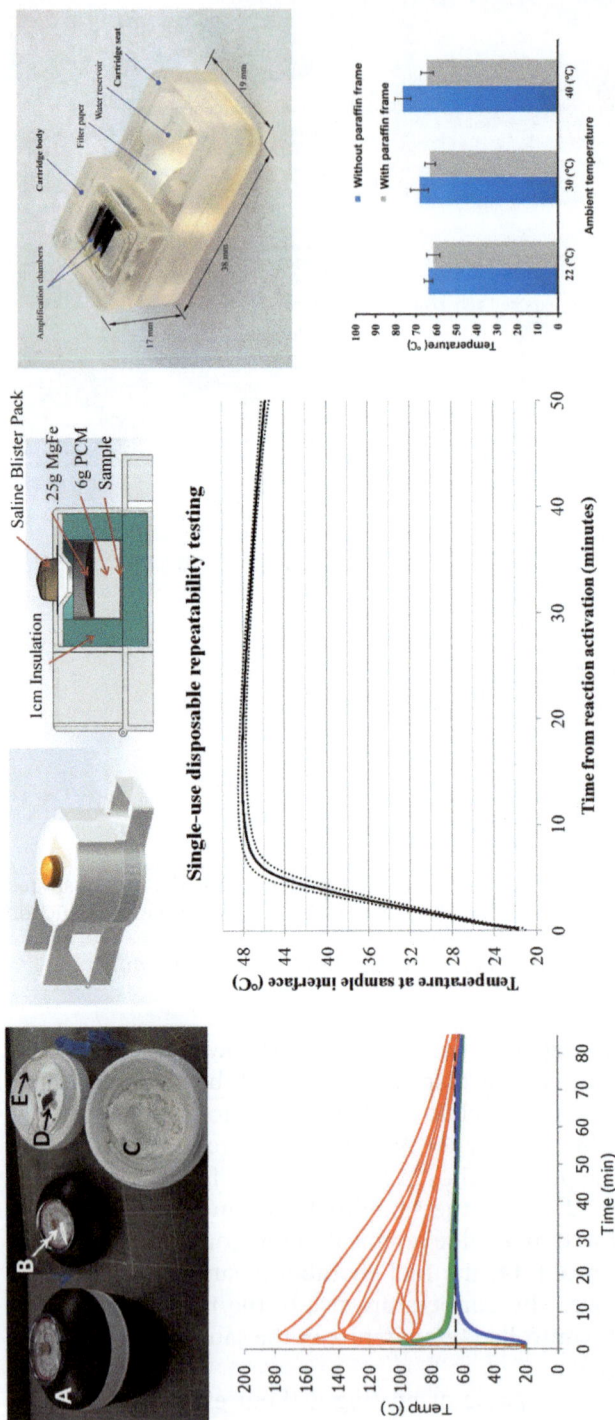

Figure 8.14 Comparison of different chemical heaters that can be used to implement point-of-care nucleic acid amplification assays. (LEFT) Thermos-based incubator demonstrated by LaBarre *et al.*[113] A: assembled incubator. B: incubator lid with integrated timer. C: Calcium oxide chamber provides heat from exothermic reaction with water. D: Assay tubes. E: thermocouple wires (not required for device operation, only for monitoring temperature). Temperature plot shows reliable achievement of a 63 °C assay temperature (blue) over 10 replicate runs. Adapted with permission from *PLOS One.* (CENTER) A single-use disposable design from Singleton *et al.*[114] Isometric and cross-sectional views of device model. Temperature plot shows assay temperature average ± one standard deviation for 12 runs over time. Adapted with permission from Singleton *et al.*, *Proc SPIE*, Microfluidics, BioMEMS, and Medical Microsystems XI, 86150R, 2013. (RIGHT) Cartridge heater from Liu *et al.*[115] The addition of a paraffin phase change material regulates the assay temperature to 63 °C for three ambient temperatures. Reprinted with permission from the Royal Society of Chemistry.

fuel chamber was provided water by a filter paper wick, which restricted the delivery of water and hence the reaction rate of the fuel. Temperature was again regulated using a phase change material, and the supplied heat is used to amplify *E. coli* DNA using LAMP.

In developing world applications, the infrastructure necessary to provide external sources of electricity is often not present.[5] In addition to on-device commercially available batteries (AA, "button" Li-ion batteries, *etc.*), "fluidic batteries" can be constructed of galvanic cells incorporated into the device using paper microfluidic networks. Thom *et al.* showed that this type of power source can be used to power a UV LED used to conduct an on-chip fluorescence assay used for the detection of β–D-galactosidase.[81] Another great example of an on-chip power supply, demonstrated by Esquivel *et al.*, uses a microfluidic fuel cell.[116] Here, the authors demonstrate a micro direct methanol fuel cell (mDMFC), which has been incorporated into a microfluidic platform. In addition to providing electrical power, the CO_2 produced by the electrochemical reaction is utilized to provide a pressure for pumping liquids through the microfluidic channels.

8.4.5 Signal Read-out

The type of signal that is generated by an assay and the corresponding read-out method require special attention by developers of diagnostics for low-resource settings. The first decision point is centered on whether the test needs to be quantitative (*i.e.* measures the amount of the target analyte present) or if it can be qualitative (*i.e.* determines whether or not the analyte is present at all). For many infectious disease tests, the output can usually be qualitative, since the patient either does or does not have the given disease. For this reason, conventional lateral flow tests use qualitative evaluation of visual signal intensity to provide a simple yes-or-no answer. This type of read-by-eye signal can be achieved using any type of particle that absorbs light in the visible range of wavelengths, such as gold nanoparticles, latex beads, carbon black, or colorimetric enzyme substrates.[117]

Tests that require a quantitative measure of the target analyte, such as glucose meters or HIV viral load tests, are necessarily more sophisticated than qualitative tests, and often require the use of an external reader to determine the result. A variety of methodologies exist to quantify signal intensity, and handheld readers[118] have been specifically designed to measure electrochemical, absorbance, reflectance, transmittance, fluorescence, or chemiluminescence signals. Not all quantitative tests require a reader, however. As an alternative to needing an instrument to quantify signal, Pollack *et al.* have shown a multiplexed assay capable of reporting rapid, semi-quantitative measurement of two serum transaminases that are used to monitor for drug-induced liver injury.[83] Here, the authors optimize the device for users to "bin" the transaminase concentration into clinically relevant ranges visually using a color-reading guide. Multiple indicators for the same analyte can also be used to help the user visually determine analyte concentrations.[119] Additionally, quantification of analyte concentration through the use of fluidic timing elements that are incorporated into the device is another reader-less method that has been demonstrated.[120] Here, interaction of the analyte of interest and reagents patterned in the paper network cause a change in contact angle of the solution and paper network, which affects capillary flow rate. This flow rate can be compared to a reference and quantified, indicating analyte concentration.

Finally, cellular phones represent an increasingly useful tool to aid the read-out of both qualitative and quantitative assays. Cell phones have achieved extremely rapid market penetration in low-resource settings, which can largely be attributed to unreliable land-line infrastructure in these areas. In fact, Africa is now the second largest mobile phone market after Asia.[10] Phones equipped with cameras and fairly sophisticated processing capabilities are becoming increasingly present in low-resource settings, which could be a great asset for diagnostic tests.

Much interest has been shown in using mobile phone cameras for a variety of purposes in low-resource settings, from digitizing medical records (covered in the following section) to the quantification of diagnostic assays.[121] For cell phone-based diagnostics, it is important for the developer to determine whether it is most appropriate for the signal read-out algorithm to be performed on the phone itself, or for the phone to transmit the image to the cloud or other server for remote analysis and subsequent notification of the result. While cell phones may provide the most utility for enabling quantitative assays, they can also be used to automate the read-out of qualitative assays and remove the error associated with manual user interpretation.[118,122,123] The cell phone also provides a useful means of connecting with the given health system as whole, which is discussed in the next section.

8.4.6 Data Integration into Health Systems

While achieving an actionable answer is the first order of business for any diagnostic test, integrating this data into the appropriate health system is an important next step, and one that is often overlooked by test developers. Connecting a diagnostic result with the health system can improve the management of a patient's condition, especially in remote areas where there is a lack of healthcare providers.[22,122,123] Integrating results of diagnostic tests into the health system is also important for epidemiological surveillance and effective public health management.[122,123]

A prime example of the need for better health systems integration is the state of medical records in the developing world. Physical paper forms remain an integral role in health systems data in low-resource settings, with this media being well understood, trusted, easy to use, and relatively low cost. The data contained within these paper records is not as easily aggregated, shared, or analyzed as digital data.[124] Digitizing paper forms is not a new design challenge, but the specialized systems that have been successful in developed regions are generally costly, and require maintenance and

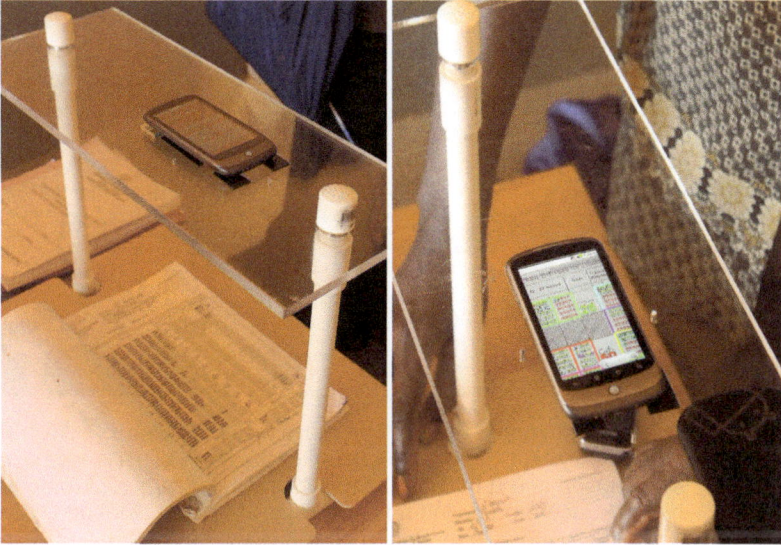

Figure 8.15 Medical record digitization using a mobile phone, demonstrated by Dell *et al.*[124]
Courtesy of Nicola Dell.

reliable electrical infrastructure. Recent work from the Open Data Kit team at the University of Washington demonstrates a smartphone application, ODK Scan (previously mScan), that uses computer vision to capture data from multiple-choice or bubble-format paper forms.[124] The authors used a low-cost stand fashioned from polyvinylchloride (PVC) pipes and sheets of plastic, as shown in Figure 8.15, holding the smartphone steady and aligning it with the form. The ODK Scan software was demonstrated in this study to capture data with remarkable accuracy; even when the form was crumpled and filled out messily, accuracy was still over 80%. The accuracy of 19 other conditions tested was always over 90%, and usually (15/19) over 99%. Much data can be stored on a smartphone, which can be uploaded to a remote database when a network becomes available.

Finally, the fully integrated version of the mChip from the Sia group[59] represents an excellent example of a microfluidic diagnostic system capable of communicating with the overarching health system. Not only does this device generate an HIV diagnosis from 1 µL of whole blood by automating an ELISA-type assay, but it also transmits the result to a cloud-based medical record system (Figure 8.16). This data integration allows the patient's data to be stored and accessed easily and enables the health system to track HIV incidence. While the device developed here is still fairly expensive (estimated <$1000) for some resource-poor settings, this work represents a great step forward for fully integrated diagnostic testing and may serve as a blueprint for other less expensive, potentially cell-phone-based systems.

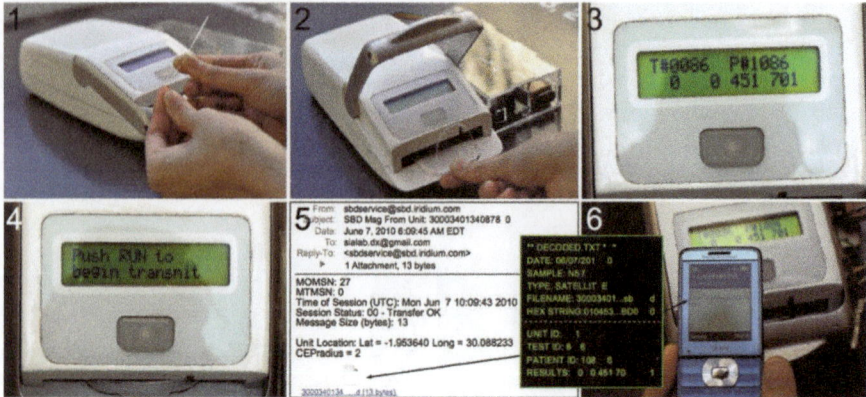

Figure 8.16 The fully integrated mChip system from the Sia group.[59] This device-plus-chip system performs ELISA-like diagnosis of HIV from 1 µL of whole blood and transmits the result to the patient's electronic health record stored on the cloud. Figure from Chin *et al.*, Mobile Device for Disease Diagnosis and Data Tracking in Resource-Limited Settings, *Clin. Chem.*, 2013, **59**, 629–640.
Figure reproduced with permission from the American Association for Clinical Chemistry.

Figure 8.17 Photographs of medical waste in low-resource settings, presented in review by Yager *et al.*[2] (a) Medical waste in an open incinerator in Senegal (PATH). (b) Incinerator in Tanzania overflowing with medical waste (PATH). (c) Open incinerator in Nigeria (PATH). (d) Unsafe sharps disposal in a public waste dump in India (Mark Koska).
Reprinted with permission.

8.4.7 Disposal

Responsible design of diagnostics for any setting includes planning for disposal. Open outdoor incinerators are often utilized by hospitals in developing countries, and needle handling and disposal remain major challenges. Biosafety protocols, which protect workers and others in the community from infectious agents, are not always implemented.[2] Diagnostic devices need to be designed appropriately for the settings in which they will be used, and should minimize waste.[2] Several representative images of disposal issues in low-resource settings are shown in Figure 8.17, highlighting the importance of planning for waste impact.[2]

8.5 Conclusions

Many strides have been made in developing microfluidic technology in a way that could potentially improve the lives of people worldwide, but only a small number of lab-on-a-chip-based devices have been successfully commercialized.[98] The long-term impact of recently introduced instrumented systems like the Cepheid GeneXpert remains to be studied, but these technologies are expected to greatly expand access to TB diagnostics to vulnerable populations. Success stories, such as the home pregnancy test and portable glucometers, illustrate the transformative potential of low-cost, easy-to-use diagnostic technologies. Future technologies under development, including more accessible nucleic acid amplification assays, have the opportunity to expand the reach of microfluidics within global health. Creating standalone devices capable of these advanced assays requires that developers fully understand not only what it takes to transform a biological analyte into a human-understandable readout, but also the market and societal forces that will dictate whether such technology has a chance of being commercially viable. Finally, almost all microfluidic technologies being pursued for developing-world applications are also well-suited to tackle diagnostic challenges in the developed world, as low-cost, point-of-care testing represents a tremendously valuable capability in any country. Microfluidics-based diagnostics are therefore poised to improve health around the world, truly advancing the state of *global* health.

References

1. P. Yager, T. Edwards, E. Fu, K. Helton, K. Nelson, M. R. Tam and B. H. Weigl, *Nature*, 2006, **442**, 412–418.
2. P. Yager, G. J. Domingo and J. Gerdes, *Annu. Rev. Biomed. Eng.*, 2008, **10**, 107–144.
3. The value of diagnostics innovation, adoption and diffusion into health care, Advanced Medical Technology Association, 2005, http://www.lewin.com/~/media/Lewin/Site_Sections/Publications/ValueofDiagnostics.pdf, (Accessed February 12, 2013).

4. K. Lorenson, M. Mvundura and World Health Organization, *Delivering Vaccines: A Cost Comparison of In-Country Vaccine Transport Container Options*, 2013.
5. B. H. Weigl, D. S. Boyle, T. de los Santos, R. B. Peck and M. S. Steele, *Expert Rev. Med. Devices*, 2009, **6**, 461–464.
6. R. McNerney and P. Daley, *Nat. Rev. Microbiol.*, 2011, **9**, 204–213.
7. D. Mabey, R. W. Peeling, A. Ustianowski and M. D. Perkins, *Nat. Rev. Microbiol.*, 2004, **2**, 231–240.
8. V. Gubala, L. F. Harris, A. J. Ricco, M. X. Tan and D. E. Williams, *Anal. Chem.*, 2012, **84**, 487–515.
9. C. D. Chin, V. Linder and S. K. Sia, *Lab. Chip*, 2007, 7, 41.
10. A. K. Yetisen, M. S. Akram and C. R. Lowe, *Lab. Chip*, 2013, **13**, 2210–2251.
11. Institute for Health Metrics and Evaluation, *The Global Burden of Disease: Generating Evidence, Guiding Policy*, IHME, University of Washington, Seattle, WA, 2013.
12. C. J. L. Murray, T. Vos, R. Lozano, M. Naghavi, A. D. Flaxman, C. Michaud, M. Ezzati, K. Shibuya, J. A. Salomon, S. Abdalla, V. Aboyans, J. Abraham, I. Ackerman, R. Aggarwal, S. Y. Ahn, M. K. Ali, M. A. AlMazroa, M. Alvarado, H. R. Anderson, L. M. Anderson, K. G. Andrews, C. Atkinson, L. M. Baddour, A. N. Bahalim, S. Barker-Collo, L. H. Barrero, D. H. Bartels, M.-G. Basáñez, A. Baxter, M. L. Bell, E. J. Benjamin, D. Bennett, E. Bernabé, K. Bhalla, B. Bhandari, B. Bikbov, A. B. Abdulhak, G. Birbeck, J. A. Black, H. Blencowe, J. D. Blore, F. Blyth, I. Bolliger, A. Bonaventure, S. Boufous, R. Bourne, M. Boussinesq, T. Braithwaite, C. Brayne, L. Bridgett, S. Brooker, P. Brooks, T. S. Brugha, C. Bryan-Hancock, C. Bucello, R. Buchbinder, G. Buckle, C. M. Budke, M. Burch, P. Burney, R. Burstein, B. Calabria, B. Campbell, C. E. Canter, H. Carabin, J. Carapetis, L. Carmona, C. Cella, F. Charlson, H. Chen, A. T.-A. Cheng, D. Chou, S. S. Chugh, L. E. Coffeng, S. D. Colan, S. Colquhoun, K. E. Colson, J. Condon, M. D. Connor, L. T. Cooper, M. Corriere, M. Cortinovis, K. C. de Vaccaro, W. Couser, B. C. Cowie, M. H. Criqui, M. Cross, K. C. Dabhadkar, M. Dahiya, N. Dahodwala, J. Damsere-Derry, G. Danaei, A. Davis, D. D. Leo, L. Degenhardt, R. Dellavalle, A. Delossantos, J. Denenberg, S. Derrett, D. C. Des Jarlais, S. D. Dharmaratne, M. Dherani, C. Diaz-Torne, H. Dolk, E. R. Dorsey, T. Driscoll, H. Duber, B. Ebel, K. Edmond, A. Elbaz, S. E. Ali, H. Erskine, P. J. Erwin, P. Espindola, S. E. Ewoigbokhan, F. Farzadfar, V. Feigin, D. T. Felson, A. Ferrari, C. P. Ferri, E. M. Fèvre, M. M. Finucane, S. Flaxman, L. Flood, K. Foreman, M. H. Forouzanfar, F. G. R. Fowkes, M. Fransen, M. K. Freeman, B. J. Gabbe, S. E. Gabriel, E. Gakidou, H. A. Ganatra, B. Garcia, F. Gaspari, R. F. Gillum, G. Gmel, D. Gonzalez-Medina, R. Gosselin, R. Grainger, B. Grant, J. Groeger, F. Guillemin, D. Gunnell, R. Gupta, J. Haagsma, H. Hagan, Y. A. Halasa, W. Hall, D. Haring, J. M. Haro, J. E. Harrison, R. Havmoeller, R. J. Hay,

H. Higashi, C. Hill, B. Hoen, H. Hoffman, P. J. Hotez, D. Hoy, J. J. Huang, S. E. Ibeanusi, K. H. Jacobsen, S. L. James, D. Jarvis, R. Jasrasaria, S. Jayaraman, N. Johns, J. B. Jonas, G. Karthikeyan, N. Kassebaum, N. Kawakami, A. Keren, J.-P. Khoo, C. H. King, L. M. Knowlton, O. Kobusingye, A. Koranteng, R. Krishnamurthi, F. Laden, R. Lalloo, L. L. Laslett, T. Lathlean, J. L. Leasher, Y. Y. Lee, J. Leigh, D. Levinson, S. S. Lim, E. Limb, J. K. Lin, M. Lipnick, S. E. Lipshultz, W. Liu, M. Loane, S. L. Ohno, R. Lyons, J. Mabweijano, M. F. MacIntyre, R. Malekzadeh, L. Mallinger, S. Manivannan, W. Marcenes, L. March, D. J. Margolis, G. B. Marks, R. Marks, A. Matsumori, R. Matzopoulos, B. M. Mayosi, J. H. McAnulty, M. M. McDermott, N. McGill, J. McGrath, M. E. Medina-Mora, M. Meltzer, Z. A. Memish, G. A. Mensah, T. R. Merriman, A.-C. Meyer, V. Miglioli, M. Miller, T. R. Miller, P. B. Mitchell, C. Mock, A. O. Mocumbi, T. E. Moffitt, A. A. Mokdad, L. Monasta, M. Montico, M. Moradi-Lakeh, A. Moran, L. Morawska, R. Mori, M. E. Murdoch, M. K. Mwaniki, K. Naidoo, M. N. Nair, L. Naldi, K. M. V. Narayan, P. K. Nelson, R. G. Nelson, M. C. Nevitt, C. R. Newton, S. Nolte, P. Norman, R. Norman, M. O'Donnell, S. O'Hanlon, C. Olives, S. B. Omer, K. Ortblad, R. Osborne, D. Ozgediz, A. Page, B. Pahari, J. D. Pandian, A. P. Rivero, S. B. Patten, N. Pearce, R. P. Padilla, F. Perez-Ruiz, N. Perico, K. Pesudovs, D. Phillips, M. R. Phillips, K. Pierce, S. Pion, G. V. Polanczyk, S. Polinder, C. A. Pope, S. Popova, E. Porrini, F. Pourmalek, M. Prince, R. L. Pullan, K. D. Ramaiah, D. Ranganathan, H. Razavi, M. Regan, J. T. Rehm, D. B. Rein, G. Remuzzi, K. Richardson, F. P. Rivara, T. Roberts, C. Robinson, F. R. De Leòn, L. Ronfani, R. Room, L. C. Rosenfeld, L. Rushton, R. L. Sacco, S. Saha, U. Sampson, L. Sanchez-Riera, E. Sanman, D. C. Schwebel, J. G. Scott, M. Segui-Gomez, S. Shahraz, D. S. Shepard, H. Shin, R. Shivakoti, D. Singh, G. M. Singh, J. A. Singh, J. Singleton, D. A. Sleet, K. Sliwa, E. Smith, J. L. Smith, N. J. Stapelberg, A. Steer, T. Steiner, W. A. Stolk, L. J. Stovner, C. Sudfeld, S. Syed, G. Tamburlini, M. Tavakkoli, H. R. Taylor, J. A. Taylor, W. J. Taylor, B. Thomas, W. M. Thomson, G. D. Thurston, I. M. Tleyjeh, M. Tonelli, J. A. Towbin, T. Truelsen, M. K. Tsilimbaris, C. Ubeda, E. A. Undurraga, M. J. van der Werf, J. van Os, M. S. Vavilala, N. Venketasubramanian, M. Wang, W. Wang, K. Watt, D. J. Weatherall, M. A. Weinstock, R. Weintraub, M. G. Weisskopf, M. M. Weissman, R. A. White, H. Whiteford, N. Wiebe, S. T. Wiersma, J. D. Wilkinson, H. C. Williams, S. R. Williams, E. Witt, F. Wolfe, A. D. Woolf, S. Wulf, P.-H. Yeh, A. K. Zaidi, Z.-J. Zheng, D. Zonies and A. D. Lopez, *The Lancet*, 2012, **380**, 2197–2223.

13. C. H. Chew, Y. A. L. Lim, P. C. Lee, R. Mahmud and K. H. Chua, *J. Clin. Microbiol.*, 2012, **50**, 4012–4019.

14. F. S. Younai, *Int. J. Oral Sci.*, 2013, **5**, 191–199.

15. J. M. Binley, T. Wrin, B. Korber, M. B. Zwick, M. Wang, C. Chappey, G. Stiegler, R. Kunert, S. Zolla-Pazner, H. Katinger, C. J. Petropoulos and D. R. Burton, *J. Virol.*, 2004, **78**, 13232–13252.

16. World Health Organization, *Global Tuberculosis Report*, 2013.
17. UNITAID, *Tuberculosis: Diagnostic Technology Landscape*, World Health Organization, 2012.
18. World Health Organization, About the Diabetes Programme, http://www.who.int/diabetes/goal/en/index.html, (Accessed 16 January 2014).
19. World Health Organization, *Use of glycated haemoglobin (HbA1c) in the diagnosis of diabetes mellitus*, 2011.
20. B. H. Weigl, G. Zwisler, R. Peck and E. Abu-Haydar. Proc. SPIE 8615, Microfluidics, BioMEMS, and Medical Microsystems XI, 86150L (March 13, 2013).
21. World Health Organization, *Diabetes Fact Sheet*, 2013.
22. S. Brown and T. X. Brown, *Proceedings of the Sixth International Conference on Information and Communication Technologies and Development: Full Papers - Volume 1 (ICTD '13)*, 2013, **1**, 267–273.
23. D. M. Smith and R. T. Schooley, *Clin. Infect. Dis.*, 2008, **46**, 1598–1600.
24. A. Calmy, N. Ford, B. Hirschel, S. J. Reynolds, L. Lynen, E. Goemaere, F. G. de la Vega, L. Perrin and W. Rodriguez, *Clin. Infect. Dis.*, 2007, **44**, 128–134.
25. M. Usdin, M. Guillerm and A. Calmy, *J. Infect. Dis.*, 2010, **201**, S73–S77.
26. W. S. Stevens, L. E. Scott and S. M. Crowe, *J. Infect. Dis.*, 2010, **201**, S16–S26.
27. Department of Health and Human Services, Adult and Adolescent Treatment Guidelines, http://aidsinfo.nih.gov/guidelines/html/1/adult-and-adolescent-treatment-guidelines/0/, (Accessed 2 February 2012).
28. X. Wei, S. K. Ghosh, M. E. Taylor, V. A. Johnson, E. A. Emini, P. Deutsch, J. D. Lifson, S. Bonhoeffer, M. A. Nowak, B. H. Hahn, M. S. Saag and G. M. Shaw, *Nature*, 1995, **373**, 117–122.
29. UNAIDS, *Global report: UNAIDS report on the global AIDS epidemic 2010*, 2010.
30. World Health Organization, Antiretroviral therapy, http://www.who.int/hiv/topics/treatment/en/index.html, (Accessed 2 February 2012).
31. C.-C. Li, K. D. Seidel, R. W. Coombs and L. M. Frenkel, *J. Clin. Microbiol.*, 2005, **43**, 3901–3905.
32. S. A. Fiscus, B. Cheng, S. M. Crowe, L. Demeter, C. Jennings, V. Miller, R. Respess, W. Stevens and the Forum for Collaborative HIV Research Alternative Viral Load Assay Working Group, *PLoS Med.*, 2006, **3**, e417.
33. S. Wang, F. Xu and U. Demirci, *Biotechnol. Adv.*, 2010, **28**, 770–781.
34. H. E. Rawizza, B. Chaplin, S. T. Meloni, G. Eisen, T. Rao, J.-L. Sankale, A. Dieng-Sarr, O. Agbaji, D. I. Onwujekwe, W. Gashau, R. Nkado, E. Ekong, P. Okonkwo, R. L. Murphy, P. J. Kanki and for the APIN PEPFAR Team, *Clin. Infect. Dis.*, 2011, **53**, 1283–1290.
35. World Health Organization, *Antiretroviral therapy for HIV infection in adults and adolescents: Recommendations for a public health approach: 2010 revision*, 2010.
36. M. Badri, S. D. Lawn and R. Wood, *BMC Infect. Dis.*, 2008, **8**, 89.
37. M. Piatak, M. Saag, L. Yang, S. Clark, J. Kappes, K. Luk, B. Hahn, G. Shaw and J. Lifson, *Science*, 1993, **259**, 1749–1754.

38. World Health Organization, *Medicines: spurious/falsely-labelled/ falsified/counterfeit (SFFC) medicines–Fact Sheet*, 2012.
39. A. A. Weaver, H. Reiser, T. Barstis, M. Benvenuti, D. Ghosh, M. Hunckler, B. Joy, L. Koenig, K. Raddell and M. Lieberman, *Anal. Chem.*, 2013, **85**, 6453–6460.
40. J. C. Jokerst, J. M. Emory and C. S. Henry, *The Analyst*, 2012, **137**, 24–34.
41. Pan American Health Organization (PAHO). Epidemiological update. Cholera, 23 August 2013, Washington, DC; PAHO; 2013. Available from: http://www.paho.org/hq/index.php?option = com_docman&task = doc_view&gid = 22751&Itemid = .
42. P. Farmer, C. P. Almazor, E. T. Bahnsen, D. Barry, J. Bazile, B. R. Bloom, N. Bose, T. Brewer, S. B. Calderwood, J. D. Clemens, A. Cravioto, E. Eustache, G. Jérôme, N. Gupta, J. B. Harris, H. H. Hiatt, C. Holstein, P. J. Hotez, L. C. Ivers, V. B. Kerry, S. P. Koenig, R. C. Larocque, F. Léandre, W. Lambert, E. Lyon, J. J. Mekalanos, J. S. Mukherjee, C. Oswald, J.-W. Pape, A. Gretchko Prosper, R. Rabinovich, M. Raymonville, J.-R. Réjouit, L. J. Ronan, M. L. Rosenberg, E. T. Ryan, J. D. Sachs, D. A. Sack, C. Surena, A. A. Suri, R. Ternier, M. K. Waldor, D. Walton and J. L. Weigel, *PLoS Negl. Trop. Dis.*, 2011, **5**, e1145.
43. G. Madico, W. Checkley, R. H. Gilman, N. Bravo, L. Cabrera, M. Calderon and A. Ceballos, *J. Clin. Microbiol.*, 1996, **34**, 2968–2972.
44. S. Richardson and T. Ternes, *Anal. Chem.*, 2005, **81**, 4645–4677.
45. S. C Terry, J. H. Jerman and J. B. Angell, *IEEE Trans. Electron. Devices*, 1979, **26**, 1880–1886.
46. A. Manz, N. Graber and H. M. Widmer, *Sensor. Actuator. B Chem.*, 1990, **1**, 244–248.
47. D. C. Duffy, J. C. McDonald, O. J. Schueller and G. M. Whitesides, *Anal. Chem.*, 1998, **70**, 4974–4984.
48. T. Thorsen, S. J. Maerkl and S. R. Quake, *Science*, 2002, **298**, 580–584.
49. L. Lafleur, D. Stevens, K. McKenzie, S. Ramachandran, P. Spicar-Mihalic, M. Singhal, A. Arjyal, J. Osborn, P. Kauffman, P. Yager and B. Lutz, *Lab Chip*, 2012, **12**, 1119–1127.
50. S. Choi, M. Goryll, L. Y. M. Sin, P. K. Wong and J. Chae, *Microfluid. Nanofluid.*, 2010, **10**, 231–247.
51. C. C. Boehme, M. P. Nicol, P. Nabeta, J. S. Michael, E. Gotuzzo, R. Tahirli, M. T. Gler, R. Blakemore, W. Worodria, C. Gray, L. Huang, T. Caceres, R. Mehdiyev, L. Raymond, A. Whitelaw, K. Sagadevan, H. Alexander, H. Albert, F. Cobelens, H. Cox, D. Alland and M. D. Perkins, *The Lancet*, 2011, **377**, 1495–1505.
52. Cepheid, Xpert® MTB/RIF, http://www.cepheid.com/us/cepheid-solutions/clinical-ivd-tests/critical-infectious-diseases/xpert-mtb-rif, (Accessed September 26, 2014).
53. A. Vassall, S. van Kampen, H. Sohn, J. S. Michael, K. R. John, S. den Boon, J. L. Davis, A. Whitelaw, M. P. Nicol, M. T. Gler, A. Khaliqov, C. Zamudio, M. D. Perkins, C. C. Boehme and F. Cobelens, *PLoS Med.*, 2011, **8**, e1001120.

54. Price for Xpert® MTB/RIF and FIND country list, http://www.finddiagnostics.org/about/what_we_do/successes/find-negotiated-prices/xpert_mtb_rif.html, (Accessed September 25, 2014).

55. N. P. Pai, C. Vadnais, C. Denkinger, N. Engel and M. Pai, *PLoS Med.*, 2012, **9**, e1001306.

56. A. Pantoja, C. Fitzpatrick, A. Vassall, K. Weyer and K. Floyd, *Eur. Respir. J.*, 2013, **42**, 708–20.

57. I. R. Lauks, *Acc. Chem. Res.*, 1998, **31**, 317–324.

58. C. D. Chin, T. Laksanasopin, Y. K. Cheung, D. Steinmiller, V. Linder, H. Parsa, J. Wang, H. Moore, R. Rouse, G. Umviligihozo, E. Karita, L. Mwambarangwe, S. L. Braunstein, J. van de Wijgert, R. Sahabo, J. E. Justman, W. El-Sadr and S. K. Sia, *Nat. Med.*, 2011, **17**, 1015–1019.

59. C. D. Chin, Y. K. Cheung, T. Laksanasopin, M. M. Modena, S. Y. Chin, A. A. Sridhara, D. Steinmiller, V. Linder, J. Mushingantahe, G. Umviligihozo, E. Karita, L. Mwambarangwe, S. L. Braunstein, J. van de Wijgert, R. Sahabo, J. E. Justman, W. El-Sadr and S. K. Sia, *Clin. Chem.*, 2013, **59**, 629–640.

60. Microfluidics 2.0, http://www.mf20.org/, (Accessed September 2014).

61. J. P. Rolland and D. A. Mourey, *MRS Bull.*, 2013, **38**, 299–305.

62. X. Li, D. R. Ballerini and W. Shen, *Biomicrofluidics*, 2012, **6**, 011301.

63. B. O'Farrell, in *Lateral Flow Immunoassay*, ed. R. C. Wong and H. Y. Tse, Humana Press, New York, NY, 2009, pp. 1–33.

64. J. L. Osborn, *Novel Paper Networks for Point-of-Care Sample Preparation and Indirect IgM Detection*, University of Washington, 2011.

65. A. W. Martinez, S. T. Phillips, G. M. Whitesides and E. Carrilho, *Anal. Chem.*, 2010, **82**, 3–10.

66. S. Rosen, in *Lateral Flow Immunoassay*, ed. R. C. Wong and H. Y. Tse, Humana Press, New York, NY, 2009, pp. 35–49.

67. G. E. Fridley, C. A. Holstein, S. B. Oza and P. Yager, *MRS Bull.*, 2013, **38**, 326–330.

68. T. Chard, *Hum. Reprod. Oxf. Engl*, 1992, **7**, 701–710.

69. A. B. Bonner, K. W. Monroe, L. I. Talley, A. E. Klasner and D. W. Kimberlin, *Pediatrics*, 2003, **112**, 363–367.

70. G. A. Posthuma-Trumpie, J. Korf and A. van Amerongen, *Anal. Bioanal. Chem.*, 2009, **393**, 569–582.

71. M. H. Craig, B. L. Bredenkamp, C. H. V. Williams, E. J. Rossouw, V. J. Kelly, I. Kleinschmidt, A. Martineau and G. F. J. Henry, *Trans. R. Soc. Trop. Med. Hyg.*, 2002, **96**, 258–265.

72. Centers for Disease Control and Prevention, *Morb. Mortal. Wkly. Rep*, 2012, **61**, 873–876.

73. S. D. Blacksell, J. A. Doust, P. N. Newton, S. J. Peacock, N. P. J. Day and A. M. Dondorp, *Trans. R. Soc. Trop. Med. Hyg.*, 2006, **100**, 775–784.

74. OraQuick ADVANCE® Rapid HIV-1/2 Antibody Test, http://www.orasure.com/products-infectious/products-infectious-oraquick.asp, (Accessed September 26, 2014).

75. A. Carballo-Diéguez, T. Frasca, I. Balan, M. Ibitoye and C. Dolezal, *AIDS Behav.*, 2012, **16**, 1753–60.

76. A. W. Martinez, S. T. Phillips, M. J. Butte and G. M. Whitesides, *Angew. Chem. Int. Ed.*, 2007, **46**, 1318–1320.
77. E. Carrilho, A. W. Martinez and G. M. Whitesides, *Anal. Chem.*, 2009, **81**, 7091–7095.
78. A. W. Martinez, S. T. Phillips and G. M. Whitesides, *Proc. Natl. Acad. Sci.*, 2008, **105**, 19606–19611.
79. A. W. Martinez, S. T. Phillips, Z. Nie, C.-M. Cheng, E. Carrilho, B. J. Wiley and G. M. Whitesides, *Lab. Chip*, 2010, **10**, 2499–2504.
80. K. M. Schilling, D. Jauregui and A. W. Martinez, *Lab. Chip*, 2013, **13**, 628.
81. N. K. Thom, K. Yeung, M. B. Pillion and S. T. Phillips, *Lab Chip*, 2012, **12**, 1768–1770.
82. M. M. Mentele, J. Cunningham, K. Koehler, J. Volckens and C. S. Henry, *Anal. Chem.*, 2012, **84**, 4474–4480.
83. N. R. Pollock, J. P. Rolland, S. Kumar, P. D. Beattie, S. Jain, F. Noubary, V. L. Wong, R. A. Pohlmann, U. S. Ryan and G. M. Whitesides, *Sci. Transl. Med.*, 2012, **4**, 152ra129–152ra129.
84. Diagnostics For All, Liver Function, http://www.dfa.org/projects/liver-function.php, (Accessed September 26, 2014).
85. S. T. Phillips and G. G. Lewis, *MRS Bull.*, 2013, **38**, 315–319.
86. E. J. Maxwell, A. D. Mazzeo and G. M. Whitesides, *MRS Bull.*, 2013, **38**, 309–314.
87. E. Fu, S. A. Ramsey, P. Kauffman, B. Lutz and P. Yager, *Microfluid. Nanofluid.*, 2010, **10**, 29–35.
88. E. Fu, B. Lutz, P. Kauffman and P. Yager, *Lab. Chip*, 2010, **10**, 918–920.
89. B. R. Lutz, P. Trinh, C. Ball, E. Fu and P. Yager, *Lab. Chip*, 2011, **11**, 4274.
90. E. Fu, P. Kauffman, B. Lutz and P. Yager, *Sensor. Actuator. B Chem.*, 2010, **149**, 325–328.
91. E. Fu, T. Liang, J. Houghtaling, S. Ramachandran, S. A. Ramsey, B. Lutz and P. Yager, *Anal. Chem.*, 2011, **83**, 7941–7946.
92. E. Fu, T. Liang, P. Spicar-Mihalic, J. Houghtaling, S. Ramachandran and P. Yager, *Anal. Chem.*, 2012, **84**, 4574–4579.
93. J. L. Osborn, B. Lutz, E. Fu, P. Kauffman, D. Y. Stevens and P. Yager, *Lab. Chip*, 2010, **10**, 2659.
94. P. Kauffman, E. Fu, B. Lutz and P. Yager, *Lab. Chip*, 2010, **10**, 2614.
95. B. A. Rohrman and R. R. Richards-Kortum, *Lab. Chip*, 2012, **12**, 3082.
96. A. Apilux, Y. Ukita, M. Chikae, O. Chailapakul and Y. Takamura, *Lab. Chip*, 2013, **13**, 126.
97. M. Funes-Huacca, A. Wu, E. Szepesvari, P. Rajendran, N. Kwan-Wong, A. Razgulin, Y. Shen, J. Kagira, R. Campbell and R. Derda, *Lab. Chip*, 2012, **12**, 4269–4278.
98. C. D. Chin, V. Linder and S. K. Sia, *Lab. Chip*, 2012, **12**, 2118–2134.
99. Y. Thakker and S. Woods, *BMJ*, 1992, **304**, 756–758.
100. D. Y. Stevens, C. R. Petri, J. L. Osborn, P. Spicar-Mihalic, K. G. McKenzie and P. Yager, *Lab. Chip*, 2008, **8**, 2038–2045.
101. G. E. Fridley, H. Q. Le, E. Fu and P. Yager, *Lab. Chip*, 2012, **12**, 4321–4327.

102. N. J. Cira, J. Y. Ho, M. E. Dueck and D. B. Weibel, *Lab. Chip*, 2012, **12**, 1052–1059.
103. S. Begolo, F. Shen and R. F. Ismagilov, *Lab. Chip*, 2013, **13**, 4331–4342.
104. W. Du, L. Li, K. P. Nichols and R. F. Ismagilov, *Lab. Chip*, 2009, **9**, 2286–2292.
105. F. Shen, B. Sun, J. E. Kreutz, E. K. Davydova, W. Du, P. L. Reddy, L. J. Joseph and R. F. Ismagilov, *J. Am. Chem. Soc.*, 2011, **133**, 17705–17712.
106. K. G. McKenzie, L. K. Lafleur, B. R. Lutz and P. Yager, *Lab. Chip*, 2009, **9**, 3543–3548.
107. L. A. Marshall, L. L. Wu, S. Babikian, M. Bachman and J. G. Santiago, *Anal. Chem.*, 2012, **84**, 9640–9645.
108. M. Mahalanabis, H. Al-Muayad, M. D. Kulinski, D. Altman and C. M. Klapperich, *Lab. Chip*, 2009, **9**, 2811–2817.
109. J. Siegrist, R. Gorkin, M. Bastien, G. Stewart, R. Peytavi, H. Kido, M. Bergeron and M. Madou, *Lab. Chip*, 2010, **10**, 363–371.
110. P. E. Vandeventer, K. M. Weigel, J. Salazar, B. Erwin, B. Irvine, R. Doebler, A. Nadim, G. a Cangelosi and A. Niemz, *J. Clin. Microbiol.*, 2011, **49**, 2533–2539.
111. OmniLyse® Ultra-Rapid Cell Lysis Kits, http://www.claremontbio.com/OmniLyse_Cell_Lysis_Kits_s/56.htm, (Accessed September 25, 2014).
112. R. Kubota, P. Labarre, J. Singleton, A. Beddoe, B. H. Weigl, A. M. Alvarez and D. M. Jenkins, *Biol. Eng. Trans.*, 2011, **4**, 69–80.
113. P. LaBarre, K. R. Hawkins, J. Gerlach, J. Wilmoth, A. Beddoe, J. Singleton, D. Boyle and B. Weigl, *PloS One*, 2011, **6**, e19738.
114. J. Singleton, C. Zentner, J. Buser, P. Yager, P. LaBarre and B. H. Weigl, *Proc. SPIE*, 2013, **8615**, 86150R.
115. C. Liu, E. Geva, M. Mauk, X. Qiu, W. R. Abrams, D. Malamud, K. Curtis, S. M. Owen and H. H. Bau, *The Analyst*, 2011, **136**, 2069–2076.
116. J. P. Esquivel, M. Castellarnau, T. Senn, B. Löchel, J. Samitier and N. Sabaté, *Lab. Chip*, 2012, **12**, 74–79.
117. P. Chun, in *Lateral Flow Immunoassay*, ed. R. C. Wong and H. Y. Tse, Humana Press, New York, NY, 2009, pp. 75–93.
118. K. Faulstich, R. Gruler, M. Eberhard, D. Lentzsch and K. Haberstroh, in *Lateral Flow Immunoassay*, ed. R. C. Wong and H. Y. Tse, Humana Press, New York, NY, 2009, pp. 157–183.
119. W. Dungchai, O. Chailapakul and C. S. Henry, *Anal. Chim. Acta*, 2010, **674**, 227–233.
120. G. G. Lewis, M. J. DiTucci and S. T. Phillips, *Angew. Chem. Int. Ed Engl.*, 2012, **51**, 12707–12710.
121. D. J. You, T. S. Park and J.-Y. Yoon, *Biosens. Bioelectron.*, 2012, **40**, 180–185.
122. O. Mudanyali, S. Dimitrov, U. Sikora, S. Padmanabhan, I. Navruz and A. Ozcan, *Lab. Chip*, 2012, **12**, 2678.
123. A. W. Martinez, S. T. Phillips, E. Carrilho, S. W. Thomas, H. Sindi and G. M. Whitesides, *Anal. Chem.*, 2008, **80**, 3699–3707.
124. N. Dell, N. Breit, T. Chaluco, J. Crawford and G. Borriello, *Proc. 2nd ACM Symp. Comput. Dev. – ACM DEV 12*, 2012, *1*.

CHAPTER 9

Isolation and Characterization of Circulating Tumor Cells

YOONSUN YANG AND LEON W. M. M. TERSTAPPEN*

Medical Cell BioPhysics Group, MIRA Institute, University of Twente, Hallenweg 23, 7522 NH, Enschede, The Netherlands
*Email: l.w.m.m.terstappen@utwente.nl

9.1   Introduction

In 1869 Thomas Ashworth, an Australian physician, after observing micro-scopically circulating tumor cells (CTCs) in the blood of a man with meta-static cancer, postulated "... cells identical with those of the cancer itself being seen in the blood may tend to throw some light upon the mode of origin of multiple tumors existing in the same person".[1] After comparing the morphology of the circulating cells to tumor cells from different lesions Ashworth concluded that "One thing is certain, that if they [CTCs] came from an existing cancer structure, they must have passed through the greater part of the circulatory system to have arrived at the internal saphena vein of the sound leg". Over the last 140 years, cancer research has indeed dem-onstrated the critical role circulating tumor cells play in the metastatic spread of carcinomas. In most cases death from cancer is not caused by expansion of the primary tumor, but through dissemination of the disease. To settle in distant sites tumor cells must travel through the peripheral blood, but at what frequency is still unknown. In blood of most patients with metastatic disease CTCs are extremely rare (<1 CTC mL^{-1} blood $= \sim 1$ in 10^7 white blood cells (WBCs, leukocytes) and 1 in 10^{10} red blood cells (RBCs, erythrocytes)) creating a large challenge for the development of technology to

RSC Nanoscience & Nanotechnology No. 36
Microfluidics for Medical Applications
Edited by Albert van den Berg and Loes Segerink
© The Royal Society of Chemistry 2015
Published by the Royal Society of Chemistry, www.rsc.org

reliably detect and characterize CTCs.[2,3] To date, various technologies have been developed and with the arrival of the validated CellSearch® System the diagnostic potential of these rare cells could be exploited.[3] The CellSearch® System has been developed to enumerate CTCs from 7.5 mL of venous blood. Evaluation of CTCs at any time during the course of disease of metastatic breast, colorectal, and prostate demonstrated that CTC was the strongest independent predictor of progression free and overall survival.[4–6] These studies as well as those demonstrating that treatment targets can be assessed on CTCs[7–14] has spurred extensive interest and brought many novel technologies into the CTC research field. Among these technologies microfluidic devices have been explored to identify and characterize CTCs. This includes the use of diverse materials and smart geometries in microchannels to manipulate and capture CTCs as well as the integration of optical and electrical detection systems.

9.2   CTC Definition in CellSearch System

Definitions of CTCs vary widely between studies and have led to a large range of reported CTC frequencies. The CellSearch system is, however, the only clinically validated and FDA cleared technology for enumeration of CTCs. In the CellSearch System, antibodies targeting epithelial cell adhesion molecule (EpCAM) are conjugated with ferrofluids and an automated system enriches cells expressing EpCAM from 7.5 mL of blood by immunomagnetic separation. The CellTracks Autoprep system also fluorescently labels the enriched cells with DAPI (nucleus), CD45-APC (leukocyte marker), and Cytokeratin-PE (epithelial cell marker) and resuspends the cells into a cartridge placed in the CellTracks Magnest. The design of the magnets guides the magnetically labeled cells to the analysis surface and the Cell-Tracks Magnest containing the chamber is placed on the CellTracks® Analyzer II. This semi-automated fluorescence-based microscope system acquires images to cover the complete surface area of the analysis chamber. A computer identifies objects staining with DAPI and PE in the same location and generates images for the DAPI, FITC, PE, and APC filters. A reviewer selects the CTC defined as nucleated DAPI + cells, lacking CD45 and expressing Cytokeratin-PE from the gallery of objects. The final classification of CTCs is performed by the operator and is the main contributor of the error of the assay.[15] Consistency in CTC classification among operators is therefore important.[16–18] To illustrate how challenging this classification can be, examples of CTC candidates images are shown in Figure 9.1. Green represents the Cytokeratin-PE staining and purple the DAPI staining. The corresponding APC images of the candidates showed no staining. The operators can only classify objects larger than 4 μm as CTCs. For that purpose, they are provided with a 4×4 μm box to determine whether or not the cell is at least as large as the box. The objects shown in panel A will all be classified as CTCs whereas none of the objects shown in panel C will be classified as

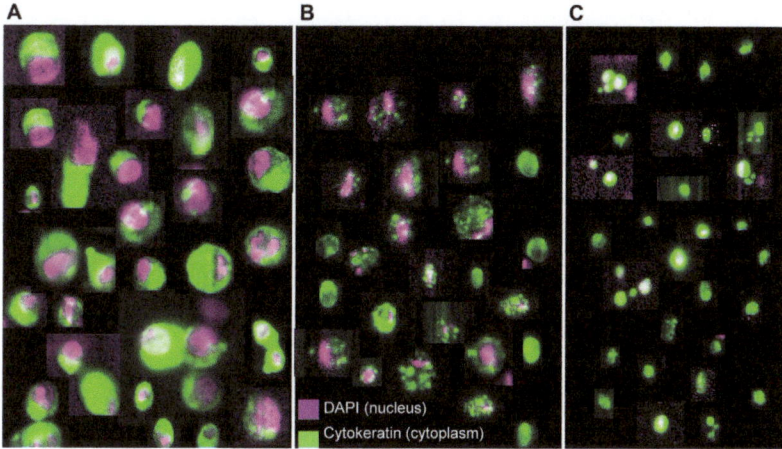

Figure 9.1 Images of CTC candidates identified by the CellTracks Analyzer II. The images show an overlay of DAPI (purple) and CK-PE (green). In the corresponding CD45-APC image no staining was observed. Operators trained to review CellSearch CTC data will all classify the objects in Panel A as CTCs, a more discordant assignment of objects as CTCs by trained operators will occur when classifying the objects displayed in Panel B and most likely represent CTCs undergoing apoptosis and all operators will not classify the objects in Panel C as CTCs and these likely represent tumor microparticles.

CTCs. Many of the small cytokeratin-PE positive objects can be considered as tumor microparticles (TMP).[19] More difficult to assign are the objects shown in panel B. The granulated objects in panel B clearly show features consistent with apoptosis.[20] The definition of a CTC greatly influences the determined frequency of CTCs. In general the more strict the definition the better the relation with clinical outcome.[18,19,21] Therefore, reliable standards for a CTC definition are required and clinical studies are needed to determine the clinical utility of the CTC identified by alternative approaches.

9.3 Clinical Relevance of CTCs

Clinical studies in metastatic breast, colorectal, and prostate cancer patients demonstrated that the presence of CTCs was a strong predictor of poor outcome.[6,17,22–26] Even after a few weeks of therapy, patients with a higher number of CTCs (Unfavorable group) had much shorter overall survival than did patients with a few or without CTCs (Favorable group). Cutoff value of $5\,CTC\,(7.5\,mL)^{-1}$ was used to stratify patients into those with Favorable outcomes (CTC < 5) and those with Unfavorable outcomes (CTC ≥ 5). Effective therapies – ones that result in elimination of CTCs – can prolong survival, irrespective of the line of therapy, as shown by the improvement in overall survival in patients that converted from an Unfavorable to a Favorable CTC count. Kaplan–Meier plots can be generated based on CTCs

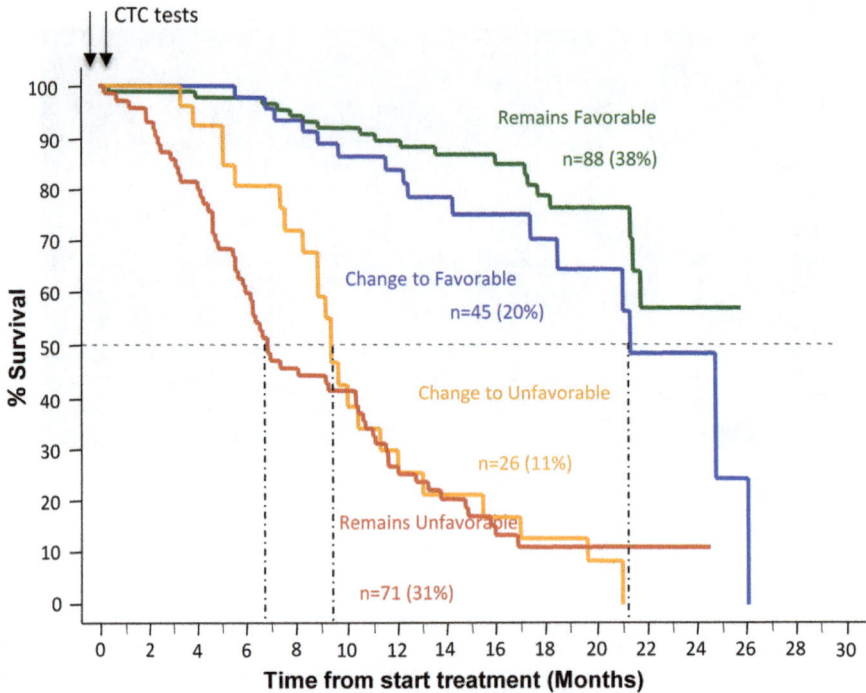

Figure 9.2 Kaplan–Meier analysis of overall survival based on CTC changes after initiation of treatment in 233 castration-resistant prostate cancer patients. Favorable CTCs is defined as <5 CTC $(7.5 \text{ mL})^{-1}$ and unfavorable CTCs as ≥ 5 CTC $(7.5 \text{ mL})^{-1}$ blood. Blood was taken before administration of a new line of therapy (first arrow) and 3–6 weeks after initiation of therapy (second arrow). Median overall survival of 88 (38%) patients that remained with Favorable CTCs after 3–6 weeks of therapy was more than 26 months (green curve), in contrast with a median overall survival of only 9.3 months for the 71 (31%) patients that remained with Unfavorable CTCs. In 45 (20%) patients the CTCs converted to Favorable CTCs and their survival chances significantly increased (21.3 months) whereas the survival significantly decreased for the 26 (11%) patients that converted to Unfavorable CTCs (6.8 months).

at baseline and follow-up blood collections for survival analyses. The Kaplan–Meier curves in Figure 9.2 show the probability of overall survival (OS) for 233 castration-resistant prostate cancer patients with Favorable and Unfavorable CTC counts after treatment to evaluate whether treatment can alter survival prospects.[27] Patients were divided into four groups: 1) CTCs remain Favorable (green lines); 2) CTCs remain Unfavorable (red lines); 3) CTCs change from Favorable to Unfavorable (orange lines); and 4) CTCs change from Unfavorable to Favorable (blue lines) during the course. CTCs in 88 (38%) patients remained Favorable with a median OS of more than 26 months. CTCs in 71 (31%) patients remained Unfavorable with a median OS of 6.8 months. CTCs in 45 (20%) patients converted to Favorable CTC with a median OS of 21.3 months and CTCs converted to Unfavorable in 26 (11%) patients median with an OS of 9.3 months. Those patients with Favorable CTCs after the first weeks of therapy benefited from the therapy whereas those with Unfavorable CTCs are on a futile therapy and may benefit from alternative therapy.[6,21,22,24,25]

Today, the current standard of care calls for routine assessments of a patient's clinical status at or about one-month intervals and depends on the type of therapy. Imaging studies are an exception as they are usually performed at some intermediate time point and at the end of a given line of therapy. Consequently most clinicians do not entertain a change in treatment until several cycles of drug have been administered. It is also felt that a certain minimum time, usually two to three cycles of therapy, are often needed before clinical benefit may be evident. Thus, the advantages associated with changing treatment at a significantly earlier time point must be first demonstrated before the persistence of CTCs will indeed be used widely in the clinic to change therapy. Studies are ongoing aiming to prove this hypothesis.

9.4 Identification of Treatment Targets on CTCs

Detailed characterization of CTCs may reveal the presence or absence of targets for specific therapies. For instance, trastuzumab is a HER2 targeted therapy blocking signaling pathways of proliferation in breast cancer patients overexpressing the HER2 receptor. However, HER2 expression status can change during the course of the disease.[9] Tumor biopsies are, however, difficult if not impossible to obtain during the course of the disease and identification of the HER2 status on CTCs may solve this issue. At present expression of the HER2 receptor is determined on tissue biopsies either by assessment of the cell surface receptor by immunohistochemistry or by the amplification of the HER2 proto-oncogene by fluorescence *in situ* analysis (FISH). Figure 9.3 illustrates that expression of HER2 can be assessed at both the protein and the gene level on CTCs. In panel A three objects are shown that were detected in a patient with metastatic breast cancer after 7.5 mL was processed with the CellSearch System to which FITC-labeled HER2 antibody was added to the staining cocktail. Two of the three objects were classified as CTCs and both expressed the HER2 protein (HER2 +). Quantitative assessment of the HER2 expression on CTCs of breast cancer patients has been reported recently.[14] The cartridges containing the enriched CTCs can be stored and exposed to a FISH staining procedure.[12] The CTCs can now be relocated and assessed for amplification of the HER2 gene. Panel B shows such a CTC with the DAPI staining nucleus (blue), the centromere of chromosome 17 (CEP-17) shown in green, and the HER2 gene in red. Whereas only two copies of the centromere of chromosome 17 are visible clearly, multiple copies of the HER2 gene are present demonstrating that this CTC has an amplification of the HER2 gene. Feasibility of detection of a variety of treatment targets on CTCs such as HER2, epidermal growth factor

A

B

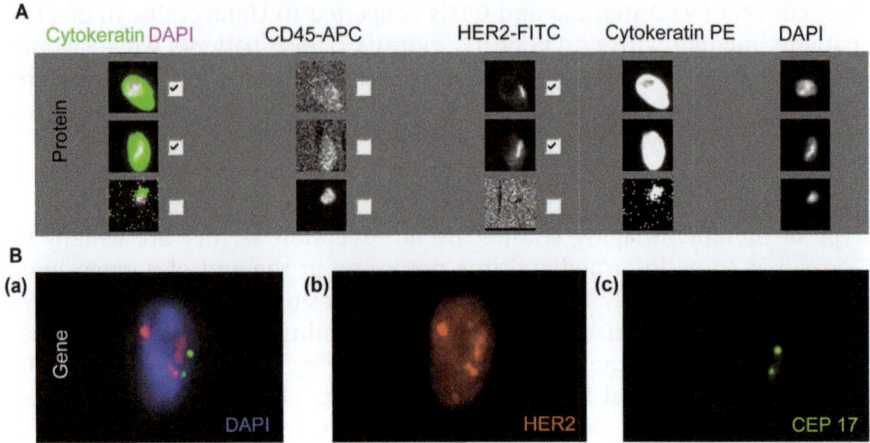

Figure 9.3 HER2 Protein & Gene expression of CTCs. CTCs were enriched with the CellSearch System and fluorescently labeled with the nucleic acid dye DAPI and antibodies recognizing Cytokeratin labeled with PE, CD45 labeled with APC, and HER2 labeled with FITC. After image acquisition the sample was hybridized with probes recognizing the centromere of chromosome 17 and the HER2 gene.[12] Panel A shows images of three objects. The top rows show images corresponding to CTCs expressing HER2 and the bottom row shows images corresponding to an object not classified as a CTC. Panel B shows fluorescence *in situ* hybridization (FISH) analysis of a CTC. (a) A merged fluorescence image of DAPI (blue), HER2 (red), and CEP 17 (green). (b) A fluorescence image showing more than two copies of the HER2 gene probes. (c) A fluorescence image showing two copies of the centromere of chromosome 17 (CEP 17).

receptor (EGFR), ERG gene, androgen receptor (AR), phosphatase and tensin homolog protein encoded by the PTEN gene, insulin-like growth factor-1 receptor (IGF-1R), urokinase plasminogen activator receptor (uPAR) have been demonstrated.[7,9–11,13,20] Expression of treatment targets in tumor cells within the individual patient is heterogeneous and this heterogeneity is also present in CTCs.[14] To evaluate the extent of this heterogeneity requires analysis of treatment targets at the individual tumor cell level. Preferably this should be done during the course of the disease to monitor the dynamic of cancer biology and have the ability to alter treatment in a timely and rational fashion.

9.5   Technologies for CTC Enumeration

The demonstration that CTCs have clinical relevance and can potentially be used as a liquid biopsy has led to the development of a variety of technologies for CTC enumeration and characterization. Table 9.1 lists some of these technologies and compares important parameters such as flow rate, type of enrichment method, typical blood volume analyzed, time for sample preparation and acquisition, recovery rate, and enrichment folds over WBCs.

Table 9.1 CTC isolation techniques.

	Fixed volume	Methods	Flow rate [mL hr^{-1}]	Analyzed blood volume [mL]	Sample preparation time [hr]	Sample acquisition time [hr]/SA [mm^2]	Total time [hr]	Recovery rate [%] (cell line)	Enrichment fold
CellSearch[3]	Yes	I	NA	7.5	2.8	~0.2/89.1	~3	≥85 (SKBR-3)	~10^4
MagSweeper[31]	Yes	I	NA	9.0	2^a	~1.8/950	~3.8	62 ± 7 (MCF-7)	~10^5
AccuCyte[33]	Yes	I/P	NA	3.0	2.5^a	~3.4/~1833	~5.9	90 (MDA-MB-453)	NR
FAST[34]	Yes	I	NA	1.0	3.5	1/NR	~4.5	95 (HT29)	~10^6
Micro Sieve[29]	No	P	100	1~7.5	~1.4^a	~0.1/6.2	~1.5	25~80^c	~10^3
Parylene[30,35]	No	P	90~100	1~7.5	~1.3^a	~0.2/100	~1.5	~90 (LNCaP)	10^3~10^4
TrackEtch[29]	No	P	100	1~7.5	~1.4^a	~0.2/102.1	~1.6	25~80^c	10^3~10^4
CTC-chip[64]	No	I	1~2	0.9~5.1	5.3	~1.4^e/970	~6.7	>60 (NCI-H1650)	~10^4
HB-chip[63]	No	I	1.5~2.5	~4	5.3	~2.5^{d,e}/~1658	~7.8	91.8 ± 5.2 (PC3)	~10^4
Spiral[61]	No	P	3	6	3^a	FACS/NA	~4.5^b	>85 (MCF-7, MDA-MB-231)	10^2~10^3

NA: Not Applicable, NR: Not Reported, SA: Surface Area, I: Immunological property-based isolation method, P: Physical property-based isolation method.
^{a}Include estimated staining time from standard protocol as 1 hr.
^{b}Process time of FACS is estimated as 1.5 hr.
^{c}Range of capturing efficiency of nine different cell lines.
^{d}Images are acquired at three different Z points.
^{e}Scanning time is estimated based on three different fluorescence filters instead of four (used for the rest of the techniques) according to the reference protocol.

For sample preparation time, whole process time for isolation and staining were included. To compare acquisition time of the technologies, we based the scanning time on the surface area at which the cells are present and imaged with a 10X/NA 0.45 objective as used in the CellTracks Analyzer. Analysis with a higher NA objective will result in a better resolution to identify the cells but is accompanied by a longer data acquisition time. The exception is the FAST system/platform in which an area of 0.90×0.67 mm is imaged with a much faster analysis system. Regarding recovery rates, it is, however, difficult if not impossible to compare these values, as for positive selection targeting the EpCAM molecule the density of this antigen on the cell surface will be the critical factor and varies greatly among cell lines.[28] For instance, breast cancer SKBR-3 and MCF-7 cells have >455 000 EpCAM molecules per cell whereas prostate cancer PC-3 cells express ~50 000 molecules per cell and the bladder cancer T-24 cells only about 2000 EpCAM molecules per cell. Same variations are expected to influence results on separation by physical characteristics, such as differences in size and deformability.[29] For example the recovery rates between nine different cell lines filtered under the same conditions varied between 25% and 80%.[29] Median diameter of LnCaP, MCF-7, MDA-MB-231 cell lines, isolated with higher recovery rates by size-based filtration methods in Table 9.1, are bigger than 15 μm.[29,30] In Table 9.1 we listed the names of the cells that were spiked and in case more cell lines were used the cell line with the lowest and highest recovery were indicated. Although fold enrichment and tumor cell purity are frequently reported the numbers provided between platforms are hampered by lack of uniformity making it difficult to make a fair comparison. For enrichment methods the important values are the number of WBCs (and/or RBCs when they are still present) that are retained per mL of blood that is processed and for the enumeration method the number of cells that are identified in blood of healthy donors that stain non-specifically or are mistakenly identified as tumor cells per mL of blood. Spiking experiments with well-defined cell lines can then be used to assess the potential sensitivity and the linearity of the system. A sufficient number of control samples is a basic requirement to report on potential sensitivity of a CTC detection platform.

In the CellSearch system sample preparation is automated and performed in the CellTracks Autoprep that can process 7.5 mL blood samples from eight patients. Data/image acquisition is performed automatically in the CellTracks Analyzer.

In the MagSweeper blood is incubated with 4.5 μm paramagnetic beads functionalized with antibodies against EpCAM and CTCs are immuno-magnetically enriched by sweeping a magnetic bar in a well containing a 9 mL blood sample.[31] Repeated capture, wash, and release steps in buffer-filled wells resulted in the isolation of cells with higher purity. Capture efficiency of EpCAM positive cells from the breast cancer cell line MCF-7 spiked into the blood is reported to be 62% ± 7%, with a purity of 51% ± 18%. After the captured CTCs are released from the magnetic bar they are identified based on the stained probes under fluorescence microscope and can be further characterized.

RareCyte, Inc. developed a platform called Accucyte, a modification of the Quantitative Buffy Coat analysis platform in which a float is placed in a small tube. After whole blood is added the tube is centrifuged and the white blood cells and platelets layer is expended permitting the differential count of the blood cells.[32] In the AccuCyte platform the volume of the blood is increased from 55–65 μl to 3 mL and blood is incubated with the fluorescently labeled antibodies and magnetic particles coated with antibodies.[33] After centrifugation the CTCs can be identified by fluorescent microscopy in the expanded buffy coat that is confined to a 35 μm thick space. Cells of interest can be moved to a desired position by placing a magnet over the cells for further characterization.

A Fiber-optic Array Scanning Technology (FAST) was introduced to scan a relatively large number of cells (100 million cells) in a relatively short period of time (5 min.).[34] To scan this amount of cells with an automatic digital microscope (ADM) would take ~18 hrs. This enables 1 mL of whole blood to be scanned after RBC lysis and staining with fluorescently labeled antibodies (~10^7 cells). The FAST cytometer reported to enrich ~10^5-fold with an average sensitivity of 98% while scanning at a rate of 300 000 cells per second. A tandem approach, FAST prescan of 100 million cells (1500 objects) followed by ADM rescanning (300 objects), improves an enrichment-fold to 10^6 and takes approximately 1 hr (Table 9.1).

Filtration of blood cells has been introduced by several investigators and is based on the assumption that the majority of CTCs do have a size that is larger as compared to the RBCs and WBCs and their deformability is smaller as compared to the hematopoietic cells.[30,35–41] Filter material, thickness and flatness, pore size, density and distribution are all important factors that have to be taken into consideration when designing the filter. Blood dilution, fixation and pressure will also influence the passage of blood through the filters and thereby the effectiveness of CTC isolation.[29,42] The advantage of the filtration technique is that it can easily handle large blood volumes, which is needed because of the rarity of CTCs, to find sufficient numbers of CTCs.[2] Moreover, filtration can be run with relatively higher flow rate (100 mL hr^{-1}) resulting in a 10^3-fold enrichment and can concentrate cells on the surface of the filter and due to the filter's flatness, microscopic detection is relatively simple.[29,42]

9.6 Isolation and Identification of CTCs in Microfluidic Devices

The use of microfluidic devices is being explored for a variety of biomedical applications and, more recently, for the isolation, identification and characterization of CTCs.[43–45] Microfluidic devices are constructed to isolate CTCs from hematopoietic cells based on differences in their physical

characteristics such as size, deformability, and conductivity or in their immunological properties, such as the expression of the EpCAM antigen.

9.6.1 Microfluidic Devices for CTC Isolation Based on Physical Properties

Deterministic lateral displacement (DLD) arrays[46] in microfluidic devices
have been exploited to separate CTCs from blood cells.[47] An array of
microposts guides particles bigger than a critical size at different angles to
the fluid flow to permit the separation of blood cells based on size while the
blood passes through the device. This principle is depicted in Figure 9.4,
panel A, where the flow paths of the RBCs and WBCs (smaller than 7 μm) are
barely influenced by an array of microposts while larger CTCs flow along the
tilted axis of the array. Maximum flow rate was determined at 10 mL min^{-1}
with more than 85% capture efficiency without damaging the cells. In addition, the use of triangle posts improved clogging problems.[47]

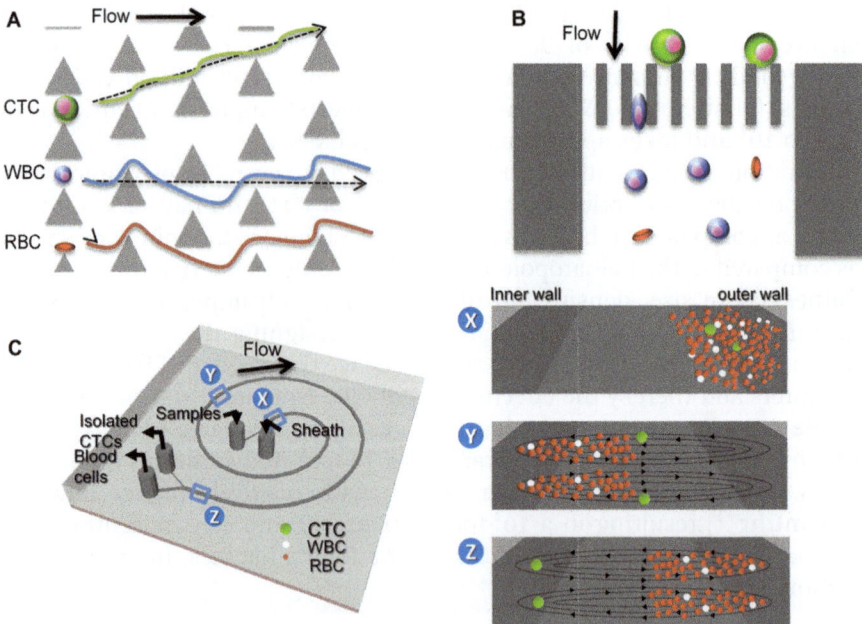

Figure 9.4 Size-based isolation of CTCs in microfluidic devices. Panel A, Deterministic lateral displacement (DLD) arrays.[47] Movement paths are indicated
for CTCs with green, WBCs with blue, and RBCs with red. Panel B,
Differential size separation by filtration through pores.[48] Cross-section
view of microsieve-integrated lab-chip. Panel C, Spiral micro-channel
with centrifugal force (Dean flow fractionation) for CTC enrichment.[61]
3D view of spiral channels on the left and side view of X, Y, Z regions on
the right. CTCs, WBCs, and RBCs are represented with green, white, and
red, respectively.

Membrane-based filtration methods were integrated into microfluidic devices as depicted in panel B of Figure 9.4.[48,49] WBCs and RBCs are smaller than the pore size or are more easily deformed to flow through the pores of the membrane, while the larger and more rigid CTCs remain on the membrane. As microfabrication techniques have advanced, pores of the membrane can be made with more precision to control size, shape, and spacing on various materials with high reproducibility. To prevent clogging problems a sufficient number of pores are needed and, for instance, silicon microsieves with 10^5 pores were used to recover 80% of tumor cells spiked within blood at $1 \, \text{mL} \, \text{min}^{-1}$ flow rate.[48] In membrane-based filtration, enumeration and further characterization of the captured cells are feasible and will permit the characterization of both EpCAM negative and EpCAM positive cells. The microfluidic unit conducts flow control, washing, and staining steps on the membrane.[48] Hosokawa and colleagues reported a nickel microcavity array integrated PDMS device. 10 000 cavities with $8.4 \, \mu\text{m}$–$9.1 \, \mu\text{m}$ diameter captured approximately 90% of tumor cells spiked into the blood.[49] From inlet to outlet of flow chamber samples are horizontally injected into a PDMS chamber and cells are captured by negative pressure from the chamber underneath the cavity array. Therefore, washing and staining steps are easily assessed from the top channel after cell capture. Zheng and colleagues reported electrode-embedded 2D microfilter device and 3D microfilter device to improve accessibility for downstream analysis.[30,50] To analyze genetic information, an electrode-integrated membrane demonstrated electrical lysis of captured cells on a parylene C microfilter. The same group presented 3D microfilters to evaluate living cells. Two layers of parylene membranes were overlapped to support the membrane and minimize the stress of the cells captured on the membrane. Captured cells in the 3D microfilter were kept viable for two weeks and accessible for molecular characterization.

Within microfluidic devices microfilters can be replaced by structures inside the channels that form obstacles for the cells to pass. Tan and colleagues presented multiple arrays of crescent-shaped micro-cups in the channels with gaps of $5 \, \mu\text{m}$ to capture CTCs. Cells from tumor cell lines were captured with an efficiency of approximately 80%.[51] Due to the similar size of the smaller CTCs and WBCs, the filtration process will miss the smaller CTCs. Reducing the pore size will prohibitively increase the contamination of WBCs.

Electrical properties and size of CTCs have been utilized to separate breast cancer cells from blood cells in a dielectric filed column by Becker and colleagues.[52–54] Difference of response to the electric field meant that cancer cells were retained in the electric field applied column while WBCs passed through. After removal of WBCs, deactivation of the electric field retrieved cancer cells at a recovery rate of 95%.[54]

Differences in size of CTCs as compared to hematopoietic cells were also explored by measurement of impedance in a microcytometer.[55] Polyelectrolytic gel electrodes (PGEs) were embedded in both sides of a micro-channel and transport ions between Ag/AgCl electrodes. When cells pass by the PGEs, transport of ions is hindered as the cell membrane behaves like an insulator. The impedance signal is proportional to the volume of the cell and using label-free DC impedance signals cells from ovarian cancer line were captured with an efficiency of 88% within blood cells. Similarly, Adams and colleagues enumerated CTCs within blood cells based on conductivity.[56,57] Differential conductometric responses between CTCs and blood cells determined the number of spiked MCF-7 breast cancer cells and detected near 100% of 10–250 cells mL^{-1} without labeling.

The Dean Flow Fractionation principle[58–60] has been explored to separate CTCs from WBCs and RBCs and also is based on size differences. Figure 9.4C shows a spiral microfluidic channel that consists of two inlets and two outlets.[61] In curvilinear channels the spiral shape drags cells back and forth along the channel width as indicated by the black dotted arrows. Larger cells experience additional inertial lift forces. When blood sample and sheath flow is pumped into the outer and inner inlets respectively, differences in the movements of CTCs as compared to blood cells occur as illustrated on right panel of Figure 9.4C. WBCs and RBCs are sorted to the outer outlet by Dean vortices while larger CTCs are influenced by inertial lift force and are sorted to the inner outlet as illustrated in the side view of the channel at position Z. Repeating the sorting process with another spiral channel (2nd cascade) is used to improve the purity with a capture efficiency of spiked tumor cells of 85%. Enrichment fold increases from 10^2 (1st spiral) to 10^3 (2nd cascade) fold over WBCs.

9.6.2 Microfluidic Devices to Isolate CTCs Based on Immunological Properties

Microfluidic channels can be chemically activated by flowing reagent solutions through the device. After functionalization of activated surfaces with antibodies or nucleic acids that are targeting molecules expressed on the surface of the CTCs and not on hematopoietic cells, blood can flow through the channels and the CTCs can be captured.[62–66] Examples of capture reagents are monoclonal antibodies or nucleic acid such as DNA aptamers recognizing EpCAM or markers overexpressed on tumor cells such as PSMA.[63,64,67] Coating of surfaces is frequently done with streptavidin followed by biotinylated antibodies (Figure 9.5A(a)) or biotinylated DNA aptamers (Figure 9.5A(b)).[57,62] The high-affinity, rapid, and stable reaction of streptavidin to biotin makes it attractive to use for coating of surfaces. In addition this combination can be used to selectively release the captured cells by conjugation of the antibodies to desbiotin that has a lower affinity for streptavidin and its binding can be broken by the addition of biotin.[68] Figure 9.5A(c) shows the use of protein G to immobilize antibodies.[69] Protein G binds preferably to the Fc region of antibodies and F(ab') region will be exposed to target cells. Enhancement of surface area for cell capture in

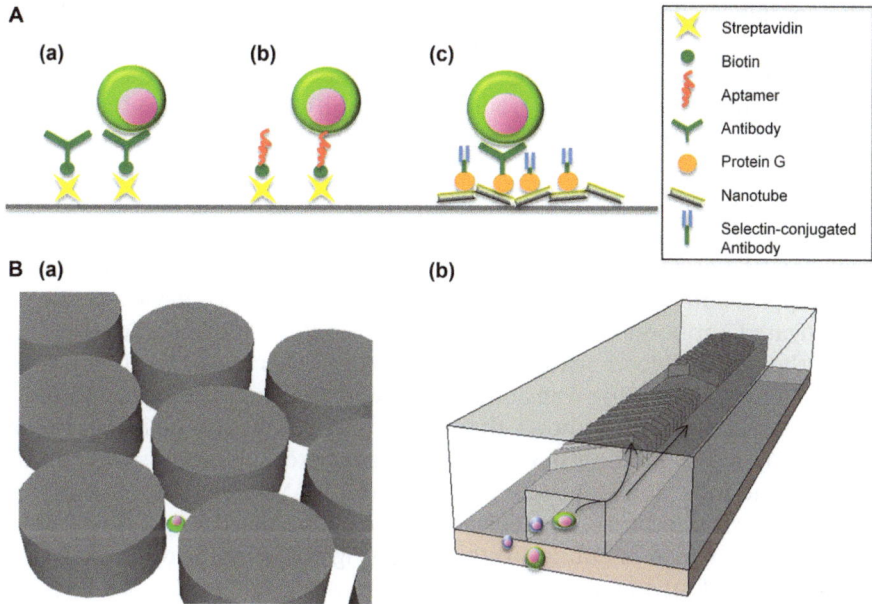

Figure 9.5 Affinity-based capture of CTCs in microfluidic devices. Panel A, antibody immobilization methods described in microfluidic devices. (a) Streptavidin coated surface with biotinylated antibodies (*e.g.* anti-EpCAM,[64] anti-PSMA[78]); (b) Streptavidin coated surface with biotin conjugated aptamers (*e.g.* KCHA10 aptamer, KDED2a-3 aptamer[67]); (c) Protein G-mediated antibody immobilization. Fc region of antibodies bind to protein G. Selectin-conjugated antibodies are immobilized together with EpCAM antibodies to enhance the purity of CTC isolation.[69] Panel B, representative examples of approaches to increase the surface area for CTC capture in microfluidic devices. (a) Microposts.[64] (b) Herringbone-structures.[63]

microfluidic devices is increased by the introduction of micro-/nanostructures such as antibody coated microposts, herringbone structured channels, hallocyte nanotubes or silicon nanowires.[63,64,69–71] One of these configurations, the CTC chip, uses 78 000 silicon microposts as a support structure to coat anti-EpCAM antibodies as illustrated in Figure 9.5B. 116 blood samples from patients with metastatic lung, prostate, pancreatic, breast, and colon cancer were processed and in 115 of 116 samples CTCs were detected between 5 and 1281 CTC mL^{-1}.[64] Purity of isolated CTCs was approximately 50%. The number of CTCs and the percentage of patients in which CTCs were detected was significantly larger compared to the numbers reported with the CellSearch System, questioning whether or not the same strict definition of a CTC was used in this system. To improve CTC isolation with higher flow rate for high-throughput 2nd-generation baptized Herringbone chip (HB-chip) was developed. This employed micro-vortex mixing based on a herringbone structure of micro-channels.[63,72] It enhanced the number of interactions between antibodies and cells compared to flat channel without the complex geometry of the CTC-chip as depicted in Figure 9.5C. With a flow rate of $1.5\,mL\,hr^{-1}$–$2.5\,mL\,hr^{-1}$, the reported capture efficiency of spiked cells from PC-3 cell lines was 26% higher compared to the CTC-chip. In addition, transparent materials – PDMS and glass – are used to observe cells inside the channel. In the microtube platform, hallocyte nanotubes and E-selectin are used to facilitate substrate to capture EpCAM positive cells as illustrated in Figure 9.5A(c).[69] Hallocyte nanotube coating provides a large surface area to immobilize capturing reagents as well as a rough surface to enhance the purity (>50%) of captured CTCs compared to a smooth surface by suppression of leukocyte (WBC) spreading. In this device E-selectin conjugated antibodies are co-immobilized with EpCAM antibodies to mimic endothelium in blood vessels.[69] Flowing leukocytes start to roll slowly on endothelium by binding transiently to selectin molecules for extravasation.[73] Experimental evidence shows that CTCs may extravasate using a similar mechanism.[74,75] Binding selectin molecules makes CTCs roll slowly on the wall of the tube until they strongly bind to EpCAM antibodies and this results in 50% capturing efficiency of spiked tumor cells at a flow rate of $4.8\,mL\,hr^{-1}$, which is similar to the capturing efficiency of other platforms processed at $1\,mL\,hr^{-1}$.

9.6.3 Microfluidic Devices to Isolate CTCs Based on Physical as well as Immunological Properties

A combination of affinity-based selection and size-based filtration has clear advantages to achieve high CTC capture efficiency with less contamination of WBCs. To combine filtration based on size with affinity separation the size of EpCAM expressing CTCs was increased by incubation of the blood with EpCAM antibody coated polystyrene beads.[76] The size of the CTCs expressing EpCAM increases and can now be captured by the multiple obstacles in the channel with higher purity. Another example is geometrically enhanced differential immunocapture (GEDI) in which microposts arrays are designed to enhance the collision rate of larger CTCs (>14 μm) and structures are coated with antibodies targeting prostate cancer cells.[77,78] High collision rate facilitates interaction between CTCs and antibodies on the posts while WBCs flow through with lower collision rate, which causes less interaction with antibodies on the posts. It results in higher capture efficiency of CTCs in GEDI devices compared to straight posts arrayed in channels. Enumeration of CTCs (PSMA+/DAPI+/CD45-) in blood samples from castration-resistant prostate cancer patients (CRPC) found a 2–400-fold higher number of CTCs compared to CellSearch System.

9.6.4 Characterization of CTCs in Microfluidic Devices

After cells are captured inside the microfluidic devices protein and gene identification can be performed by immunostaining and FISH analysis.

For downstream analysis, isolated cells are exposed to lysing solutions to isolate DNA, RNA, or proteins. Reports with the CTC-chip, Herringbone chip, and GEDI chip have identified genetic alterations of captured CTCs by qPCR or sequencing results.[64,78,79] Organ specific markers (*e.g.* PSA in prostate cancer) and well-known molecular characterization of cancer cells (*e.g.* TMPRSS2:ERG fusion protein) were shown to be present on the captured CTCs. The lysates, however, still contain the content of the non-specific or unwanted captured WBCs, which reduces the specificity and makes it impossible to reveal the heterogeneity of the cancer cells. Accutase and Trypsin have been used to detach cells from substrates in these devices to collect the captured cells.[66,69] One of those is the high-throughput microsampling unit (HTMSU) platform integrated with electrokinetic manipulation.[66] Colorectal cancer cells are captured on EpCAM antibody coated surfaces in HTMSU and detached by Trypsin 0.25% for genetic analysis. Detached cells are enumerated based on conductivity and directed to the separate chamber by applying an electric field to perform polymerase chain reaction (PCR) and ligase detection reaction (LDR). Point mutations of *KRAS* gene were only detected in blood spiked with SW620 cells and not in blood spiked with HT29 cells, which is a consistent result with known genotype. In the microtube platform, Accutase detached captured CTCs from the surface and viable CTCs were retrieved and can be maintained in cell culture for 15 days.[69] For functional characterization of CTCs, long-term observation of CTCs was reported on the GEDI device. With this GEDI chip, a study was performed to monitor the response of captured CTCs to the drug candidates for 24 hours.[78]

9.7 Summary and Outlook

Clinical relevance of CTCs has been shown in many studies and stimulated the development of a variety of technologies to isolate and characterize CTCs. The use of microfluidics has also been explored in this quest for CTCs. One of the largest challenges for microfluidic devices is their relatively low throughput. CTCs are extremely rare and in approximately 50% of metastatic cancer patients <1 CTC in 7.5 mL of blood are present.[2] Specific markers to identify all CTCs have not yet been found. Use of epithelial cell markers and epithelial tumor origin markers might not be sufficient to identify all CTCs. Most approaches target the EpCAM antigen and indeed not all cancers and cancer cells will express a sufficient number of EpCAM antigens on the cell surface to select the CTCs. During the epithelial mesenchymal transition (EMT) process for example epithelial markers can be lost from the cell surface and prohibit their isolation. To overcome this limitation several technologies have been reported that attempt to separate the CTCs from hematopoietic cells based on differences in physical characteristics. More recently the combination of both immunological properties and physical properties is being explored to increase the CTC yield and detect

CTCs in those patients not expressing EpCAM or other cell surface molecules.

Another large challenge the CTC field is facing is the large variation in the reported CTC frequency, which can in most cases be traced to the definition used for CTC. This is not surprising as reports from the early studies in the development of the CellSearch System[80,81] show the number of CTCs and the frequency at which they were detected was also significantly higher as compared to the FDA cleared CellSearch System.[3] The apparent differences in numbers can be attributed to the different detection platform (flowcytometry *versus* fluorescent microscopy), the strict size ($4\,\mu m \times 4\,\mu m$), and morphological features such as "intact morphology" that are being implied, which are necessary to avoid detection of CTCs in controls and to maintain a significant relation with clinical outcome. To advance the CTC field it is therefore of utmost importance to clearly define a CTC and the demonstration that they are indeed relevant in controlled prospective studies.

The largest expectation for the CTC field is their ability to provide a liquid biopsy for all patients and at all time points during the course of the disease when it has not been eliminated from the body. Not only will we need a technology that can detect CTCs in all patients, but they will also need to be preserved in such a manner that their protein, RNA, and DNA content can be analyzed. The heterogeneity of the cancer cells and their acquired resistance to certain therapies during the course of the disease imposes the need for individual tumor cell characterization. The first requirement to achieve this goal is that tumor cells must be present in the blood.[2] The difficulty that arises is that for a given patient the physical characteristics and the cell surface antigens expressed on the CTCs are not known and a tumor biopsy taken years prior may no longer be representative. Key features for a future platform are illustrated in Figure 9.6 and contain the following critical elements: 1) a blood volume containing a sufficient number of CTCs for further characterization to choose the most appropriate therapy, 2) identification of CTCs from whole blood and their individual isolation, 3) direction of CTCs for single cell genotyping and phenotyping. Manipulation of a single cell in microfluidic platforms has been reported with numerous methods such as dielectrophoresis, optical fiber guidance, and smart geometry.[82–88] The inherent capability of microfluidic systems to handle small volumes in parallel together with the ability of integration with highly sensitive detection systems make them very well suited for single cell analysis. However, difficulties in integrating high-throughput analysis with single cell analysis have discouraged the investigation of single CTCs in a microfluidic platform. A high-throughput, micromechanical valve integrated platform is depicted in Figure 9.6 as one of the promising methods for a future single CTC platform for molecular characterization.[89–92] Transparent materials such as PDMS enable the visualization of the captured single cell before proceeding for further analysis. Genetic analysis of single CTCs such as mutation analysis and expression level comparison by RT-qPCR or sequencing will enable the exploration of treatment targets. For protein

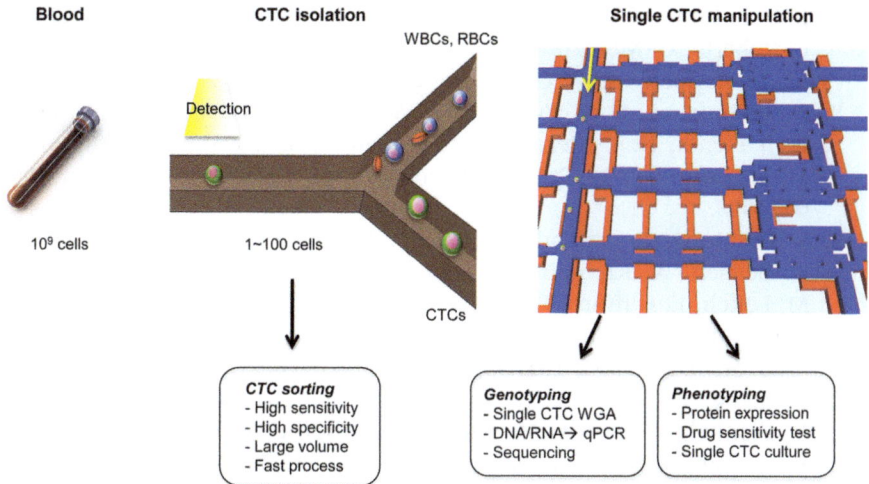

Figure 9.6 Outlook for future microfluidic platforms for CTC isolation and characterization. Basic requirements are: a blood volume that contains a sufficient number of tumor cells for enumeration and characterization, a CTC isolation system that identifies and sorts the individual tumor cells, a single cell manipulation system for genotyping and phenotyping the single CTC.

expression and observation of cellular response to drug candidates viable single CTCs that can be maintained will be needed. Technology will improve and one day a liquid biopsy for all cancer patients will be realized. The CTC field will, however, only move forward if indeed researchers presenting technologies to isolate and characterize CTCs validate their data in controlled clinical studies.

References

1. T. R. Ashworth, *Med. J. Aust.*, 1869, **14**, 146.
2. F. A. Coumans, S. T. Ligthart, J. W. Uhr and L. W. Terstappen, *Clin. Cancer Res.*, 2012, **18**, 5711.
3. W. J. Allard, J. Matera, M. C. Miller, M. Repollet, M. C. Connelly, C. Rao, A. G. Tibbe, J. W. Uhr and L. W. Terstappen, *Clin. Cancer Res.*, 2004, **10**, 6897.
4. M. Cristofanilli, L. Terstappen and D. F. Hayes, *New Engl. J. Med.*, 2004, **351**, 2453.
5. S. J. Cohen, L. W. Terstappen, C. J. Punt, E. P. Mitchell, T. M. Fynan, T. Li, J. Matera, G. V. Doyle and N. J. Meropol, *J. Clin. Oncol.*, 2006, **24**, 153S.
6. J. S. de Bono, H. I. Scher, R. B. Montgomery, C. Parker, M. C. Miller, H. Tissing, G. V. Doyle, L. W. W. M. Terstappen, K. J. Pienta and D. Raghavan, *Clin. Cancer Res.*, 2008, **14**, 6302.

7. J. S. de Bono, G. Attard, A. Adjei, M. N. Pollak, P. C. Fong, P. Haluska, L. Roberts, C. Melvin, M. Repollet, D. Chianese, M. Connely, L. W. M. M. Terstappen and A. Gualberto, *Clin. Cancer Res.*, 2007, **13**, 3611.

8. D. A. Smirnov, D. R. Zweitzig, B. W. Foulk, M. C. Miller, G. V. Doyle, K. J. Pienta, N. J. Meropol, L. M. Weiner, S. J. Cohen, J. G. Moreno, M. C. Connelly, L. Terstappen and S. M. O'Hara, *Cancer Res.*, 2005, **65**, 4993.

9. S. D. Meng, D. Tripathy, S. Shete, R. Ashfaq, B. Haley, S. Perkins, P. Beitsch, A. Khan, D. Euhus, C. Osborne, E. Frenkel, S. Hoover, M. Leitch, E. Clifford, E. Vitetta, L. Morrison, D. Herlyn, L. Terstappen, T. Fleming, T. Fehm, T. Tucker, N. Lane, J. Q. Wang and J. Uhr, *Proc. Natl Acad. Sci. USA*, 2004, **101**, 9393.

10. S. Meng, D. Tripathy, S. Shete, R. Ashfaq, H. Saboorian, B. Haley, E. Frenkel, D. Euhus, M. Leitch, C. Osborne, E. Clifford, S. Perkins, P. Beitsch, A. Khan, L. Morrison, D. Herlyn, L. Terstappen, N. Lane, J. Wang and J. Uhr, *Proc. Natl Acad. Sci. USA*, 2006, **103**, 17361.

11. D. F. Hayes, T. M. Walker, B. Singh, E. S. Vitetta, J. W. Uhr, S. Gross, C. Rao, G. V. Doyle and L. Terstappen, *Int. J. Oncol.*, 2002, **21**, 1111.

12. J. F. Swennenhuis, A. G. J. Tibbe, R. Levink, R. C. J. Sipkema and L. W. M. M. Terstappen, *Cytometry Part A*, 2009, **75A**, 520.

13. G. Attard, J. F. Swennenhuis, D. Olmos, A. H. Reid, E. Vickers, R. A'Hern, R. Levink, F. Coumans, J. Moreira, R. Riisnaes, N. B. Oommen, G. Hawche, C. Jameson, E. Thompson, R. Sipkema, C. P. Carden, C. Parker, D. Dearnaley, S. B. Kaye, C. S. Cooper, A. Molina, M. E. Cox, L. W. Terstappen and J. S. de Bono, *Cancer Res.*, 2009, **69**, 2912.

14. S. T. Ligthart, F. C. Bidard, C. Decraene, T. Bachelot, S. Delaloge, E. Brain, M. Campone, P. Viens, J. Y. Pierga and L. W. Terstappen, *Ann. Oncol.*, 2012, DOI: 10.1093/annonc/mds625.

15. A. G. J. Tibbe, M. C. Miller and L. W. M. M. Terstappen, *Cytometry Part A*, 2007, **71A**, 154.

16. J. Kraan, S. Sleijfer, M. H. Strijbos, M. Ignatiadis, D. Peeters, J. Y. Pierga, F. Farace, S. Riethdorf, T. Fehm, L. Zorzino, A. G. J. Tibbe, M. Maestro, R. Gisbert-Criado, G. Denton, J. S. de Bono, C. Dive, J. A. Foekens and J. W. Gratama, *Cytometry B Clin. Cytometry*, 2011, **80B**, 112.

17. S. Riethdorf, H. Fritsche, V. Muller, T. Rau, C. Schindlbeck, B. Rack, W. Janni, C. Coith, K. Beck, F. Janicke, S. Jackson, T. Gornet, M. Cristofanilli and K. Pantel, *Clin. Cancer Res.*, 2007, **13**, 920.

18. S. T. Ligthart, F. A. W. Coumans, G. Attard, A. Mulick Cassidy, J. S. de Bono and L. W. M. M. Terstappen, *PloS One*, 2011, **6**, e27419.

19. F. A. W. Coumans, C. J. M. Doggen, G. Attard, J. S. de Bono and L. W. M. M. Terstappen, *Ann. Oncol.*, 2010, **21**, 1851.

20. C. J. Larson, J. G. Moreno, K. J. Pienta, S. Gross, M. Repollet, S. M. O'Hara, T. Russell and L. W. M. M. Terstappen, *Cytometry*, 2004, **62A**, 46.

21. F. A. W. Coumans, S. T. Ligthart and L. Terstappen, *Transl. Oncol.*, 2012, **5**, 486.

22. G. T. Budd, M. Cristofanilli, M. J. Ellis, A. Stopeck, E. Borden, M. C. Miller, J. Matera, M. Repollet, G. V. Doyle, L. W. M. M. Terstappen and D. F. Hayes, *Clin. Cancer Res.*, 2006, **12**, 6403.
23. M. Cristofanilli, G. T. Budd, M. J. Ellis, A. Stopeck, J. Matera, M. C. Miller, J. M. Reuben, G. V. Doyle, W. J. Allard, L. W. Terstappen and D. F. Hayes, *N. Engl. J. Med.*, 2004, **351**, 781.
24. D. F. Hayes, *Clin. Cancer Res.*, 2006, **12**, 4218.
25. S. J. Cohen, C. J. A. Punt, N. Iannotti, B. H. Saidman, K. D. Sabbath, N. Y. Gabrail, J. Picus, M. Morse, E. Mitchell, M. C. Miller, G. V. Doyle, H. Tissing, L. W. M. M. Terstappen and N. J. Meropol, *J. Clin. Oncol.*, 2008, **26**, 3213.
26. S. J. Cohen, C. J. A. Punt, N. Iannotti, B. H. Saidman, K. D. Sabbath, N. Y. Gabrail, J. Picus, M. A. Morse, E. Mitchell, M. C. Miller, G. V. Doyle, H. Tissing, L. W. M. M. Terstappen and N. J. Meropol, *Ann. Oncol.*, 2009, **20**, 1223.
27. M. C. Miller, G. V. Doyle and L. W. M. M. Terstappen, *J. Oncol.*, 2010, **2010**, 617421.
28. C. G. Rao, D. Chianese, G. V. Doyle, M. C. Miller, T. Russell, R. A. Sanders, Jr. and L. W. Terstappen, *Int. J. Oncol.*, 2005, **27**, 49.
29. F. A. Coumans, G. van Dalum, M. Beck and L. W. Terstappen, *PloS One*, 2013, **8**, e61770.
30. S. Zheng, H. Lin, J.-Q. Liu, M. Balic, R. Datar, R. J. Cote and Y.-C. Tai, *J. Chrom.*, 2007, **1162**, 154.
31. A. H. Talasaz, A. A. Powell, D. E. Huber, J. G. Berbee, K. H. Roh, W. Yu, W. Xiao, M. M. Davis, R. F. Pease, M. N. Mindrinos, S. S. Jeffrey and R. W. Davis, *Proc. Natl Acad. Sci. USA*, 2009, **106**, 3970.
32. S. C. Wardlaw and R. A. Levine, *J. Am. Med. Assoc*, 1983, **249**, 617.
33. http://rarecyte.com, 2012.
34. R. T. Krivacic, A. Ladanyi, D. N. Curry, H. B. Hsieh, P. Kuhn, D. E. Bergsrud, J. F. Kepros, T. Barbera, M. Y. Ho, L. B. Chen, R. A. Lerner and R. H. Bruce, *Proc. Natl Acad. Sci. USA*, 2004, **101**, 10501.
35. H. K. Lin, S. Zheng, A. J. Williams, M. Balic, S. Groshen, H. I. Scher, M. Fleisher, W. Stadler, R. H. Datar, Y. C. Tai and R. J. Cote, *Clin. Cancer Res.*, 2010, **16**, 5011.
36. H. W. Hou, Q. S. Li, G. Y. H. Lee, A. P. Kumar, C. N. Ong and C. T. Lim, *Biomed. Microdevices*, 2009, **11**, 557.
37. R. Rosenberg, R. Gertler, J. Friederichs, K. Fuehrer, M. Dahm, R. Phelps, S. Thorban, H. Nekarda and J. R. Siewert, *Cytometry*, 2002, **49**, 150.
38. H. J. Kahn, A. Presta, L. Y. Yang, J. Blondal, M. Trudeau, L. Lickley, C. Holloway, D. R. McCready, D. Maclean and A. Marks, *Breast Canc. Res. Treat*, 2004, **86**, 237.
39. I. Desitter, B. S. Guerrouahen, N. Benali-Furet, J. Wechsler, P. A. Janne, Y. Kuang, M. Yanagita, L. Wang, J. A. Berkowitz, R. J. Distel and Y. E. Cayre, *Anticancer Res.*, 2011, **31**, 427.
40. S. Zheng, H. K. Lin, B. Lu, A. Williams, R. Datar, R. J. Cote and Y.-C. Tai, *Biomed. Microdevices*, 2010, **13**, 203.

41. V. J. Hofman, M. I. Ilie, C. Bonnetaud, E. Selva, E. Long, T. Molina, J. M. Vignaud, J. F. Flejou, S. Lantuejoul, E. Piaton, C. Butori, N. Mourad, M. Poudenx, P. Bahadoran, S. Sibon, N. Guevara, J. Santini, N. Venissac, J. Mouroux, P. Vielh and P. M. Hofman, *Am. J. Clin. Pathol.*, 2010, **135**, 146.

42. F. A. Coumans, G. van Dalum, M. Beck and L. W. Terstappen, *PloS One*, 2013, **8**, e61774.

43. G. M. Whitesides, *Nature*, 2006, **442**, 368.

44. P. Yager, T. Edwards, E. Fu, K. Helton, K. Nelson, M. R. Tam and B. H. Weigl, *Nature*, 2006, **442**, 412.

45. J. El-Ali, P. K. Sorger and K. F. Jensen, *Nature*, 2006, **442**, 403.

46. L. R. Huang, E. C. Cox, R. H. Austin and J. C. Sturm, *Science*, 2004, **304**, 987.

47. K. Loutherback, J. D'Silva, L. Liu, A. Wu, R. H. Austin and J. C. Sturm, *AIP Adv*, 2012, **2**, 42107.

48. L. S. Lim, M. Hu, M. C. Huang, W. C. Cheong, A. T. L. Gan, X. L. Looi, S. M. Leong, E. S.-C. Koay and M.-H. Li, *Lab on a Chip*, 2012, **12**, 4388.

49. M. Hosokawa, T. Hayata, Y. Fukuda, A. Arakaki, T. Yoshino, T. Tanaka and T. Matsunaga, *Anal. Chem.*, 2010, **82**, 6629.

50. S. Y. Zheng, H. K. Lin, B. Lu, A. Williams, R. Datar, R. J. Cote and Y. C. Tai, *Biomed. Microdevices*, 2011, **13**, 203.

51. S. J. Tan, L. Yobas, G. Y. Lee, C. N. Ong and C. T. Lim, *Biomed. Microdevices*, 2009, **11**, 883.

52. P. R. C. Gascoyne, J. Noshari, T. J. Anderson and F. F. Becker, *Electrophoresis*, 2009, **30**, 1388.

53. P. R. C. Gascoyne, *Anal. Chem.*, 2009, **81**, 8878.

54. F. F. Becker, X.-B. Wang, Y. Huang, R. Pethig, J. Vykoukal and P. Gascoyne, *Proc. Natl Acad. Sci*, 1995, **92**, 860.

55. H. Choi, K. B. Kim, C. S. Jeon, I. Hwang, S. Lee, H. K. Kim, H. C. Kim and T. D. Chung, *Lab Chip*, 2013, **13**, 970.

56. A. A. Adams, P. I. Okagbare, J. Feng, M. L. Hupert, D. Patterson, J. Göttert, R. L. McCarley, D. Nikitopoulos, M. C. Murphy and S. A. Soper, *J. Am. Chem. Soc.*, 2008, **130**, 8633.

57. U. Dharmasiri, S. Balamurugan, A. A. Adams, P. I. Okagbare, A. Obubuafo and S. A. Soper, *Electrophoresis*, 2009, **30**, 3289.

58. D. Di Carlo, *Lab Chip*, 2009, **9**, 3038.

59. S. S. Kuntaegowdanahalli, A. A. S. Bhagat, G. Kumar and I. Papautsky, *Lab Chip*, 2009, **9**, 2973.

60. D. Di Carlo, D. Irimia, R. G. Tompkins and M. Toner, *Proc. Natl Acad. Sci*, 2007, **104**, 18892.

61. H. W. Hou, M. E. Warkiani, B. L. Khoo, Z. R. Li, R. A. Soo, D. S.-W. Tan, W.-T. Lim, J. Han, A. A. S. Bhagat and C. T. Lim, *Sci. Rep*, 2013, **3**, 1259.

62. Y. Xu, J. A. Phillips, J. Yan, Q. Li, Z. H. Fan and W. Tan, *Anal. Chem.*, 2009, **81**, 7436.

63. S. L. Stott, C. H. Hsu, D. I. Tsukrov, M. Yu, D. T. Miyamoto, B. A. Waltman, S. M. Rothenberg, A. M. Shah, M. E. Smas, G. K. Korir,

F. P. Floyd, Jr., A. J. Gilman, J. B. Lord, D. Winokur, S. Springer, D. Irimia, S. Nagrath, L. V. Sequist, R. J. Lee, K. J. Isselbacher, S. Maheswaran, D. A. Haber and M. Toner, *Proc. Natl Acad. Sci. USA*, 2010, **107**, 18392.

64. S. Nagrath, L. V. Sequist, S. Maheswaran, D. W. Bell, D. Irimia, L. Ulkus, M. R. Smith, E. L. Kwak, S. Digumarthy, A. Muzikansky, P. Ryan, U. J. Balis, R. G. Tompkins, D. A. Haber and M. Toner, *Nature*, 2007, **450**, 1235.

65. J. A. Phillips, Y. Xu, Z. Xia, Z. H. Fan and W. Tan, *Anal. Chem.*, 2009, **81**, 1033.

66. U. Dharmasiri, S. K. Njoroge, M. A. Witek, M. G. Adebiyi, J. W. Kamande, M. L. Hupert, F. Barany and S. A. Soper, *Anal. Chem.*, 2011, **83**, 2301.

67. W. Sheng, T. Chen, R. Kamath, X. Xiong, W. Tan and Z. H. Fan, *Anal. Chem.*, 2012, **84**, 4199.

68. L. W. M. M. Terstappen, G. C. Rao, D. L. Gohel, B. Feeley, S. Gross, E. Church and P. Liberti, *US Pat.*, 5,646,001, 1997.

69. A. D. Hughes, J. Mattison, L. T. Western, J. D. Powderly, B. T. Greene and M. R. King, *Clin. Chem.*, 2012, **58**, 846.

70. S. Wang, K. Liu, J. Liu, Z. T.-F. Yu, X. Xu, L. Zhao, T. Lee, E. K. Lee, J. Reiss, Y.-K. Lee, L. W. K. Chung, J. Huang, M. Rettig, D. Seligson, K. N. Duraiswamy, C. K.-F. Shen and H.-R. Tseng, *Angew. Chem. Int. Ed. Engl.*, 2011, **50**, 3084.

71. S. Wang, K. Liu, J. Liu, Z. T. Yu, X. Xu, L. Zhao, T. Lee, E. K. Lee, J. Reiss, Y. K. Lee, L. W. Chung, J. Huang, M. Rettig, D. Seligson, K. N. Duraiswamy, C. K. Shen and H. R. Tseng, *Angew. Chem. Int. Ed. Engl.*, 2011, **50**, 3084.

72. A. D. Stroock, S. K. Dertinger, A. Ajdari, I. Mezić, H. A. Stone and G. M. Whitesides, *Science*, 2002, **295**, 647.

73. A. D. Hughes and M. King, *Bioengineering Conference (NEBEC)*, 2011 IEEE 37th Annual Northeast, 2011.

74. S. Gout, P.-L. Tremblay and J. Huot, *Clin. Exp. Metastasis*, 2008, **25**, 335.

75. F. W. Orr, H. H. Wang, R. M. Lafrenie, S. Scherbarth and D. M. Nance, *J. Pathol.*, 2000, **190**, 310.

76. M. S. Kim, T. S. Sim, Y. J. Kim, S. S. Kim, H. Jeong, J.-M. Park, H.-S. Moon, S.-I. Kim, O. Gurel, S. S. Lee, J.-G. Lee and J. C. Park, *Lab on a Chip*, 2012, **12**, 2874.

77. J. P. Gleghorn, E. D. Pratt, D. Denning, H. Liu, N. H. Bander, S. T. Tagawa, D. M. Nanus, P. A. Giannakakou and B. J. Kirby, *Lab Chip*, 2010, **10**, 27.

78. B. J. Kirby, M. Jodari, M. S. Loftus, G. Gakhar, E. D. Pratt, C. Chanel-Vos, J. P. Gleghorn, S. M. Santana, H. Liu, J. P. Smith, V. N. Navarro, S. T. Tagawa, N. H. Bander, D. M. Nanus and P. Giannakakou, *PloS One*, 2012, **7**, e35976.

79. S. L. Stott, C.-H. Hsu, D. I. Tsukrov, M. Yu, D. T. Miyamoto, B. A. Waltman, S. M. Rothenberg, A. M. Shah, M. E. Smas, G. K. Korir, F. P. Floyd, A. J. Gilman, J. B. Lord, D. Winokur, S. Springer, D. Irimia,

S. Nagrath, L. V. Sequist, R. J. Lee, K. J. Isselbacher, S. Maheswaran, D. A. Haber and M. Toner, *Proc. Natl Acad. Sci. USA*, 2010, **107**, 18392.

80. L. Terstappen, C. Rao, S. Gross, V. Kotelnikov, E. Racilla, J. Uhr and A. Weiss, *Vox Sang.*, 1998, **74**, 269.

81. E. Racila, D. Euhus, A. J. Weiss, C. Rao, J. McConnell, L. Terstappen and J. W. Uhr, *Proc. Natl Acad. Sci. USA*, 1998, **95**, 4589.

82. M. Kirschbaum, C. R. Guernth-Marschner, S. Cherré, A. de Pablo Peña, M. S. Jaeger, R. A. Kroczek, T. Schnelle, T. Mueller and C. Duschl, *Lab on a Chip*, 2012, **12**, 443.

83. S. Le Gac and A. van den Berg, *Trends Biotechnol.*, 2010, **28**, 55.

84. D. Psaltis, S. R. Quake and C. Yang, *Nature*, 2006, **442**, 381.

85. V. Sanchez-Freire, A. D. Ebert, T. Kalisky, S. R. Quake and J. C. Wu, *Nat. Protoc.*, 2012, 7, 829.

86. T. Sun and H. Morgan, *Microfluidics and Nanofluidics*, 2010, **8**, 423.

87. J. S. Marcus, W. F. Anderson and S. R. Quake, *Anal. Chem.*, 2006, **78**, 3084.

88. D. Wang and S. Bodovitz, *Trends Biotechnol.*, 2010, **28**, 281.

89. T. Thorsen, S. J. Maerkl and S. R. Quake, *Science*, 2002, **298**, 580.

90. M. A. Unger, H. P. Chou, T. Thorsen, A. Scherer and S. R. Quake, *Science*, 2000, **288**, 113.

91. Y. Marcy, T. Ishoey, R. S. Lasken, T. B. Stockwell, B. P. Walenz, A. L. Halpern, K. Y. Beeson, S. M. D. Goldberg and S. R. Quake, *PLoS Genet.*, 2007, **3**, e155.

92. Y. Yang, J. F. Swennenhuis, H. S. Rho, S. Le Gac and L. W. M. M. Terstappen, *PLoS One*, 2014, **9**, e107958.

CHAPTER 10

Microfluidic Impedance Cytometry for Blood Cell Analysis

HYWEL MORGAN* AND DANIEL SPENCER

Faculty of Physical and Applied Sciences, and Institute for Life Sciences, University of Southampton, UK
*Email: hm@ecs.soton.ac.uk

10.1 Introduction

Counting and discriminating small particles such as cells, viruses, and bacteria is an important and widely used measurement tool in medical science. In particular, the enumeration of different blood cells is widely used to aid the diagnosis of a wide variety of medical conditions including anemia, pneumonia, and thrombocytopenia, or to indicate underlying health problems such as leukemia. This so-called Full Blood Count (FBC) or Complete Blood Count (CBC) is currently performed using large, expensive but high-throughput equipment which requires the patient's sample to be sent to a hospital or centralized laboratory for analysis. A decentralized, low-cost Point of Care (PoC) FBC would clearly have many benefits, including treating patients suffering from chronic disease at home, aiding rapid diagnosis in emergencies and for companion diagnostics. A relatively cheap, robust and portable PoC device would also benefit developing countries where the costs of traditional tests are prohibitively high or the facilities to run these tests do not exist.

RSC Nanoscience & Nanotechnology No. 36
Microfluidics for Medical Applications
Edited by Albert van den Berg and Loes Segerink
Published by the Royal Society of Chemistry, www.rsc.org

The FBC is performed using hematology analyzers, which are automated flow cytometers that perform sample processing, and cell identification and counting. Flow cytometry is a technique that counts individual small particles, for example cells suspended in an electrolyte. Particles pass through a detection region where their optical and/or electrical properties are measured. The first flow cytometer was the Coulter counter,[1] a technology that went on to completely revolutionize the field of hematology. Prior to the invention of the Coulter counter all blood counting was performed manually, but the Coulter counter enabled automated high-throughput cell counting for the first time. The working principle of the Coulter counter is shown schematically in Figure 10.1. It consists of two liquid reservoirs separated by a small aperture approximately 0.1 mm diameter. Particles, suspended in an electrolyte, pass through the aperture from one reservoir to the other. A DC or low-frequency AC voltage is applied between the measurement electrodes in each reservoir and the current flowing between the electrodes is monitored. Cells pass through the aperture one-by-one and momentarily change the current flow. In a DC field, a cell behaves as an insulating particle, therefore, as each cell moves through the aperture the current is reduced. The size of the current change is proportional to the volume of the cell.[2,3] Thus single particles can be both counted and sized.

The Coulter counter performs high-speed cell counting, but a full blood count also requires some means of accurately metering sample volume. The Coulter counter therefore includes metering electrodes to start and stop

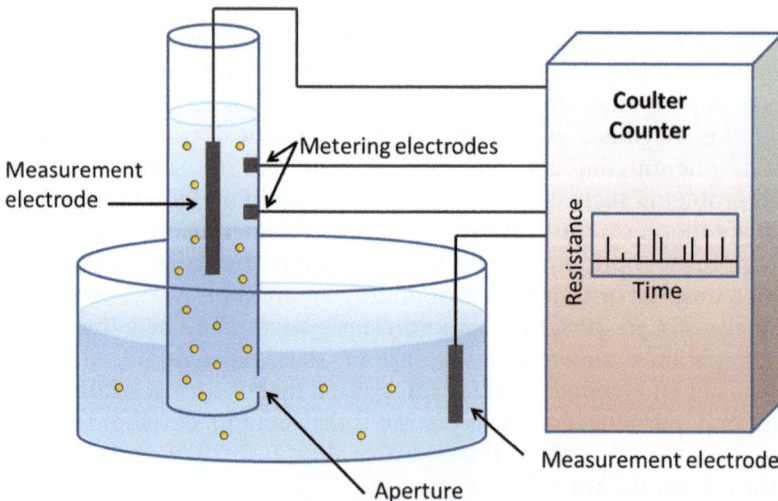

Figure 10.1 Overview of a Coulter counter. A small aperture separates two reservoirs with the sample loaded into one reservoir. Electrodes measure the current flow through the aperture, which is reduced as a cell passes through. Metering electrodes start and stop the measurement when a fixed volume of sample has been dispensed, so that the absolute cell particle concentration can be determined.

the counting when a fixed volume of liquid has been measured. The system therefore determines the number of cells per unit volume of blood. Modern day hematology analyzers also include a number of proprietary chemistry steps as part of the process, used for various sample pretreatment steps such as dilution, red cell lysis or selective labeling of white cell sub-populations.

Cells can also be distinguished using optical methods, usually with a combination of scattered light and fluorescence, and the first fluorescent-based flow cytometers were developed in the late 1960s.[4] The suspension of cells is sheathed in a liquid (usually saline) and passed through a nozzle to reduce the sample stream width such that only one particle passes through the interrogating laser beam at any one time as shown in Figure 10.2. This sheath flow is required because the intensity profile of a laser beam is Gaussian and thus particles must all transit the same point within the beam to keep the interrogation power constant. Commercial flow cytometers can measure up to 100 000 cells per second and simultaneously detect scattered light (to measure cell size and granularity) and multiple fluorescent markers – for a full review see ref. 4. Although extremely powerful, cytometers have traditionally been quite large and complex due to the optical components such as lenses and filters required for the sheath flow system, but lower cost, smaller devices are now available, such as the BD Accuri C6, Millipore Guava EasyCyte, and Life Technologies Attune.

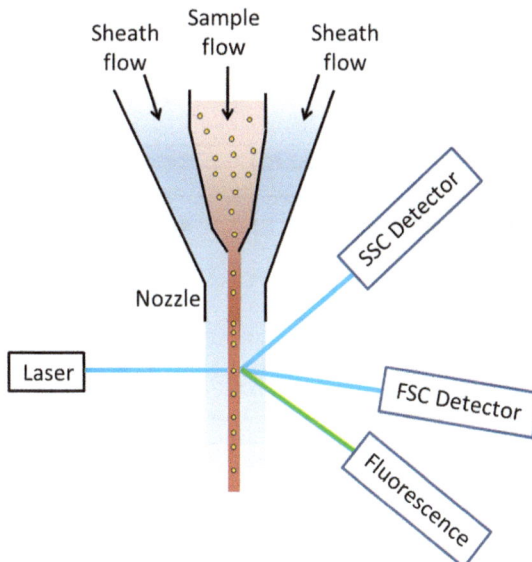

Figure 10.2 Illustration of an optical flow cytometer. A concentric sheath flow is used to surround the sample (hydrodynamic focusing) before it passes through a nozzle, which generates a tight sample stream that flows through the center of a laser beam. Scattered light (SSC, Side scattered, and FSC, Forward scattered) and fluorescence is detected using photo-multiplier tubes.

Microfluidics is the obvious way of reducing both the cost and size of these devices and chip-based optical flow cytometers are now being developed for portable operation.[5,6] These devices vary in complexity, and often include integrated fiber-optics, micro-lenses and waveguides. However, the lasers, optical filters and detectors are bulky and remain to be integrated in a cost-effective manner.

Over the last ten years, all-electrical impedance-based systems have also been developed. An all-electrical analytical system has many advantages over an optical system, principally from the perspective of the ease of mass manufacture, low cost of electronics and the robustness of such a device. The Microfluidic Impedance Cytometer (MIC) is a type of Coulter counter without an aperture. Instead miniature electrodes are fabricated within a microfluidic channel, as shown in Figure 10.3. The channel is usually a few hundred microns long and takes the place of the orifice in the Coulter counter. The dimensions of the channel and electrodes are similar to the diameters of the particles to be measured. For instance, to measure blood cells (2 to 15 μm in diameter), the height and width of the channel is around 30 to 40 μm. The limit of detection is approximately proportional to channel dimensions, thus if the channel is too large very small particles cannot be measured (just like the Coulter counter). If the channel is too small, larger particles will block the channel. Advanced signal processing can be used to enable detection of the very smallest particles such as platelets in large channels that do not become blocked with large cells.

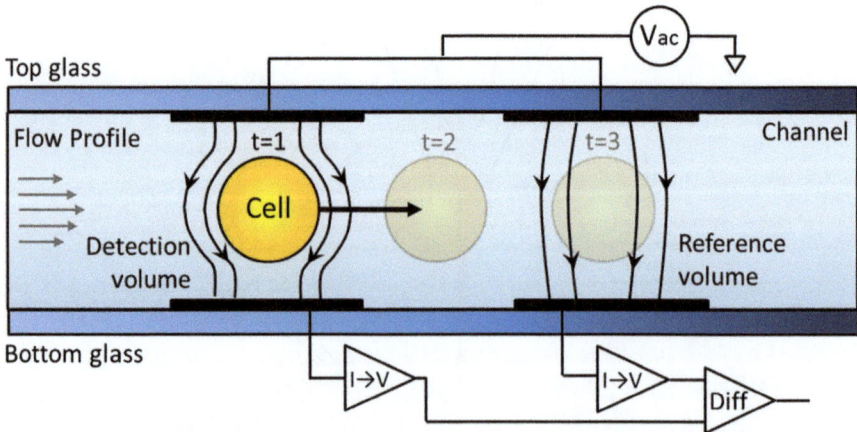

Figure 10.3 Diagram of a miniature microfluidic impedance cytometer. Cells are suspended in an electrolyte and flow through a miniature channel (typical cross-sectional dimensions of 40×40 μm). The impedance is measured as cells between micro-electrodes fabricated within the channel.

Adapted with permission from Gawad *et al.*, *Rev. Sci. Instrum.*, 2007, **78**, 054301. Copyright © AIP Publishing LLC.

The Coulter counter uses very large electrodes in order to produce a high current density in the channel and provide the best SNR. However, this limits the frequency bandwidth of the system. MIC uses miniature electrodes fabricated directly in the channel, which means that a very wide range of AC frequencies can be used, up to 500 MHz, providing much more information on cell properties, including the electrical properties of the membrane and cytoplasm. Combining two or more simultaneous interrogation frequencies provides information on both cell size and membrane capacitance. This enables the discrimination of many more types of similarly sized cells than is possible with the Coulter counter.

Impedance cytometry is inherently smaller and much simpler than conventional optical detection and therefore offers a novel solution to low-cost cell analysis and discrimination technology for PoC diagnostics. In addition the miniature size of a MIC means that it can be easily integrated with microfluidic sample preparation systems. Miniaturized microfluidic technologies also require less blood (*e.g.* a fingerprick) and are therefore suitable for elderly patients and infants.

10.2   The Full Blood Count

A typical adult has approximately 6 liters of blood, and of this volume about 55% is plasma (consisting of about 90% water), with the remaining 45% blood cells. These are, in order of abundance, red blood cells whose role is to transport oxygen around the body, platelets which cause blood to clot, and white blood cells which form the immune system. The FBC measures the numbers of these cells and the test is widely used to diagnose many forms of disease. Tables 10.1 and 10.2 list the parameters measured and typical ranges of the FBC test.

10.2.1   Clinical Diagnosis and the Full Blood Count

The demand for PoC testing in hematology has increased significantly over recent years. For example, the International Council for Standardization in Haematology (ICSH) guidelines[7] have identified a clear need for a PoC FBC

Table 10.1   Normal WBC and platelet ranges for the FBC. Adapted with permission from Hoffbrand and Moss, *Essential Haematology*, 6th edn. Copyright © Wiley-Blackwell.

| White blood cells | Normal range | | Platelets | Normal range |
	Count per nL	% of total		
Total	4–11	100	Mean volume	7.5–11.5 fL
Basophils	0.01–0.1	0–1	Count per nL	150–400
Eosinophils	0.04–0.44	1–6		
Lymphocytes	1.5–3.5	20–45		
Monocytes	0.2–0.8	2–10		
Neutrophils	2.5–7.5	40–75		

Table 10.2 Normal RBC ranges for the FBC. Adapted with permission from Hoffbrand and Moss, *Essential Haematology*, 6th edn. Copyright © Wiley-Blackwell.

Red Blood Cells	Normal Range	
	Male	Female
Count per pL	4.5–6.5	3.9–5.6
Packed cell volume in L/L	0.40–0.52	0.36–0.48
Volume in fL	80–95	80–95
Hemoglobin in g dL^{-1}	13.5–17.5	11.5–15.5
RBC size distribution (CV in %)	11–15	11–15
Mean cell hemoglobin in pg	27–34	27–34
Mean cell hemoglobin concentration in g dL^{-1}	20–35	20–35

in many areas including GP surgeries, pharmacies, and residential settings. Such a device needs to be portable, the test must be minimally invasive, and the system simple to use. For self-testing devices, capillary blood can be collected from a finger puncture, such as with glucose tests or anti-coagulation therapy.

10.2.1.1   Red Blood Cells and Hemoglobin

Red blood cells (RBCs) are filled with hemoglobin, which binds to oxygen and transports it around the body. An FBC measures the number of RBCs in blood and the hemoglobin concentration. The red cell count is multiplied by the mean red cell volume to calculate the packed cell volume (hematocrit).

10.2.1.2   Platelets

Platelets are integral to the control of hemostasis and their main function is to stop bleeding from a wound. A low count, called thrombocytopenia, indicates that a patient could be susceptible to bleeding, which could for example be caused by an autoimmune disorder. A very low count (less than 20 per nanoliter) means the patient is at risk of spontaneous bleeding. High counts mean the blood is more likely to clot, however the activation level of platelets also determines how easily platelets form clots. The mean platelet volume increases when platelets are produced by the body at a higher rate and can be used to infer problems with platelet production.

10.2.1.3   White Blood Cells

The three main types of WBC (in order of decreasing abundance) are as follows:

Neutrophils account for up to 75% of the total WBC count and are responsible for ingesting bacteria, fungi, and damaged cells. An increased count is called neutrophilia and can be an indication of a bacterial infection. Other causes of a high neutrophil count are stress (for instance trauma or burns) or inflammation (such as after a heart attack). A reduction in the neutrophil count means the body is less able to defend itself and is common in patients undergoing chemotherapy. It could also indicate a viral infection, severe bacterial infection (such as from typhoid), or bone marrow failure.

Lymphocytes are the second most numerous WBC population and are responsible for immunity and antibody protection. A high count is an indication of viral or chronic infection such as tuberculosis or hepatitis. The cells are further divided into three types: T-lymphocytes, B-lymphocytes, and natural killer (NK) cells depending on the surface marker expression (for example T-cells express CD3 whereas B-lymphocytes express CD19 and NK cells express CD16). A low helper T-lymphocyte (CD4) count is commonly associated with HIV infection.

Monocytes are a sub-population of WBC which differentiate into macrophages to ingest pathogens and dead cells. An increased count can be seen post-chemotherapy, but can also be an indication of a chronic infection such as malaria or tuberculosis.

In addition there are many rarer cells including Eosinophils that defend against infection from protozoa (parasites) and form part of the allergic response; and Basophils, one of the rarest where increased numbers are present in some inflammatory disorders such as rheumatoid arthritis.

10.2.2 Commercial FBC Devices

The FBC is the most widely performed test in hospitals, normally undertaken in centralized laboratories using high-throughput automated hematology analyzers. Flow cytometers are complex and require trained laboratory technicians. They also require large volumes of expensive reagents and since they are a centralized resource they cannot be used to provide rapid continuous small sample testing that may be required to monitor certain diseases or trauma patients. Heikali and Di Carlo[8] recently reviewed the field of portable hematology analyzers, clearly outlining the need for microfluidic solutions. They argue that developments in microfluidic cytometry will ultimately cause hematology to change from a niche laboratory science to become a low-cost tool with widespread use in medical diagnostics. One example of a hematology analyzer is the Sysmex XE-2100, which can measure 150 samples per hour, costs around $100 000, requires nine different reagents, and weighs more than 100 kg. Some of these reagents are used to dilute the RBCs for accurate counting, others to lyse the red cells for accurate WBC counting, and yet others to stain the different white cell populations.

Several portable blood counting devices are being developed, however none can perform the full range of FBC measurements in a single platform. A recent device made by HemoCue was launched in 2012. It measures the five WBC sub-populations using a camera to count stained white cells after chemically lysing the RBCs. Discrimination between the WBC sub-populations is performed using image processing techniques that examine the shape of each cell. This device does not measure RBCs, platelets, or hemoglobin and thus has limited diagnostic use. Another device is the Chempaq XBC.[9] It measures RBCs, hemoglobin, and the three most numerous WBC sub-populations. The device is a miniature Coulter counter, with electrodes and a laser-drilled aperture built into a single-use plastic cartridge. Hemoglobin is measured using a miniature photometric method. The device only performs one measurement at a time (WBCs, RBCs, or hemoglobin), each requiring a separate cartridge. In addition the device does not measure platelets.

10.3   Microfluidic Impedance Cytometry (MIC)

The first high-throughput microfluidic single cell impedance technology was described by Gawad.[10] This paper demonstrated clear differentiation of beads, erythrocytes, and ghost cells. The paper demonstrated quantitative analysis of particles, and the system discriminated between two cell populations. Interestingly the idea of dual frequency impedance measurements goes back to the time of Coulter,[11] and in the 1970s Hoffman and Britt[12,13] developed a miniature system for simultaneous low- and high-frequency impedance analysis of single cells using a flow cell with small integrated electrodes.

The device described by Gawad *et al.*[10] consisted of a glass chip with a pair of electrodes fabricated on the bottom of the channel only. This design is easy to fabricate and was adopted for single particle/cell impedance sensing and counting by many authors.[14–19] Unfortunately because of the geometry of the electric field, this design suffers from a high variation in the impedance signal with particle position in the channel. Improved designs adopted parallel facing electrode pairs as shown in Figure 10.3, which significantly improve on the performance,[20] but are more demanding to fabricate (for further details see Section 10.3.6).

Measurement of the frequency-dependent dielectric properties of single cells can be used to determine the heterogeneity of a population. This measurement can be performed by sweeping a stimulation signal over a range of frequencies while cells flow through the device.[10,21–23] Simultaneous multi-frequency techniques such as Maximum Length Sequences (MLS) can also be used[24] or electronic hardware that generates multiple discrete frequencies.[25]

The first demonstration of WBC differentiation in human blood using MIC was by Holmes *et al.*[23] The system was used to discriminate the three main types of leukocyte using a few microliters of blood. RBCs were lysed using a solution of saponin and formic acid, followed by addition of a quench solution to restore solution osmolarity and pH. Lymphocytes were distinguished from granulocytes using a low-frequency size measurement, as in the Coulter counter. A high-frequency signal measures the membrane capacitance of the cells and enables differentiation of monocytes from the similarly sized granulocytes.

10.3.1 Measurement Principle

Single cell impedance cytometry is performed using a microfluidic chip as shown in Figure 10.4. A glass chip contains a microfluidic channel, with two pairs of electrodes fabricated within the channel. An AC voltage at one or more discrete frequencies is applied to the top pair of electrodes, creating an electric field within the channel (Figure 10.4a) and the bottom electrodes are connected to amplifiers that measure the current. The two electrode pairs form a differential measurement. As the cell passes between the first pair it changes the electrical current. This is measured by the amplifier with respect to the current passing through the second electrode (through the pure medium), which acts as a reference; likewise, when the cell transits the second pair of electrodes. The electrical current is converted into voltage signals and a differential amplifier and lock-in amplifiers are used to demodulate the in-phase and out-phase impedance signals at the stimulating frequency, whilst rejecting noise at other frequencies (Figure 10.4a). The impedance signal is read out as a pair of peaks, one for the in-phase (Real) and one for the out of phase signal (Imaginary) at each frequency (see Figure 10.4b). Typically the data is plotted as the magnitude impedance phasor (as shown in Figure 10.4b). At low frequencies (below 1 MHz) the magnitude of the impedance is dominated by the resistance change in the system and is an accurate measurement of particle volume (for an insulating particle). The impedance is measured as a voltage, but can be scaled to an absolute volume measurement. This differential impedance sensing scheme provides several advantages: (i) the properties of the cell are measured directly against the suspending medium; (ii) any uneven drift in the properties of the electrodes is canceled; (iii) the velocity of the flowing cells is determined from the transit time between the peaks (Figure 10.4b).

To measure the properties of cells, they are first suspended in an electrolyte (*e.g.* phosphate buffered saline, PBS) at an appropriate concentration. The sample is loaded into a syringe and pushed through the microfluidic channel. Figure 10.4b illustrates the change in the impedance signal as a particle travels between the electrodes. As the particle passes between the first pair, a peak in impedance signal is measured. The maximum of this peak corresponds to when the particle is at the mid-point of the first electrode pair (point t1). As the particle passes between the second electrode pair, a similar change in the impedance signal is observed. The time taken for the particle to pass between the electrodes from t1 to t3 is the transit time (dt).

Just as in traditional flow cytometers, if the particle concentration is too high, then there is a high probability of measuring doublets, conversely if the sample concentration is too low, then the throughput will be very low and analysis will take a long time. For example, with typical channel cross-section of 40 μm×40 μm, and electrodes of 30 μm width (40 μm spacing) a typical volumetric throughput would be up to 100 μL min^{-1}. At a particle number density of 600 per μL the system can measure up to 1000 particles per second with a coincidence level of 5% (calculated using Poisson statistics).

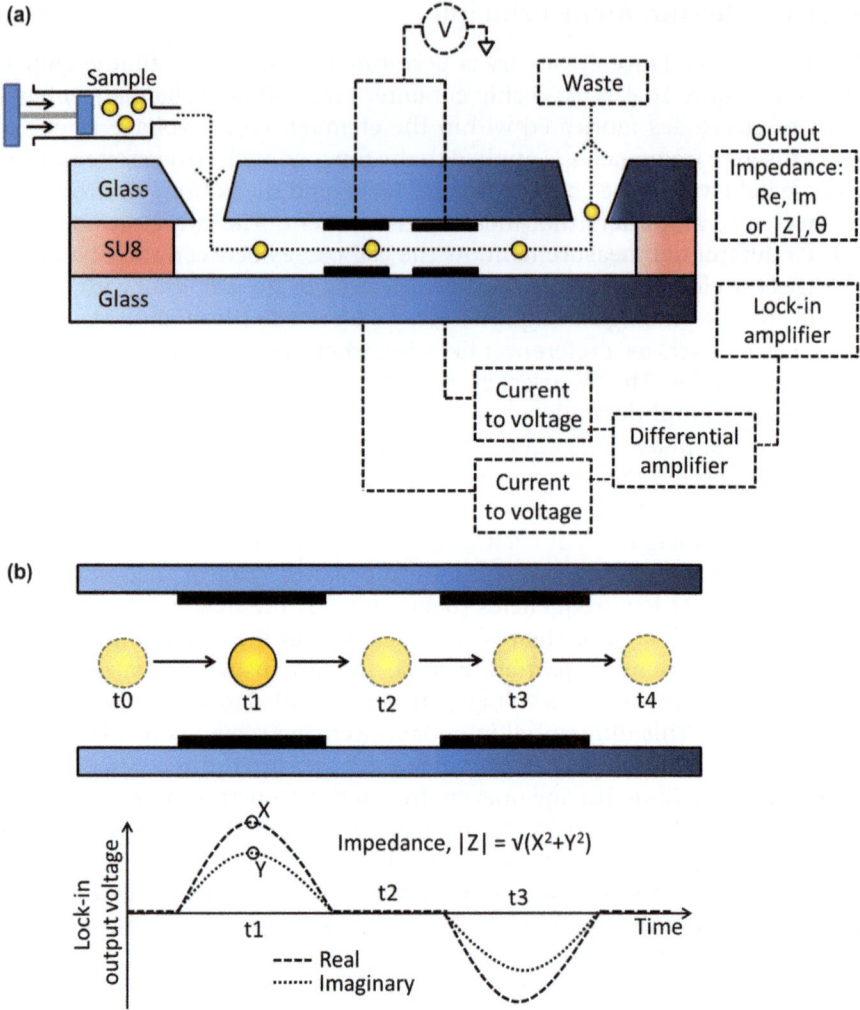

Figure 10.4 Illustration of the measurement of a sample of cells using MIC. (a) Cells are suspended in an electrolyte at physiological conductivity, *e.g.* PBS. They are pumped through the chip and the electronic signals from the electrode measured using dedicated electronic hardware. (b) Schematic representation of the signals from the system. A pair of Gaussian shaped anti-symmetric peaks are produced as the cells pass each pair of electrodes. There are two signals for each applied frequency, one for the in phase and one for the out of phase impedance. These signals are combined to give the impedance magnitude and the phase.

10.3.2   Behavior of Cells in AC fields

The frequency-dependent dielectric property of cells is well known and has been the subject of numerous studies for nearly a century. Excellent reviews can be found in Schwan,[26] Asami,[27,28] and Pethig and Kell.[29] From an electrical point of view, a cell can be considered to be a particle with a conducting core surrounded by a very thin insulating lipid membrane, suspended in a conducting medium or electrolyte as shown in Figure 10.5a.

The impedance measurement system can be approximated to a volume with a pair of electrodes as shown in Figure 10.5b. Without a cell the electric field lines are parallel and the electrical behavior is dominated by the electrical properties of the electrolyte. When a cell enters the box (Figure 10.5c) the electrical properties of the system change. At low AC frequencies the cell approximates to an insulating particle and the lines of electrical current pass around the cell. When the cell enters the measurement chamber an equivalent volume of conducting electrolyte is displaced. This increases the overall resistance of the system, and it is this change that forms the basis of the Coulter (electrical) volume measurement. Provided the membrane cell is not compromised in any way, and that the cell is viable, the electrical volume (low-frequency impedance) is an accurate measurement of the actual volume of the cell.

The very thin lipid membrane gives the cell a high capacitance, of the order of 1 μF cm^{-2}. As the frequency of the electric field increases the properties of the system change; the membrane becomes transparent to the electric field and at very high frequencies the behavior is ultimately now dominated by the cytoplasm (Figure 10.5d). The transition from these two regimes occurs in the MHz frequency region, and measurements in this frequency range enable the capacitance of the membrane to be measured. The transition between these two regimes is gradual and occurs over at least two decades in frequency; Figure 10.5e illustrates the overall change in complex impedance of the system as a function of frequency (for a cell suspended in physiological medium). As shown in the figure, the impedance is invariant with frequency up to around 500 kHz (ignoring electrode polarization effects) and here the cell volume can be measured. Cell membrane capacitance is measured around 1 to 5 MHz and at frequencies greater than 10 MHz the resistance is governed by the cell cytoplasm.

It is well known that different cells have different membrane capacitances depending on cell size and membrane morphology. For example a red cell ghost (spherical RBC) has few membrane proteins, is a perfect sphere, and has the minimum surface area, and therefore the lowest value of capacitance, at around 9 mF m^{-2}. As the membrane surface roughness increases, the membrane capacitance also increases. Features such as membrane folds, ruffles, and microvilli cause the area to increase by an effective folding factor $\varphi > 1$.[30] White blood cells have invaginated membranes and the membranes also contain protein molecules, which increases the capacitance. For example lymphocytes have a membrane capacitance in the range 10–12 mF m^{-2} and monocytes have larger values (around 15 mF m^{-2}).

Gascoyne *et al.*[31] comprehensively characterized the membrane properties of a wide number of cell types (NCI-60 panel) in terms of cell shape, dendritic projections, and cell membrane roughening. They showed a strong correlation between membrane morphology and membrane capacitance

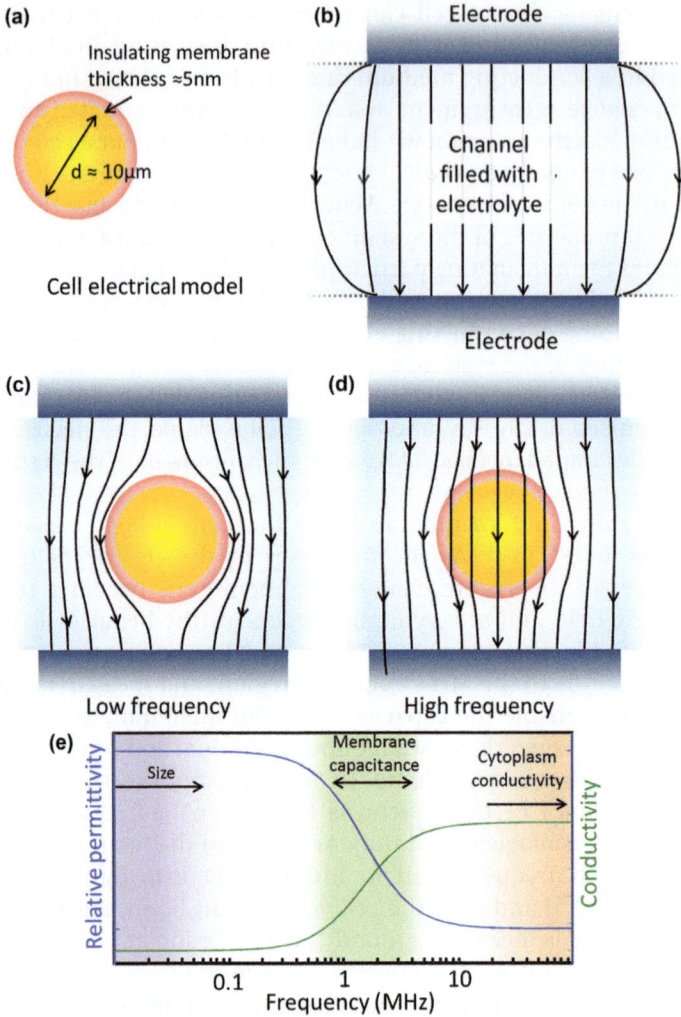

Figure 10.5 (a) Simplified representation of a cell showing the insulating membrane and conducting cytoplasm. The membrane is very thin, which gives the cell a high membrane capacitance. (b) The measurement chamber created by the micro-electrodes within the channel. Within the central region the field lines are parallel, but they diverge at the electrode edges. (c) A cell exposed to a low-frequency electric field. Viable cells approximate to insulating particles in this frequency regime and the field lines pass around the cell. The impedance in this case is proportional to cell volume. (d) The same cell at high frequencies (>10 MHz). At this frequency the membrane capacitance (reactance) is short circuited and the electric field sees the cell interior. At these frequencies the impedance of the system is governed by the combination of the cell cytoplasm and suspending electrolyte. (e) Plot of the change in the permittivity and conductivity (proportional to the real and imaginary signals) for a model cell suspended in PBS. Note the transition from low- to high-frequency behavior around 1 MHz – this is the frequency region where the cell membrane dominates the impedance.

(measured using AC electrokinetic techniques). Tumor cells are particularly large and have highly invaginated cell membranes with large folding factors ($\varphi > 3$), so that their capacitance is often much greater than white blood cells, with values up to 30 mF m^{-2}. Interestingly, it has been shown that the cell cytoplasmic membrane morphology exhibits fractal behavior, and can be used as an alternative description of membrane complexity compared with the folding factor.[32]

10.3.3 Sizing Particles

As described above, a single frequency measurement at for example 500 kHz can be used to measure the electric volume of a particle. Figure 10.6a shows data for a mixture of four different sizes of polystyrene beads that are approximately the same size as blood cells. The histogram shows experimental data together with a Gaussian fit to the populations (approximately 10 000 events). The plot in Figure 10.6b shows the linear response of the cube root of the impedance signal with particle diameter (*i.e.* impedance is proportional to particle volume). It also demonstrated the extremely high accuracy of the system. The populations are very tightly defined and the standard deviations (3σ) are much better than the manufacturer's quoted data.

Cells have a much wider distribution in size or volume and Figure 10.7 shows an impedance scatter plot for a mixture of red blood cells and platelets. This figure shows the value of the phase plotted against impedance magnitude for whole blood diluted in saline. Each point on this scatter plot is the measured impedance for one single cell. Figure 10.7b shows a histogram for these data, the difference in peak height demonstrating the different ratios in the two cell types found in human blood (around 20-fold).

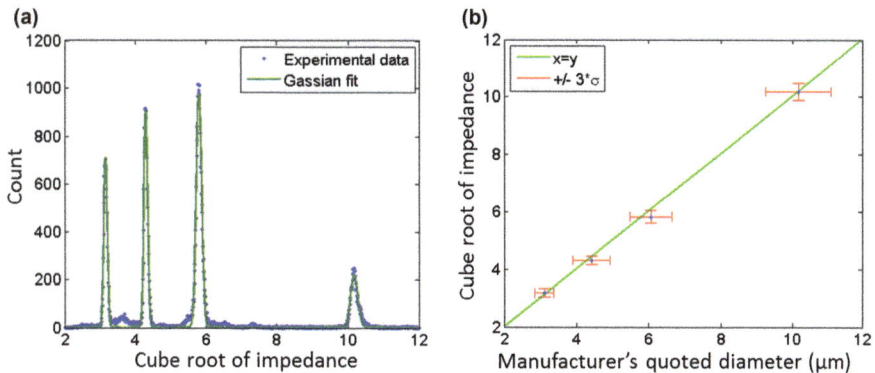

Figure 10.6 (unpublished). (a) Histogram of particle impedance for a set of four different calibration particles (note that the impedance axis is scaled with a single constant that depends on the gain in the electronic circuitry). The solid line is a Gaussian fit to the data. (b) Comparison between impedance data and the specification from the manufacturer (3 × S.D. plotted). Note the linear relationship between impedance and particle volume.

(a)

(b)

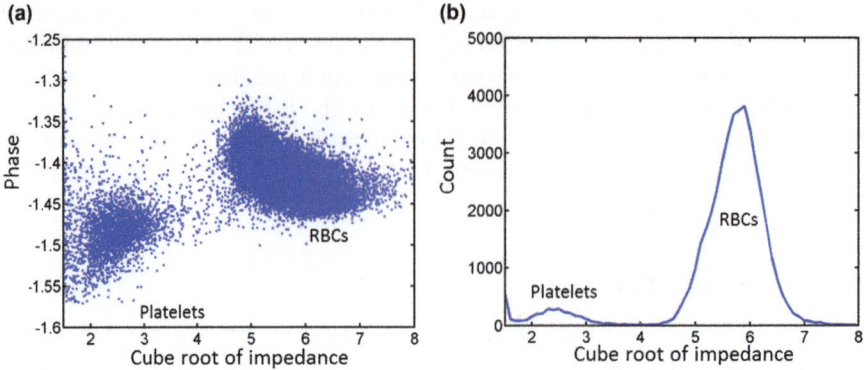

Figure 10.7   (unpublished). (a) Scatterplot and (b) histogram for human whole blood diluted in PBS $(10\,000\times)$ demonstrating discrimination based on volume differences between platelets (9 fL) and RBCs (90 fL). Impedance measured at 500 kHz.

10.3.4   Cell Membrane Capacitance Measurements

As demonstrated by Figure 10.7, discrimination of certain cell types can be performed using just electrical volume measurements. For example in a WBC count, the lymphocytes are smaller than neutrophils and therefore these two cell types can be discriminated with a single low-frequency measurement. However, an FBC also requires discrimination of monocytes from the similarly sized neutrophils. The membrane capacitance of the monocytes is higher than the neutrophils, thus these two cell types can be discriminated using a dual frequency measurement. A second frequency is used to measure the membrane capacitance, typically between 2 and 5 MHz where the impedance is dominated by the cell membrane capacitance. Characterization of the three different sub-populations of white cells is therefore performed using two discrete frequencies. Figure 10.8 shows a scatter plot for a WBC population from human blood. The *x*-axis is the cube root of the impedance at 500 kHz, the electrical radius. The *y*-axis is the ratio of the high-frequency (2 MHz) to low-frequency impedance magnitude. This metric normalizes the cell capacitance measurement to cell volume and is termed "opacity", first introduced by Coulter and Hogg.[11] A reduction in opacity correlates with an increase in the specific cell membrane capacitance, hence the capacitance of the monocytes is larger than the neutrophils.

The scatter data clearly show that low frequency impedance differentiates between the neutrophils and lymphocytes $(7-8\ \mu m$ diameter) on electrical volume but that the neutrophils and monocytes $(10-12\ \mu m)$ would be indistinguishable without the second frequency. These data are from a whole blood sample where the RBCs were first chemically lysed (see below), and the debris from this process can be seen in the upper top left of the scatter plot.

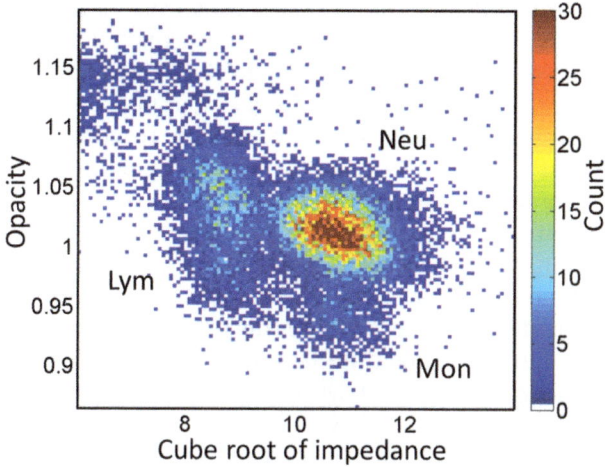

Figure 10.8 Scatter plots for WBC prepared form human blood by lysing the RBCs. The opacity (ratio of the impedance at 2 MHz to that at 500 kHz) enables discrimination of neutrophils from monocytes.

10.3.5 Microfluidic FBC Chip

In analyzing small volumes of blood, it is important to consider how the sample is processed prior to measurement. For example, a 2 μL drop of blood (smaller than a fingerprick) typically contains 10 million erythrocytes, between 300 000 and 800 000 platelets, and approximately 14 000 leukocytes, of which only approximately 700 are monocytes. In order to perform the FBC and count these different cells, whole blood must be diluted in a physiological buffer such as PBS to an appropriate concentration so that the probability of coincidence is small. Then a simple size-based measurement can distinguish the smaller platelets (9 fL) from the larger RBCs (90 fL).

The ratio of red cells to white cells is approximately 1000 to 1, therefore to enable measurement of the leukocytes in a practical time scale, the red cells must be removed so that they are not counted. If the RBCs were removed from a 4 μL blood sample, a measurement of the remaining WBCs (28 000 cells) would take about a minute at a throughput of 500 cells per second. This volume of blood contains enough WBCs to discriminate and count the three major sub-populations (neutrophil, monocyte, and lymphocyte). If the RBCs remained, the measurement would take around 17 hours at the same throughput. Therefore, microfluidic systems for sample pre-processing (lysis and dilution) are a necessary part of an integrated FBC system. To cover this very large dynamic range, two different sample preparation methods are required, one to perform dilution for RBC measurements (1 : 10 000) and another to remove the RBCs using lysis chemistry for WBC analysis.

Lysis of RBCs can be performed in a number of different ways, for example using ammonium chloride or other proprietary lysis solutions. High-quality, noise-free impedance analysis requires a lysis chemistry that minimizes the amount of debris and prevents the lysed cells from re-forming to produce ghosts. The lysis chemistry first described by Ledis[33] provides such a solution. It consists of detergent and mild acid (0.05% saponin + 0.12% formic acid) to which cells are exposed for a fixed amount of time (usually 6 s). The red cell membranes are quickly destroyed and the reaction is then quenched by the addition of base (Na_2CO_3), which recovers the original pH and osmolality.[34] Figure 10.9 illustrates this process.

The data shown in Figures 10.7 and 10.8 were obtained using blood processed according to this protocol. Such manual processing is clearly not suitable for a point of care system. A continuous flow microfluidic lysis system has been developed to replicate this process.[34,35] In microfluidic flow (low Reynolds numbers), flows are laminar and diffusive processes dominate. In this design, blood and lysis reagents flow side by side through microfluidic channels where they mix by diffusion for a specified time, which is determined by the flow velocity and the channel lengths. Figure 10.10 shows an overview of this system.

Blood and the lysis reagents are pumped through the device at a constant rate. The lysis solution diffuses into the blood stream, lysing the RBCs. After a fixed distance (time) the quench solution is introduced thus halting the

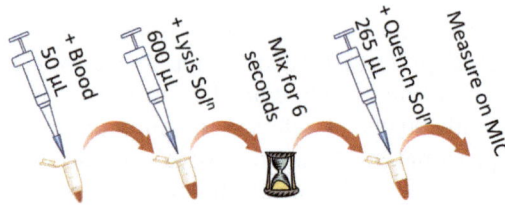

Figure 10.9   Standard RBC lysis protocol. Blood is mixed with a lysis solution for 6 seconds before a quench solution is added to restore pH and osmolality.

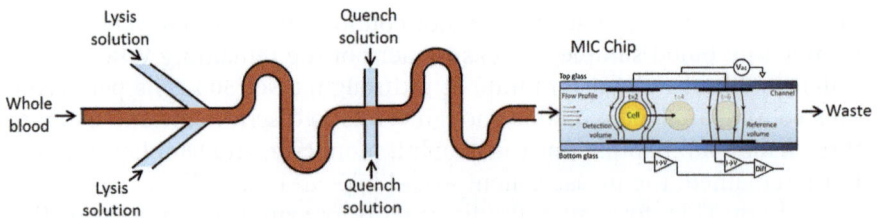

Figure 10.10   Illustration of a microfluidic lysis system. Whole blood flows along a channel where it meets a second channel containing lysis solution. Both liquids flow co-linearly and mix by diffusion. After the liquids have mixed for a predefined time (determined by the flow velocity and channel length), the mixture meets a channel containing quench solution. Mixing is performed by diffusion until the mixture flows through the MIC measurement chip and the individual cells are measured.

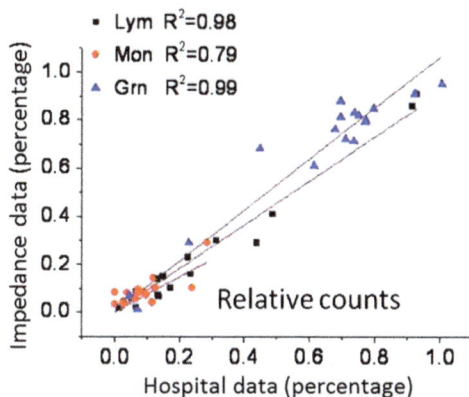

Figure 10.11 Relative counts of WBCs measured on MIC and using a centralized hospital hematology service. The samples are from clinical blood, hence the wide distribution in cell numbers.
Reprinted with permission from Han *et al.*, *Anal. Chem,*, 2012, **84**(2), 1070–1075. Copyright © 2012 American Chemical Society.

lysis process. This principle of RBC lysis using microfluidic chips is similar to that described by Sethu *et al.*[36] In cases where the desired reaction times are shorter than diffusion times, additional features can be implemented into the microfluidic channels to speed up mixing,[37] thus reducing the reaction time.

Figure 10.11 shows a comparison of the FBC data obtained using this microfluidic lysis protocol against a commercial hematology analyzer showing good correlation between the MIC results and a commercial hematology analyzer for the relative ratios of the WBC sub-populations. These data were obtained from clinical samples, hence the very low monocyte counts.[34]

10.3.6 Accuracy and Resolution

Within the channel shown in Figure 10.3, the electric field is non-homogeneous. This means that particles traveling off center see a different electric field from those traveling down the middle of the channel. This in turn influences the measured impedance signal amplitude. One way to resolve this issue is to implement sheath flow similar to that used in classical flow cytometry (Figure 10.2). However, implementing a miniature sheath flow arrangement in a microchip is difficult and does not lead to a small, compact and simple system.

Accurate mapping of the electric field distribution within the microchannel is required to model the impedance spectrum of a single cell. Gawad *et al.*[21] performed 3D finite element modeling of a pair of parallel facing electrodes to calculate the electric field. Linderholm *et al.*[38,39] and Sun *et al.*[20] used conformal mapping to analytically solve the electric field distribution for both a coplanar (not shown) and parallel facing electrode

(Figure 10.3) design. It was demonstrated that control of particle position is more significant for a coplanar design than for a parallel electrode configuration because the electric field distribution in the latter design is least divergent.[20] A full numerical analysis of the variation of impedance signal with particle position was described in Spencer and Morgan.[40] The map of the relationship between impedance and particle position showed that the variation in signal was greatest along the vertical axis between the two electrodes, rather than across the channel (side to side). Given this knowledge, complex signal processing techniques can be used to remove this positional variation leading to a simple sheath-less high-accuracy system.

The Limit of Detection (LoD) is governed by a number of factors, including the volume fraction of the particle in the channel (ratio of particle volume to the volume occupied by the electric field), the current that flows between the electrodes and the noise in the electronic amplifiers. Reducing the channel size will increase the volume fraction leading to a larger signal. However, this is at the expense of greater hydrodynamic back pressure and a significant risk of channel blockage when using real samples such as blood. Increasing the applied voltage to increase the signal is also possible but this generates electrical noise due to the production of charged species at the electrodes. For a 30 μm (height)×40 μm (width) channel the limit of detection is approximately 1 μm diameter particles as shown in the histogram of Figure 10.12.

One way of increasing the sensitivity whilst maintaining relatively large channel dimensions is to focus particles into the sensing region with the use of a sheath flow that is made of a less conducting liquid such as pure water or oil. This principle is used in Coulter counters; the sheath flow serves to concentrate the electric field lines into the sensing volume without restricting the geometry of the device. When sample focusing is achieved both vertically and horizontally along the channel axes, the size of the detection area can be precisely adjusted to the size of the particles, thus greatly increasing the sensitivity of the system. Bernabini *et al.*[41] described a micro-impedance device with a wide channel where the sample is confined on either side with an insulating oil. This sheath fluid hydrodynamically focuses the sample into the center of the chip and also confines the electric field to the conducting central stream.

Figure 10.13 illustrates how the sample stream is constrained laterally by the oil streams coming from the side inlets. The suspended particles are aligned into a single file and cross the interrogation region (two pairs of micro-electrodes located downstream) one at a time. Using this technique, the sample stream can be confined to a width of ~10 μm in a 200 μm wide channel producing stable flow and noise-free signals. Discrimination between bacteria and beads of similar sizes is shown in Figure 10.14.[41]

A similar device was recently used as the basis of a system to identify changes in platelets after activation. Platelet function testing is an important physiological test, normally performed by determining the aggregation of

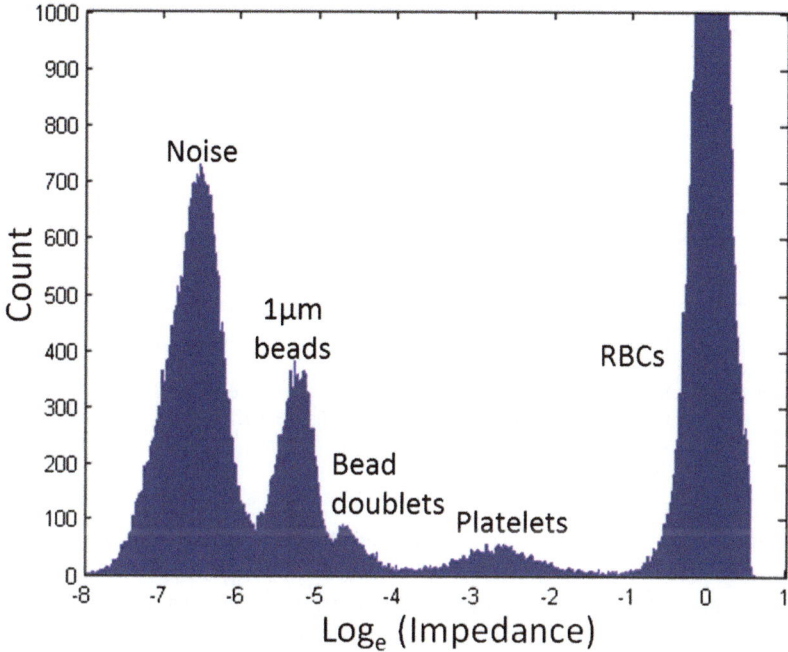

Figure 10.12 (unpublished) Histogram of RBCs, platelets and 1 μm diameter polystyrene beads demonstrating the Limit of Detection (LoD) for a typical system with a 30 μm high×40 μm wide channel. The noise is primarily electrical, and determines the LoD.

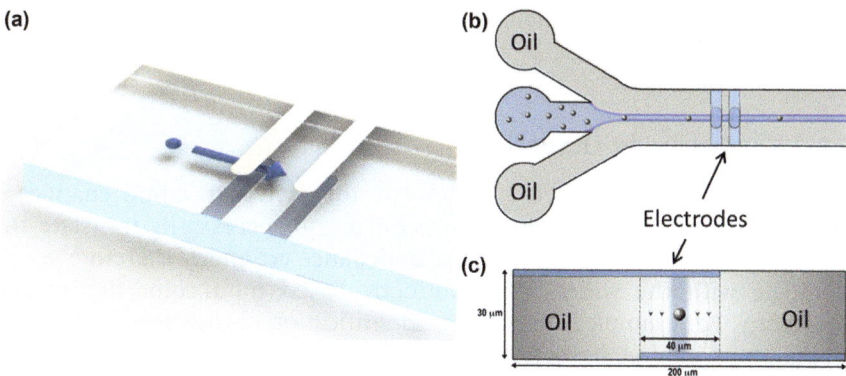

Figure 10.13 (a) Diagram of the microfluidic chip with wide channels and overlapping electrodes. This design uses oil to focus the aqueous sample stream into the center as shown in (b) and (c). Such a system provides high sensitivity in a very wide channel which has low probability of blockage.
Adapted from Bernabini *et al.*, *Lab. Chip*, 2011, **11**, 407–412.

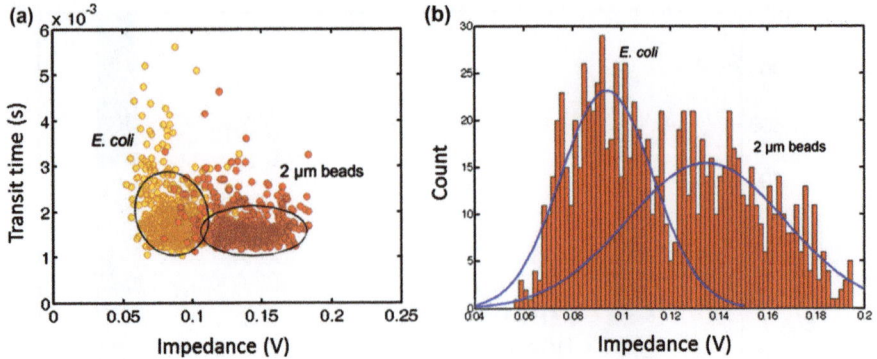

Figure 10.14   Scatter plot (a) and histogram (b) of bacteria and 2 μm beads measured with the system shown in Figure 10.13.
Adapted from Bernabini *et al.*, *Lab. Chip*, 2011, **11**, 407–412.

platelets in plasma or blood in response to an agonist. The aggregation is
normally determined optically or from bulk impedance measurements.
Evander *et al.*[42] developed an impedance analysis system to analyze platelet
activation. They used a microfluidic chip containing electrodes for impedance analysis and two phase flows to sheath the platelets, similar to that
described in ref. 41. Dielectrophoretic focusing was used to center the particles in the channel and thus avoid platelets interacting with the channel
walls and also to minimize shear forces on the platelets. The authors used
multivariate analysis to discriminate data sets containing four different
frequencies (280 kHz to 4 MHz) to distinguish activated (using Thrombin
Receptor Activating Peptide) from non-activated platelets. The physical origin of the difference was not determined, but could be a combination of
shape, size, or intrinsic dielectric properties, particularly membrane conductivity as demonstrated by Egger and Donath.[43]

10.3.7   Antibody Detection

A limitation of impedance cytometry is the fact that subtle changes in
morphology or cell antigenic expression do not necessarily translate into
changes in cell electrical properties. Impedance cytometry therefore cannot
compete with the widely used antibody based multi-color fluorescence
cytometry. For example, the accurate identification of the various T-cell sub-
populations cannot be achieved without the use of labeling antibodies.
There are several groups developing miniature cytometers that incorporate
fluorescence, however an all-electrical analysis platform would be advantageous. Therefore, an electrical analogue of a fluorescent antibody is required. In this context, it was demonstrated that a sub-population of cells
could be uniquely identified within a heterogeneous mixture using antibody-
complexed polymer beads.[44] The antibody-coated particles bind to a specific

sub-population of cells leading to a distinct and identifiable impedance signal for those sub-populations, in a manner that is analogous to fluorescence labeling in optical flow cytometry. The technique is termed impedance labeling and was demonstrated by uniquely identifying a sub-population of CD4 + T-lymphocyte in human blood. These cells are particularly important as an aid to diagnosis and management of HIV/AIDS. The absolute count of CD4 + T-lymphocytes is used to monitor disease progression and response to drug therapy in HIV-infected individuals. According to the latest WHO guidelines, antiretroviral treatment should be initiated if the CD4 count drops below 500 cells per μL.

The principle of impedance labeling is shown in Figure 10.15a. Small CD4 antibody-coated beads are mixed with whole blood. These cells bind to the T-lymphocytes, and also to the monocytes, which also express CD4. When the white cells are measured using impedance spectroscopy, the usual three-part differential is seen in the opacity *vs.* electrical volume scatter diagram as shown in Figure 10.15b. However, two further populations can be identified, both with higher than normal opacity. These are the large population of CD4 expressing T-lymphocytes and to a lesser extent the CD4 expressing monocytes.

Both the relative percentages and absolute cell numbers are often measured during diagnosis of HIV/AIDS. In conventional flow cytometry, a fixed volume of sample needs to be measured. Alternatively precise concentrations of marker beads can be added to the sample – counting the beads provides a means to determine the unknown sample concentration. Microfluidic impedance cytometry with precise volumetric flow can therefore perform both a three-part differential (total and percentage) as well as CD4 + T-cell count, enabling the CD4 count to be presented as either a

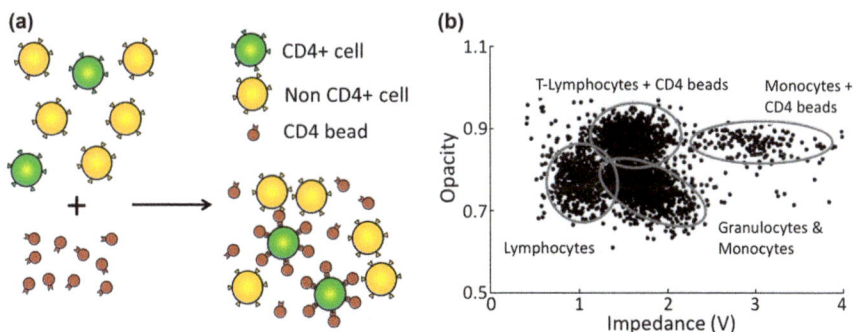

Figure 10.15 (a) Concept for labeling cells with small insulating beads to provide dielectric labels. (b) Scatter plot for blood labeled with anti-CD4 beads demonstrating discrimination of CD4 lymphocytes and monocytes. Adapted with permission from Holmes and Morgan, *Anal. Chem.*, 2001, **15**;82(4), 1455–1461. Copyright © 2010 American Chemical Society.

CD4% or absolute. This technique can provide a simple, robust, low-cost method for identifying antibody expressing cells of interest and could be used in point of care diagnosis and treatment.

10.4   Further Applications of MIC

Single cell impedance cytometry has matured over the last few years and a number of groups have explored its application in medicine and biology. A comprehensive review of the field to 2010 can be found in ref. 45. Some examples of recent developments are summarized below.

10.4.1   Cell Counting and Viability

Counting cells using impedance analysis is relatively straightforward. Segerink *et al.*[46] described a simple planar microelectrode array that is used to count spermatozoa in semen. They were able to discriminate between sperm and other cells (HL60) introduced into the sample. The technique can be used to provide a simple and quick method for determining the concentration of sperm without the need for optical analysis.

Assessment of cell viability and heterogeneity is an important aspect of many analytical techniques. David *et al.*[47] measured the impedance properties of single bacteria (*Bacillus megaterium*) using a chip with channel dimensions of 10 μm×10 μm, thereby increasing the sensitivity of the system. This organism is a large bacterium with a length of up to 4 μm and a diameter of 1.5 μm. Impedance analysis could distinguish cells in exponential *vs.* stationary phase. Differences in impedance of heat-inactivated (non-viable) cells were also shown. The authors claim that impedance cytometry can detect changes in membrane potential at a frequency of 500 kHz. However, the changes measured at this frequency are dominated by the viability of the cell membrane, which will become electrically leaky when cells lose viability. Any relationship between membrane potential and dielectric properties is predicted to occur in the very low frequency (alpha) relaxation, which is typically measured below 1 kHz[48,49] and to date measurements at this frequency have not been reported using single cell cytometry.

Impedance cytometry has also been used to for on-line quality monitoring of hybridoma cells.[50] Hybridoma cells are used for monoclonal antibody production and are grown in bioreactors. The cells were analyzed to determine viability and the data correlated with fluorescent staining with Annexin V (marker of apoptosis) and 7-AAD for necrosis. Optimum differentiation was observed in a low conductivity buffer (0.26 S m^{-1}) at a frequency of 10 MHz. At this conductivity and frequency, the impedance signal will mostly be dominated by the cytoplasmic conductivity of the cells.

10.4.2 Parasitized Cells

It has been shown that invasion of cells with protozoa changes their electrical properties.[51,52] Infection of red blood cells with the malarial parasite changes the dielectric properties in several ways, including the cell membrane conductivity, the permittivity of the interior, and the shape and volume of the cell; see Figure 10.16 for a summary.

These differences in the dielectric properties of the cells were exploited to develop a method for enriching cells using a dielectrophoretic free-flow fractionation system.[52] An example of the use of high-frequency measurements for cell analysis was reported by Küttel *et al.*,[17] who detected infection of RBCs with the parasite. They infected cells with the parasite *Babesia bovis*, a protozoan that lives in red blood cells and which is transmitted by ticks. They used a signal of 8.7 MHz and a low conductivity suspending media. The reduction in electrolyte conductivity reduces the characteristic membrane charging frequency and reduces the frequency required to probe intracellular properties. The authors demonstrated differentiation of parasitized RBCs from both uninfected RBCs and ghost RBCs due to changes in the electrical properties of the cell cytoplasm.

10.4.3 Tumor Cells and Stem Cell Morphology

There is considerable interest in developing a label-free method of identifying abnormal cells. One particular application is the identification and enumeration of circulating tumor cells (CTCs), which has been demonstrated to have prognostic value; breast cancer patients with raised numbers of CTCs (greater than 5 per 7.5 mL whole blood) on average have a decreased survival time.[53] CTCs are typically larger than most blood cells, a parameter that is utilized in label-free enrichment by size filtration; however, there is considerable variation in the size of CTCs and consequently some CTCs are missed. It is known that tumor cells have a higher membrane capacitance than other blood cells[31] and this provides a means of separating these cells using AC electrokinetic techniques such as dielectrophoresis (DEP) and DEP-Field Flow Fractionation.[54,55] The dielectric properties of a wide range of tumor cells has been measured using techniques such as DEP and electrorotation,[31,56] where analysis of a few cells can take many hours. Microfluidic impedance provides a high-throughput alternative to this.

A model cell line, representative of CTCs, is the immortalized breast cancer cell line MCF-7. These cells have a wide range in size, typically up to 20 μm in diameter, and have a very high membrane capacitance of up to 23 mF m^{-2}.[31] Figure 10.17 shows impedance scatter data for the MCF-7 cells, spiked into whole blood. The cells were labeled with EpCAM fluorescent antibody (marker for epithelial cells) represented by the false color scale. As the figure demonstrates, both the size and capacitance (opacity) are very different from the WBCs. A small number of EpCAM positive cells

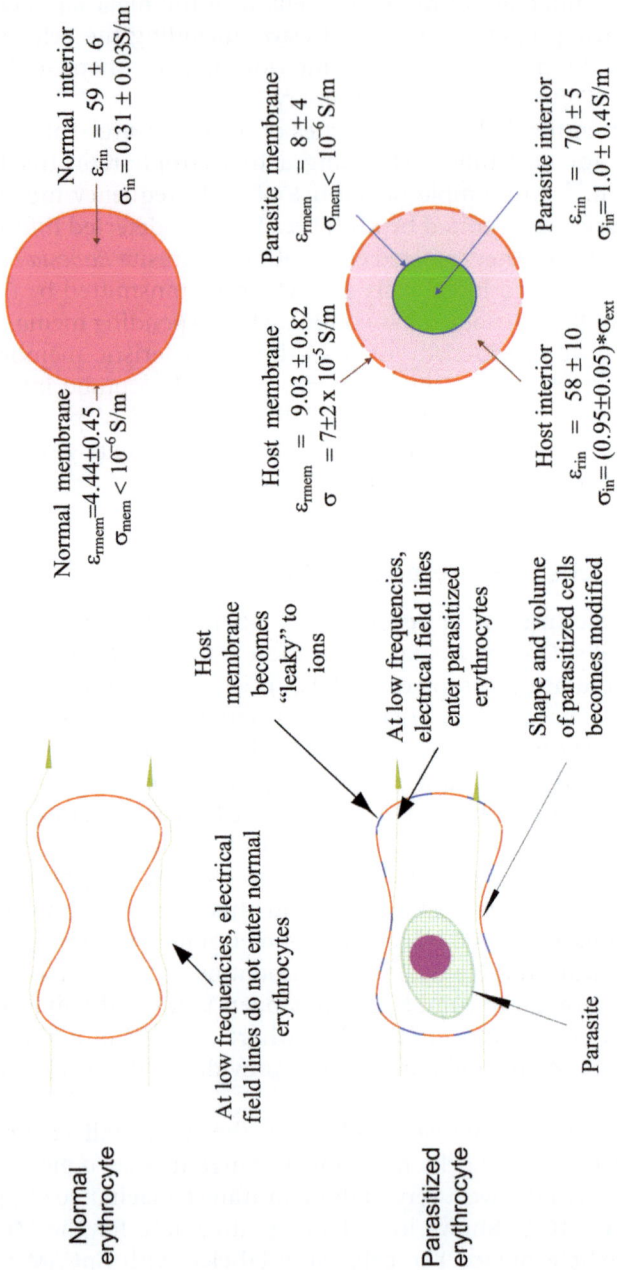

Dielectric Characteristics

Normal interior
$\varepsilon_{rin} = 59 \pm 6$
$\sigma_{in} = 0.31 \pm 0.03$ S/m

Normal membrane
$\varepsilon_{rmem} = 4.44 \pm 0.45$
$\sigma_{mem} < 10^{-6}$ S/m

Parasite membrane
$\varepsilon_{rmem} = 8 \pm 4$
$\sigma_{mem} < 10^{-6}$ S/m

Host membrane
$\varepsilon_{rmem} = 9.03 \pm 0.82$
$\sigma = 7 \pm 2 \times 10^{-5}$ S/m

Parasite interior
$\varepsilon_{rin} = 70 \pm 5$
$\sigma_{in} = 1.0 \pm 0.4$ S/m

Host interior
$\varepsilon_{rin} = 58 \pm 10$
$\sigma_{in} = (0.95 \pm 0.05) * \sigma_{ext}$

Biophysical Characteristics

Host membrane becomes "leaky" to ions

At low frequencies, electrical field lines enter parasitized erythrocytes

Shape and volume of parasitized cells becomes modified

At low frequencies, electrical field lines do not enter normal erythrocytes

Parasite

Normal erythrocyte

Parasitized erythrocyte

Figure 10.16 Summary of the electrical changes that occur in an erythrocyte after infection with a malaria parasite. The membrane conductivity increases so that the cells should appear to be smaller in electrical volume than their physical volume when measured at low frequencies. There is also a change in the internal (cytoplasmic) properties of the cell. Reprinted with permission from Gascoyne et al., *Acta Trop.*, 2004, **89**(3), 357–369. Copyright © 2004, Elsevier.

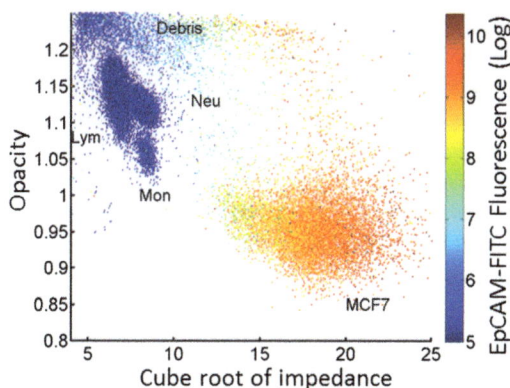

Figure 10.17 (unpublished). Impedance scatter plot showing the very large size difference between MCF-7 cells and WBCs. The capacitance of the MCF-7 is also very much larger as demonstrated by the large difference in opacity. Note the wide distribution in the properties of the MCF-7 cells compared with WBCs. Some of the MCF-7 cells are non-viable and this population is visible in the top center of the scatter plot (confirmed with PI staining).

can be seen at the top of the plot – these are non-viable cells (corroborated by flow cytometry with propidium iodide stain, data not shown). These cells express EpCAM but are electrically quite distinct from the normal cell populations.

Current MIC technology does not have a sufficiently high-throughput for the identification of extremely rare cells like CTCs and a maximum sample throughput of 100 μL min^{-1} is clearly inadequate for processing 10 mL of blood (which must be significantly diluted). Nevertheless, spiking experiments demonstrated a limit of detection of 100 tumor cells per 50 μL whole blood (without pre-concentration), flowing for 15 minutes. However, MIC provides an ideal label-free method for cell identification, which could easily be integrated into many other pre-enrichment high-throughput microfluidic platforms.

Recent work has also demonstrated that the differentiation state of stem cells may be observable using dielectric measurements. Song *et al.*[57] developed a micro-pore-based impedance analysis chip that measured the impedance of mouse embryonic carcinoma cells at different frequencies. Differentiated and undifferentiated cells could be distinguished from the differences in membrane capacitance, measured in terms of opacity.

10.4.4 High-frequency Measurements

Measurement of the dielectric properties of cells across a very wide range of frequencies up to 500 MHz should provide much more information on the internal properties of cells. Much of the information in the literature is obtained from traditional suspension impedance analysis, which provides a population average, or AC electrokinetic techniques including dielectrophoresis and electrorotation, which are accurate but have very low throughput. To date, impedance cytometry has been limited by the electronic hardware available, in terms of both particle throughput and frequency bandwidth. Recently Haandbaek *et al.*[25] demonstrated a new system that can measure from DC to 500 MHz enabling characterization of cell cytoplasm and nuclei properties. The utility of this system was demonstrated by analyzing yeast strains with and without large internal vacuoles. Differences between the cells was only apparent for frequencies above 10 MHz, where the membrane becomes transparent to the electric field and the dielectric properties are dominated by internal properties. By fitting the data with numerical simulations of the system, a range of dielectric parameters for the cells was obtained, which in general were comparable to literature values determined using traditional bulk impedance spectroscopy.[58]

10.5   Future Challenges

The technology of single cell impedance cytometry has developed rapidly over the last decade. The electronic hardware and signal processing have been optimized and are now commercially available. Chip manufacture remains a challenge. Many laboratories cannot produce the double co-planar high precision bonded glass chips that are needed to produce high-quality data, although simple cell counting and discrimination can be achieved with planar electrode pairs that can be made using simple photolithography. Throughput is typically up to 1000 cells per second, which is often adequate for most analysis, although it would be possible to increase this by a factor of 10 provided that appropriate signal acquisition hardware and software were available. Perhaps the next breakthrough would be an all electrical continuous analysis and sorting system, analogous to the optical FACS. This would enable identification of rare cells followed by isolation for further downstream processing, all in a single chip. Other challenges include onchip sample preparation. The dielectric properties of cells can be modified using chemistry to aid in discrimination, for example exposure of monocytes to formic acid increases their membrane capacitance. Binding CD4 antigen coated beads to T-lymphocytes can also be used to change the cell properties sufficient to enable measurement. Integrating automated sample preparation into the system would deliver the "sample in, answer out" devices that are so attractive for point of care applications. Such devices need to include dilution, washing, chemical processing, and antibody labeling, all in a simple, low-cost, and reliable format.

References

1. W. H. Coulter, *US Pat.*, 2,656,508, 1953.
2. H. Morgan, T. Sun, D. Holmes, S. Gawad and N. G. Green, *J. Phys. D Appl. Phys.*, 2007, **40**, 61–70.

3. T. Sun and H. Morgan, *Microfluid. Nanofluid.*, 2010, **8**, 423–443.
4. H. M. Shapiro, *Practical Flow Cytometry*, Wiley-Liss.BD, 2003.
5. F. S. Ligler and J. S. Kim, *The Microflow Cytometer*, Pan Stanford Publishing, 2010.
6. S. H. Cho, J. M. Godin, C.-H. Chen, W. Qiao, H. Lee and Y.-H. Lo, *Biomicrofluidics*, 2010, **4**, 043001.
7. C. Briggs, J. Carter, S.-H. Lee, L. Sandhaus, R. Simon-Lopez and J.-L. Vives, *Int. J. Lab. Hematol.*, 2008, **30**, 105–116.
8. D. Heikali and D. Di Carlo, *J. Assoc. Lab. Autom.*, 2010, **15**, 319.
9. L. V. Rao, B. A. Ekberg, D. Connor, F. Jakubiak, G. M. Vallaro and M. Snyder, *Clin. Chim. Acta*, 2008, **389**, 120–125.
10. S. Gawad, L. Schild and P. H. Renaud, *Lab. Chip*, 2001, **1**, 76–82.
11. W. Coulter and W. R. Hogg, *US Pat.*, 3502974 A, 1970.
12. R. A. Hoffman and W. B. Britt, *J. Histochem. Cytochem.*, 1979, **27**, 234–240.
13. R. A. Hoffman, T. S. Johnson and W. B. Britt, *Cytometry*, 1981, **1**, 377–384.
14. J. H. Nieuwenhuis, F. Kohl, J. Bastemeijer, P. M. Sarro and M. J. Vellekoop, *Sensor. Actuator. B Chem.*, 2004, **102**, 44–50.
15. D. K. Wood, S. H. Oh, S. H. Lee, H. T. Soh and A. N. Cleland, *Appl. Phys. Lett.*, 2005, **87**, 184106.
16. H. Morgan, D. Holmes and N. G. Green, *Curr. Appl. Phys.*, 2006, **6**, 367–370.
17. C. Küttel, E. Nascimento, N. Demierre, T. Silva, T. Braschler, P. H. Renaud and A. G. Oliva, *Acta Tropica*, 2007, **102**, 63–68.
18. C. Iliescu, D. P. Poenar, M. Carp and F. C. Loe, *Sensor. Actuator. B*, 2007, **123**, 168–176.
19. R. Rodriguez-Trujillo, O. Castillo-Fernandez, M. Garrido, M. Arundell, A. Valencia and G. Gomila, *Biosens. Bioelectron.*, 2008, **24**, 290–296.
20. T. Sun, N. G. Green, S. Gawad and H. Morgan, *IET Nanobiotechnol.*, 2007, **1**, 69–79.
21. S. Gawad, K. Cheung, U. Seger, A. Bertsch and P. H. Renaud, *Lab. Chip*, 2004, **4**, 241–251.
22. K. C. Cheung, S. Gawad and P. H. Renaud, *Cytometry Part A*, 2005, **65A**, 124–132.
23. D. Holmes, D. Pettigrew, C. H. Reccius, J. D. Gwyer, C. V. Berkel, J. Holloway, D. E. Davie and H. Morgan, *Lab. Chip*, 2009, **9**, 2881–2889.
24. T. Sun, S. Gawad, C. Bernabini, N. G. Green and H. Morgan, *Meas. Sci. Tech.*, 2007, **18**, 2859–2868.
25. N. Haandbæk, S. C. Bürgel, F. Heer and A. Hierlemann, *Lab. Chip*, 2014, **14**, 369.
26. H. P. Schwan, *Adv. Biol. Med. Phys.*, 1957, **5**, 147–209.
27. K. Asami, T. Yonezawa, H. Wakamatsu and N. Koyanagi, *Bioelectrochem. Bioenerg.*, 1996, **40**, 141–145.
28. K. Asami, *Prog. Polym. Sci.*, 2002, **27**, 1617–1659.

29. R. Pethig and D. B. Kell, *Phys. Med. Biol.*, 1987, **32**, 933.
30. X. B. Wang, Y. Huang, P. R. Gascoyne, F. F. Becker, R. Holzel and R. Pethig, *Biochim. Biophys. Acta*, 1994, **1193**, 330–344.
31. P. R. Gascoyne, S. Shim, J. Noshari, F. F. Becker and K. Stemke-Hale, *Electrophoresis*, 2013, **34**(7), 1042–1050.
32. X. Wang, F. F. Becker and P. R. Gascoyne, *Chaos*, 2010, **20**, 043133.
33. S. L. Ledis, H. R. Crews, T. J. Fischer and T. Sena, *US Pat.*, 5155044, 1992.
34. X. Han, C. van Berkel, J. Gwyer, L. Capretto and H. Morgan, *Anal. Chem.*, 2012, **84**, 1070–1075.
35. C. van Berkel, J. D. Gwyer, S. Deane, G. Nicolas, J. Holloway, V. Hollis and H. Morgan, *Lab on a Chip*, 2011, **11**, 1249–1255.
36. P. Sethu, M. Anahtar, L. L. Moldawer, R. G. Tompkins and M. Toner, *Anal. Chem.*, 2004, **76**, 6247–6253.
37. P. Sethu, L. L. Moldawer, M. N. Mindrinos, P. O. Scumpia, C. L. Tannahill, J. Wilhelmy, P. A. Efron, B. H. Brownstein, R. G. Tompkins and M. Toner, *Anal Chem.*, 2006, **78**, 5453–5461.
38. P. Linderholm and P. H. Renaud, *Lab Chip*, 2005, **5**, 270–279.
39. P. Linderholm, U. Seger and P. H. Renaud, *Electron. Lett. IEE*, 2006, **42**, 145–147.
40. D. Spencer and H. Morgan, *Lab on a Chip*, 2011, **11**, 1234–1239.
41. C. Bernabini, D. Holmes and H. Morgan, *Lab. Chip*, 2011, **11**, 407–412.
42. M. Evander, A. J. Ricco, J. Morser, G. T. A. Kovacs, L. K. L. Leung and L. Giovangrandi, *Lab. Chip*, 2013, **13**, 722.
43. M. Egger and E. Donath, *Biophys. J.*, 1995, **68**, 364–372.
44. D. Holmes and H. Morgan, *Anal Chem.*, 2010, **82**, 1455–1461.
45. K. C. Cheung, M. Di Berardino, G. Schade-Kampmann, M. Hebeisen, A. Pierzchalski, J. Bocsi, A. Mittag and A. Tárnok, *Cytometry A*, 2010, 77, 648–666.
46. L. I. Segerink, A. J. Sprenkels, P. M. ter Braak, I. Vermes and A. van den Berg, *Lab. Chip*, 2010, **10**, 1018–1024.
47. F. David, M. Hebeisen, G. Schade, E. Franco-Lara and M. Di Berardino, *Biotech Bioeng*, 2012, **109**, 483.
48. C. Bot and C. Prodan, *Eur. J. Biophys.*, 2009, **38**, 1049–1105.
49. C. T. Bot and C. Prodan, *Biophys. Chem.*, 2010, **46**, 2–3.
50. A. Pierzchalski, M. Hebeisen, A. Mittag, J. Bocsi, M. Di Berardino and A. Tarnok, *Lab. Chip*, 2010, **12**, 4533–4543.
51. P. Gascoyne, R. Pethig, J. Satayavivad, F. F. Becker and M. Ruchirawat, *Biochim. Biophys. Acta*, 1997, **1323**, 240–252.
52. P. Gascoyne, J. Satayavivad and M. Ruchirawat, *Acta Trop.*, 2004, **89**, 357–369.
53. M. Cristofanilli, G. T. Budd, M. J. Ellis, A. Stopeck, J. Matera, M. C. Miller, J. M. Reuben, G. V. Doyle, W. J. Allard, L. W. Terstappen and D. F. Hayes, *New Engl. J. Med.*, 2004, **351**(8), 781–791.
54. P. R. Gascoyne, J. Noshari, T. J. Anderson and F. F. Becker, *Electrophoresis*, 2009, **30**(8), 1388–1398.

55. V. Gupta, I. Jafferji, M. Garza, V. O. Melnikova, D. K. Hasegawa, R. Pethig and D. W. Davis, *Biomicrofluidics*, 2012, **6**, 024133.
56. F. F. Becker, X. B. Wang, Y. Huang, R. Pethig, J. Vykoukal and P. R. Gascoyne, *PNAS*, 1995, **92**(3), 860–864.
57. H. Song, Y. Wang, J. M. Rosano, B. Prabhakarpandian, C. Garson, K. Panta and E. A. Lai, *Lab. Chip*, 2013, **13**, 2309.
58. K. Asami and T. Yonezawa, *Biophys. J.*, 1996, **71**, 2192–2200.

CHAPTER 11

Routine Clinical Laboratory Diagnostics Using Point of Care or Lab on a Chip Technology

GÁBOR L. KOVÁCS*[a,b] AND ISTVÁN VERMES[b,c]

[a] Institute for Laboratory Medicine, Medical Faculty, University Pécs, Ifjúság u. 13, 7624 Pécs, Hungary; [b] Szentágothai Research Centre, University Pécs, Ifjúság u. 13, 7624 Pécs, Hungary; [c] BIOS, Lab in a Chip Group, MESA$^+$ Institute for Nanotechnology, University Twente, PO Box 217, 7500 AE Enschede, The Netherlands
*Email: Gabor.L.Kovacs@aok.pte.hu

11.1 Introduction

In the past decades the clinical laboratory had made significant technological advances focused on clinical efficiency and efficacy. In addition to the technical development this goal has been reached mainly *via* automation and centralization. In contrast to this centralization and increased efficiency in clinical laboratory diagnosis, there has been a recent trend towards a more decentralized diagnostic analysis near to the patients. The idea of this so-called point-of-care testing (POCT) is to bring the test immediately and in a convenient way to the patient. These devices have been developed to offer improvements in convenience, patient care, and turnaround time. POCT systems should be fast, small, and simple to use while maintaining

RSC Nanoscience & Nanotechnology No. 36
Microfluidics for Medical Applications
Edited by Albert van den Berg and Loes Segerink
© The Royal Society of Chemistry 2015
Published by the Royal Society of Chemistry, www.rsc.org

state-of-the-art performance features. Therefore it is no wonder that Lab-on-Chip technology is used in POCT.[1] Most of these devices are dedicated to glucose measurement but more and more attention is being paid to developing bedside systems capable of a wider spectrum of laboratory analysis. More than 31 000 laboratory diagnostic instruments and devices have been categorized by the FDA under CLIA (Clinical Laboratory Improvement Amendments), and POCT represents 25% of these.[2] During the last decade POCT has become an important part of laboratory medicine.[3]

11.2 Point-of-care Testing

POCT is characterized by proximity to the patient, quantitative or semi-quantitative single measurements, short turnaround time, no sample preparation, no pipetting, use of pre-made reagents, user-friendly dedicated analytical instruments, and instant, result deduced therapeutic action.[4] All outpatient or ward personnel will be expected to be able to use these analyzers. No previous knowledge in sample analysis should be required. POCT measurements are based on so-called biosensors. The biosensor is an analytical device for the detection of analytes that combines a biological component with a physicochemical-detector component. This generally occurs through the use of LOC systems, where biological components (usually antibodies) are immobilized on a solid-state surface,[4] which interacts with the analyte. POCT devices can be divided into different groups based on the biosensor used (Figure 11.1).

11.2.1 Categorization of POCT Devices

Qualitative strip-based POCT devices: These analyzers (*e.g.* pregnancy tests, urine dipstick, stool blood tests) produce qualitative results (plus or minus) *via* simple visualization of a strip which is made of a porous matrix containing the reagents.

Unit-used POCT devices: This is the most frequently and most simple quantitative POCT device, which also uses test strips. The chemical reactions are performed on a single-use test strip and the reader unit is only used to measure the end-signal of the chemical reaction. The detection principle is usually electrochemical. The measurement is performed by using whole blood which makes this type of POCT device very convenient for the user. Glucometers and multi-parameter-readers (*e.g.* iSTAT) are examples of this type of analyzer.

Bench-top POCT devices: These are more complex than unit-used instruments and apply different analytical principles (*e.g.* enzyme activity measurement, particle counting, immunoassay, sensor-based blood gas analysis). Although these are POCT analyzers, preferably only qualified personnel should operate them. Reflotron, a multi-parameter chemistry instrument (Roche), PFA

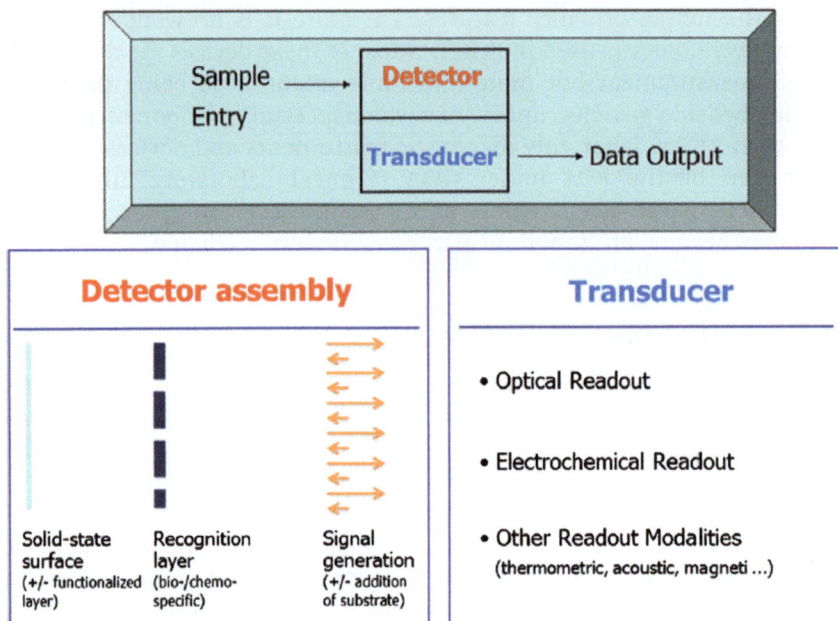

Figure 11.1 The biosensor as the basis of analysis in many point-of-care testing (POCT) instruments. Note that the recognition layer may comprise attached recognition elements (antibodies, receptors, aptamers) or immobilized enzymes. In the latter case, the addition of substrates is essential. Not illustrated is the fact that the amplified signal is finally processed by microelectronics and displayed.
Reprinted with permission from Luppa *et al. Trends Anal. Chem.*, 2011, **30**, 887. Copyright Elsevier Ltd.

100, platelet function analyzer (Siemens), and ABL blood-gas analyzers (Radiometer) are examples of this type of device.

POCT devices for continuous measurement: Real-time continuous glucose monitoring POCT systems are available for diabetes type 1 patients. A needle-based glucose-oxidase embedded sensor is placed subcutaneously and the produced hydrogen peroxide is measured electrochemically.[5] Guardian Real Time (Medtronic) or Freestyle Navigator (Abbott) are examples of the continuous glucose analyzers available commercially.

Molecular biology-based POCT analyzers: are used most frequently to detect infectious pathogens. The simple instruments use qualitative test strips to detect pathogens based on immunochromatography but nowadays rapid nucleic acid testing based on PCR technology is also used. GeneXpert (Cepheid) is one commercially available example of these instruments.

11.2.2   Role of POCT in Laboratory Medicine

The concept of POCT represents a fundamental shift in diagnostic testing where the objective is cost-effective, patient-focused testing at the site of diagnosis. This trend of decentralization is likely to assume even greater importance in the future, since financial pressures will mandate more efficient health care. The number of patients requiring in-patient treatment in a hospital, together with the length of their stay, is likely to decline further in the near future–to the advantage of outpatient treatment and patient observation in their home environment. The importance of POCT will increase in these areas. In outpatient care, it is clear what advantage immediate diagnostic and therapeutic decisions offer: the patient may not need to be further investigated. Home-care will be extended when new tests for chronic diseases become readily available. The most important trend in this context is termed "personalized medicine". The therapeutic strategies would then be tailored to each patient individually, thus improving the results of treatment. The effectiveness of home monitoring and thus patient safety are improved. In addition to home monitoring one can expect further applications of POCT devices in different parts of public life *e.g.* sport events, military service, show business, nursing homes, transport vehicles, alternative medical care, veterinary medicine, fitness studios, pharmacies, *etc.* The recent technological developments of LOC technology allow this wide opening of POCT in laboratory medicine. LOC devices are now beginning to be used in remote settings, as a result of developments in integrating fluid actuation, sample pretreatment, sample separation, signal amplification, and signal detection into a single device.

11.3 Glucometers

11.3.1 The WHO and ADA Criteria of Diabetes

According to the World Health Organization (WHO), more than 140 million people suffer from diabetes mellitus around the world.[6] WHO also estimates that by 2025 this number may well double. Patients with type 1 diabetes need to check their glucose levels several times per day. The WHO and the American Diabetes Association (ADA) define the diagnosis of diabetes mellitus by at least two measurements of fasting plasma glucose concentration of ≥ 7.0 mmol L^{-1}. As an alternative, a random venous plasma glucose concentration of ≥ 11.1 mmol L^{-1} in the presence of symptoms or a 2-h post oral glucose tolerance test result of ≥ 11.1 mmol L^{-1} suffices to make a definite diagnosis of diabetes mellitus.[7] The new classifications of "impaired fasting glycemia" have narrower intervals (6.1–6.9 mmol L^{-1}, WHO; or 5.6–6.9 mmol L^{-1}, ADA) for venous plasma glucose concentration than the previous fasting interval (5.6–7.7 mmol L^{-1}) for classifying normoglycemia and diabetes. The narrower diagnostic limits increase the need for reliable results to classify individuals correctly.

11.3.2 Plasma Glucose or Blood Glucose

Currently, various types of instruments detect and report fundamentally different glucose quantities. Biosensors for glucose are "direct reading" when they measure glucose directly, *i.e.* without prior dilution of the sample. The new generation of direct reading glucose sensors responds to the molality of glucose, which is identical in whole blood and plasma, whereas the concentration of glucose in the two systems is different. Methods requiring high sample dilution produce results equivalent to concentration when calibrated against aqueous standards, because the water concentrations of sample and calibrator are almost identical after dilution. In current clinical practice, plasma and blood glucose are used interchangeably,[8] with a consequent risk of misinterpretation. The two systems are frequently mistaken in the clinical literature, despite an average 11% difference in glucose concentration (plasma/blood). The ADA provides clinical decision limits for the concentration of glucose in venous plasma, but the WHO also provides the concentration of glucose in whole blood.[9,10]

Glucose levels in plasma are generally 10%–15% higher than glucose measurements in whole blood (and even more after eating). This is important because home blood glucose meters measure the glucose in whole blood while most lab tests measure the glucose in plasma. Currently, there are many meters on the market that give results as "plasma equivalent", even though they are measuring whole blood glucose. The plasma equivalent is calculated from the whole blood glucose reading using an equation built into the glucose meter. This allows patients to easily compare their glucose measurements in a lab test and at home. It is important for patients and their health care providers to know whether the meter gives its results as "whole blood equivalent" or "plasma equivalent". One model measures beta-hydroxybutyrate in the blood to detect ketoacidosis (ketosis). Special glucose meters for multi-patient hospital use are now used. These provide more elaborate quality control records. Their data handling capabilities are designed to transfer glucose results into electronic medical records and the laboratory computer systems.

The development of self-monitoring of blood glucose is probably the most important advance in controlling diabetes since the discovery of insulin and provides the ability for diabetes patients to test their own blood glucose and adjust insulin dosage to control their glucose needs. With the universal availability of glucose meters today, it is difficult to imagine that managing blood glucose was once considered impossible.

11.3.3   Glucometers in Medical Practice

Glucose meters are widely used in hospitals, outpatient clinics, emergency rooms, ambulatory medical care (ambulances, helicopters, cruise ships), and home self-monitoring. Glucose meters provide fast analysis of blood glucose levels and allow management of both hypoglycemic and hyperglycemic disorders with the goal of adjusting glucose to a near-normal range, depending on the patient group.

Glucose meters have two essential parts: an enzymatic reaction and a detector. The enzyme portion of the glucose meter is generally packaged in a dehydrated state in a disposable strip or reaction cuvette. Glucose in the patient's blood sample rehydrates and reacts with the enzymes to produce a product that can be detected. Some meters generate hydrogen peroxide or an intermediary that can react with a dye, resulting in a color change proportional to the concentration of glucose in solution. Other meters incorporate the enzymes into a biosensor that generates an electron that is detected by the meter. There are three principal enzymatic reactions utilized by current glucose meters: glucose oxidase, glucose dehydrogenase, and hexokinase. Each enzyme has characteristic advantages and limitations. All meters are susceptible to heat and cold, because the enzymes are proteins that can denature and become inactivated at temperature extremes. Although packaged in a dry state, exposure of the enzymes to humidity can prematurely rehydrate the proteins and limit their reactivity when utilized for patient testing. The disposable reagents for glucose meters must therefore be protected from extremes of temperature and humidity. Such conditions could occur when transporting the reagents outside in the heat of summer or cold of winter. Test strips should not be stored in closed vehicles for extended periods and must be protected from rain, snow, and other environmental elements. The detector portion of the meter is composed of electronics, so it must also be protected from extremes of temperature, humidity, moisture, and the elements. Many meters now have internal temperature checks that prevent use of the meter outside of acceptable tolerance by blocking patient results or displaying an error code if the ambient conditions of temperature and humidity are outside manufacturer ranges. Glucose meters must also not be submerged in water when cleaning and must be protected from moisture, as with any electronic device.

Accuracy of glucose meters is a common topic of clinical concern. Blood glucose meters must meet accuracy standards set by the International Organization for Standardization (ISO). According to ISO 15197 Blood glucose meters must provide results that are within 20% of a laboratory standard in 95% of the results. However, a variety of factors can affect the accuracy of a test. Factors affecting accuracy of various meters include calibration of meter, ambient temperature, pressure use to wipe off strip, size and quality of blood sample, high levels of certain substances (such as ascorbic acid) in blood, hematocrit, dirt on meter, humidity, and aging of test strips. Models vary in their susceptibility to these factors and in their ability to prevent or warn of inaccurate results with error messages.

Today, a variety of diagnostic companies offer glucose meters on the market. These include Abbott, Bayer, Suncoast, Diagnostic Devices, 77Electronics, Entra Health Systems, HealthPia, HemoCue, Home Diagnostics, Hypoguard, Lifescan, Menarini, Nova, Relion, Roche, US Diagnostics. And the list is far from being complete (Figure 11.2).

However, the quality and applicability of these instruments may vary from company to company and may also depend on the setting where used (home or hospital). For example Gijsen *et al.*[11] recently compared some glucometers generally accepted for home setting and used them in intensive care patients

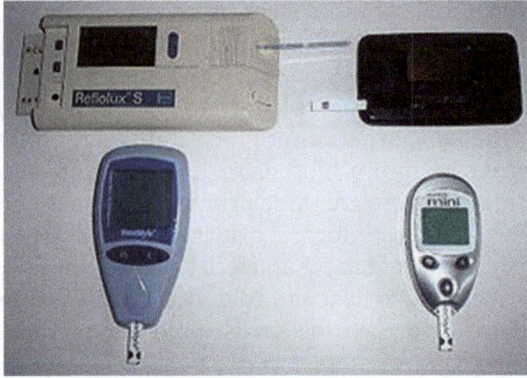

Figure 11.2   Four generations of blood glucose meter (1993–2005). Sample sizes vary from 30 to 0.3 µl. Test times vary from 5 seconds to 2 minutes (modern meters typically provide results in 5 seconds).
Reprinted from Wikipedia. No permission needed.

who were under tight glycemic control. When ISO 15197 was applied, the glucometers of Roche, HemoCue, and Abbott fulfilled the criterion in this patient population, whereas Nova and Menarini did not. When the Dutch (TNO) quality guideline or the recent British/American (NACB/ADA 2011) guideline was applied, only the meter of Roche fulfilled the criteria. Thus most home-use POC glucose meters are unsuitable for tight glycemic control in hospital settings.

The use of glucometers, based on sampling of capillary samples, is in some clinical situations and disease complications contraindicated. Such complications are severe dehydration, hypotension, shock, peripheral circulatory failure, hyperosmolar non-ketotic coma, diabetic ketoacidosis, and the unconscious patient.

There are potential interactions of different reasons that might seriously affect the interpretation of glucose measurements with glucometers. Such interferences might be of environmental nature (air, exposure of strips, altitude, humidity, temperature), due to physiological parameters (hematocrit, prandial state, hyperlipidemia, oxygenation, pH), operational reasons (hemolysis, anticoagulants, generic test strips, amniotic fluid, arterial and venous catheter, volume of sample, reuse of strips), or to various drug interactions (*e.g.* maltose, acetaminophen, ascorbate, mannitol, dopamine).

11.3.4   Glucometers in Gestational Diabetes

Diabetes is a common medical complication of pregnancy. The prevalence may range from 1 to 14% of all pregnancies. The prevalence varies with criteria and population studied. Diabetic women can be separated into those detected before pregnancy (pre-gestational or overt) and those diagnosed during pregnancy (gestational diabetes[12]). Gestational diabetes represents nearly 90% of all pregnancies complicated by diabetes. Clinical recognition of gestational diabetes is important because management and assessment (including medical nutrition therapy, insulin when necessary, and ante-partum fetal surveillance) can reduce the gestational diabetes-associated perinatal morbidity and mortality. Moreover, women whose pregnancies are complicated by this disease are prone to have diabetes in the future, approximately 50% of cases within 20 years. Gestational diabetes is defined as any degree of glucose intolerance with onset or first recognition during pregnancy.[13] Currently, a glucometer is used worldwide for self-glucose monitoring because of comfortable use and yielding a quick result. The accuracy of this equipment for the assay of capillary glucose specimens has been well justified. The ADA recommended the cutoff point of plasma glucose was 7.8 mmol L^{-1} for screening gestational diabetes. The optimal cutoff value is important for any screening test.

11.3.5 Continuous Glucose Monitoring

Diabetic patients need to check their glucose levels several times per day by puncturing their finger to fetch a droplet of blood for chemical analysis, bearing uninterrupted pain and skin injury. In order to overcome this problem, many research groups perform research aiming to develop non-invasive techniques to monitor blood glucose level. In general, experimental *in vitro* results of application of three different optical techniques to glucose sensing in highly scattering media are being used: optical coherence tomography,[14] photoacoustic spectroscopy, and time-of-flight measurement of ultra-short laser pulses. A connective factor between these techniques is the fact that variations in glucose concentration in turbid media affect the scattering properties of the media due to the change in the refractive index and, consequently, the light propagation and distribution parameters. By detecting the variations of these parameters, the glucose content in the media can be resolved. The final goal is to make low-cost individual devices for everyday non-invasive monitoring of glucose concentration in home conditions. However, there are many problems in computer modeling, device development, and integration before a commercial product is available. In the future, research will be focused on the physiological range of glucose.

11.4 i-STAT: a Multi-parameter Unit-use POCT Instrument

The company i-STAT Corporation (Canada) was founded in 1984 to develop biosensor chip technologies in order to produce POCT systems. It was the first company to develop biosensors by using semiconductor wafer technology. In 1998 the i-STAT company entered a strategic alliance with Abbott Laboratories (Abbott Park, MI, USA) for research and development of POCT. In 2003 Abbott acquired the i-STAT company and since then i-STAT systems

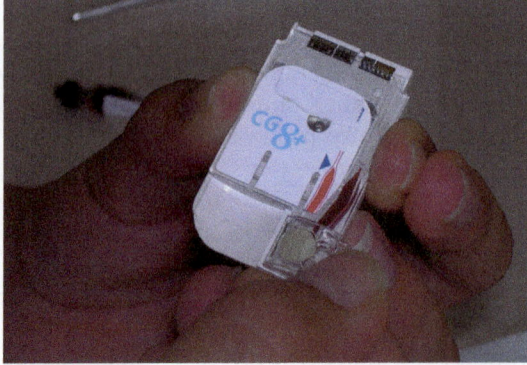

Figure 11.3 Disposable cartridge for i-STAT point of care analyzer.

have been developed, produced, and distributed by the Point of Care Division of Abbott Laboratories.

The i-STAT System performs bedside blood analysis using a single disposable cartridge, which contains microfabricated sensors, a calibrator solution, fluidics system, and a waste chamber (Figure 11.3)

The measurements of the different assays are electrochemical, using the microfabricated sensors housed in each cartridge to measure analyte concentrations directly in a single whole blood sample (Table 11.1). The analyzer used is the handheld i-STAT Portable Clinical Analyzer (Figure 11.4).

When a cartridge is filled with 2–3 drops (16–60 μL) blood and inserted into the i-STAT analyzer for analysis, it automatically controls all functions of the testing cycle including fluid movement within the cartridge, calibration, and continuous quality monitoring (Figure 11.5).

A desktop computer, the Central Data Station, provides the primary information management capabilities for the i-STAT system. An Infrared Interface Link for handheld analyzers allows for transmission of patient records from several analyzers to the Central Data Station. Data can be stored, organized, edited and can be transferred *via* the Hospital Laboratory Information System to the patient medical dossier. Currently measurements of the following analytes are available by using i-STAT cartridges (Table 11.1):

11.4.1 Clinical Chemistry

Electrolytes (sodium, potassium, chloride, ionized calcium), pH, and PCO_2 are measured by ion-selective electrode potentiometry.[15,16,17] Concentrations are calculated from the measured potential through the Nernst equation. PO_2 is measured amperometrically. The oxygen sensor is similar to a conventional Clark electrode. Oxygen permeates through a gas permeable membrane from the blood sample into an internal electrolyte solution where it is reduced at the cathode. The oxygen reduction current is proportional to the dissolved oxygen concentration. Glucose is oxidized in the presence of

Table 11.1 Diagnostic tests available on the i-STAT analyzer.

	Principle of measurement	Measurement	Range
CHEMISTRY			
Sodium	Thin film sodium selective electrode.	Potentiometric	100–180 mmol L^{-1}
Potassium	Thin film potassium selective electrode.	Potentiometric	2.0–9.0 mmol L^{-1}
Chloride	Thin film chloride selective electrode.	Potentiometric	65–140 mmol L^{-1}
Ionized calcium	Thin film calcium ion selective electrode.	Potentiometric	0.25–2.5 mmol L^{-1}
Glucose	Glucose is oxidized in the presence of glucose oxidase which is immobilized over an electrode. The oxidation of the resulting hydrogen peroxide is electro-chemically measured.	Amperometric	1.1–38.9 mmol L^{-1}
Urea	Urea is hydrolyzed in the presence of urease. The production of ammonium is measured by a micro-fabricated ammonium selective electrode coated with the immobilized enzyme.	Potentiometric	1–50 mmol L^{-1}
Creatinine	Creatinine is hydrolyzed to creatine, creatine to sarcosine, and sarcosine to glycine and hydrogen peroxide, which is electro-chemically measured.	Amperometric	17–1768 μmol L^{-1}
Lactate	Lactate oxidase selectively converts lactate to pyruvate and hydrogen peroxide.	Amperometric	0.5–17.0 mmol L^{-1}
PO$_2$	Clark type electrode, where oxygen permeates through a gas permeable membrane to a thin film oxygen reducing electrode.	Amperometric	0–800 mm Hg
PCO$_2$	Severinghaus type electrode, where CO$_2$ permeates through a gas permeable membrane to a bicarbonate solution. It dissolves and dissociates to create a pH change, which is measured at a thin film pH electrode.	Potentiometric	10–100 mm Hg
pH	Thin film hydrogen ion selective electrode.	Potentiometric	6.8–8.0
CARDIAC MARKERS			
cTnI	Two-site enzyme-linked immune-sorbent assay (ELISA) labeled with alkaline phosphatase.	Amperometric	0–50 μg L^{-1}
CK-MB	Two-site enzyme-linked immune-sorbent assay (ELISA) labeled with alkaline phosphatase.	Amperometric	0–150 μg L^{-1}
BNP	Two-site enzyme-linked immune-sorbent assay (ELISA) labeled with alkaline phosphatase.	Amperometric	15–5000 ng L^{-1}
HEMATOLOGY			
Hematocrit	A two terminal, noble metal electrodes, alternating current resistivity cell.	Conductivity	15–75%
Hemoglobin	Calculated value		2.1–15.8 mmol L^{-1}
Kaolin ACT	Complete activation of the coagulation cascade *via* conversion of a thrombin substrate other than fibrinogen.	Amperometric	50–1000 s
PT/INR	Complete activation of the extrinsic pathway of the coagulation cascade when initiated with a thromboplastin.	Amperometric	0.9–8.0 INR

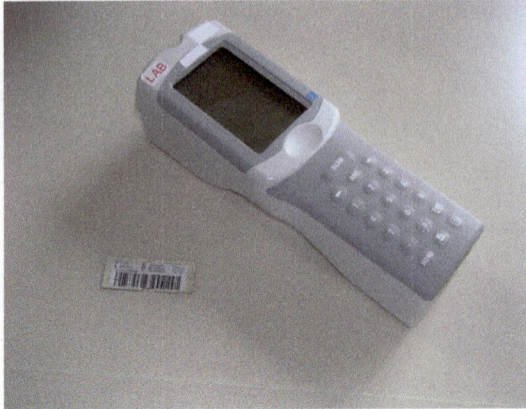

Figure 11.4   The i-STAT portable analyzer.

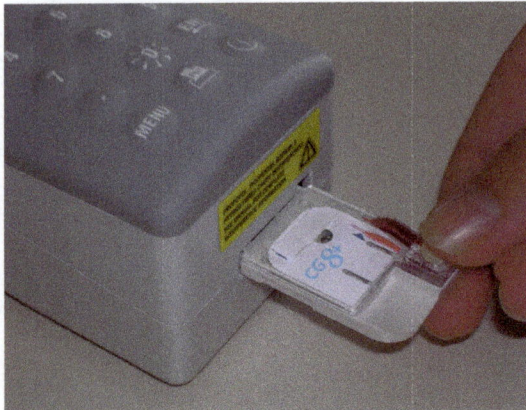

Figure 11.5   The cartridge is inserted into the i-STAT analyzer.

glucose oxidase, which is immobilized over an electrode. The oxidation of the resulting hydrogen peroxide is electro-chemically measured. Urea (BUN) is measured by a potentiometric assay. Urea is hydrolyzed in the presence of urease and the production of ammonium is measured by an ammonium selective electrode. Creatinine is analyzed by an amperometric test, which uses several enzymes to couple a series of reactions that ultimately produce hydrogen peroxide. Creatinine is hydrolyzed to creatine, creatine to sarcosine, and sarcosine to glycine and hydrogen peroxide. The hydrogen peroxide is then oxidized at a platinum electrode to produce a current that is proportional to the creatinine concentration. Lactate measurement with iSTAT is an amperometric test that measures L-lactate concentration in the plasma fraction of a whole blood sample. The enzyme lactate oxidase, immobilized in the lactate biosensor, selectively converts lactate to pyruvate and hydrogen peroxide. The liberated hydrogen peroxide is oxidized at a platinum electrode to produce a current, which is proportional to the sample lactate concentration.

11.4.2 Cardiac Markers

Human cardiac Troponin I (cTnI), creatine kinase isoenzyme MB (CK-MB), and B-type natriuretic peptide (BNP) are measured amperometrically by using microfabricated two-site enzyme-linked immune-sorbent assay (ELISA). Antibodies specific for cTnI or CK-MB or BNP are immobilized on an electrochemical sensor fabricated on a silicon chip. Antibody/alkaline phosphatase enzyme conjugate specific to a separate portion of the analyte is also immobilized in another part of the chip. The whole blood sample is brought into contact with the sensors allowing the enzyme conjugate to dissolve into the sample. The cTnI or CK-MB or BNP within the sample become labeled with alkaline phosphatase and are captured onto the surface of the chip during an incubation period of 7 minutes. The sample, as well as excess enzyme conjugate, is washed off the sensors. Within the wash fluid is a substrate for the alkaline phosphatase enzyme. The enzyme bound to the antibody/antigen/antibody sandwich cleaves the substrate releasing an electrochemically detectable product. The amperometric sensor measures this enzyme product, which is proportional to the concentration of the cardiac marker within the sample.

11.4.3 Hematology

Hematocrit is determined conductometrically. The measured conductivity, after correction for electrolyte concentration, is inversely related to the hematocrit.

Hemoglobin values are calculated based on the hematocrit measurement. The activated clotting time (ACT), described by Hattersley,[18] has been the first bedside system employed to assess coagulation during heparin therapy. This test measures the time required by whole-blood samples to clot after contact activation uses an activator, either kaolin or celite to accelerate co-agulation by activation of the contact pathway. The principles of the i-STAT ACT test are based on this assay by using kaolin (Kaolin ACT) or celite (Celite ACT) as activator.[19] The endpoint of this method is indicated by the conversion of a thrombin substrate other than fibrinogen. The substrate used in the electrogenic assay (*H*-D-phenylalanyl-pipecolyl-arginine-*p*-amino-*p*-methoxydiphenylamine) has an amide linkage that mimics the thrombin-cleaved amide linkage in fibrinogen. Thrombin cleaves the amide bond at the carboxy-terminus of the arginine residue. This reaction produces an electroactive compound that is detected amperometrically, and the time of detection is measured in seconds.

The i-STAT PT/INR test is a whole blood determination of the prothrombin (PT) time used for monitoring oral anticoagulant therapy. The test deter-mines the time required for complete activation of the extrinsic pathway of the coagulation cascade when initiated with a thromboplastin. In a PT test, coagulation is initiated by mixing the sample with tissue thromboplastin. In traditional PT tests, complete activation is indicated when activated thrombin converts fibrinogen to fibrin and extensive or localized clots are detected mechanically or optically. The i-STAT PT/INR test is similar except that the endpoint is indicated by the conversion of a thrombin substrate other than fibrinogen. An electrochemical sensor is used to detect this conversion. The added thrombin substrate is H-D-phenylalanyl-pipecolyl-arginine-p-amino-p-methoxydiphenylamine. Thrombin cleaves the amide bond at the carboxy terminus of the arginine residue because the bond structurally resembles the thrombin-cleaved amide linkage in fibrinogen. The product of the thrombin-substrate reaction is an electrochemically inert tripeptide (phenylalanyl-pipecolyl-arginine) and an electroactive compound NH_{3+}-C_6H_4-NH-C_6H_4-OCH_3. A formation of the electroactive compound is detected amperometrically and the time of detection is measured. The PT/INR test result is reported as an International Normalized Ratio (INR) and, optionally, in seconds. The INR is the recommended method of result reporting for monitoring of oral anticoagulant therapy.

11.4.4   Clinical Use and Performance

The *clinical chemical parameters* especially electrolytes and blood gas of i-STAT are most frequently used in emergency, intensive care, pediatric, and surgical departments of hospitals.[20,21,22] Serial venous POCT lactate measurements could help predict injury severity and occurrence of a bad outcome in undifferentiated trauma patients[23] and emergency department patients at increased risk of sepsis,[24] and together with blood gases help to judge the severity of tissue hypoxia as seen in severe anemia, shock, cardiac decompensation, pulmonary insufficiency, or alcohol/drug intoxication. The analytical and clinical evaluations showed very good comparisons between the results produced by the central laboratory reference instruments and i-STAT using by laboratory professionals or by non-laboratory health-care professionals.[20,21,22,25] The performance of the i-STAT in these settings indicated that this device is generally suitable in a clinical environment. The speed of the assay, which is 2 min., coupled with acceptable precision (total imprecision CV <5%) indicate that this system is appropriate for clinical use. The internal monitoring systems of i-STAT, which detect events such as insufficient sample, overfill of the sample chamber, or calibration failure provides safeguards against preanalytical errors. The small number of rejected results (<5%) shows the robust nature and simplicity of the system. Being single-use and cartridge based, the i-STAT System is not prone to persistent errors stemming from electrode drifts or repeated use of the same sensors and sample channels. In addition, each unitized i-STAT cartridge, sealed in its own separate foil pouch and having its own individual history, cannot be susceptible to events that may affect an entire reagent batch on other systems. Traditional quality control regimens such as testing a cartridge with a liquid control solution offer no benefit to this system.[16]

Biochemical markers of cardiac injury play an important role in the diagnosis, prognosis, monitoring, and risk stratification of patients with cardiovascular diseases and have become an essential part of therapeutic and intervention guidelines.[26,27] POCT in patients with ischemic heart diseases is driven by the time-critical need for fast, specific, and accurate results to initiate therapy instantly. The results of the cardiac marker testing should be available to the cardiologist within 30 min. after the patient has arrived at the emergency department.[27,28] This fact is a clear cause and an absolute indication in the use of POCT to measure cardiac markers.[29] I-STAT was the first POCT device on which measurement of cTnI, the most important cardiac biomarker, was available. It has an analytical sensitivity of 0.02 ng mL^{-1} and a measuring range to 50.0 ng mL^{-1}. The total imprecision is <10%, which seems to be sufficient for clinical use because it must be <25% according to the National Academy of Clinical Biochemistry.[28] In a consensus document from the European Society of Cardiology (ESC) and the American College of Cardiology (ACC), myocardial infarction was redefined as any amount of myocardial necrosis in the presence of myocardial ischemia, as indicated by an increased cardiac troponin above the 99th percentile of a reference population (which is <0.08 ng mL^{-1} in our hospital). The ability for cTnI to be measured at low concentrations allows therapeutic intervention to be considered at any elevation above the normal range. Acute myocardial injury is evidenced by temporal changes in troponin levels while consistent elevations of troponin may be suggestive of other chronic cardiac or non-cardiac conditions. The use of a serial protocol allows identification of temporal changes, as well as clarifying the clinical diagnosis for those patients with low-level results.

Although the cardiac-specific troponin is now considered the biochemical marker of choice in the evaluation of acute coronary syndromes, CK-MB can also be used as a secondary marker to aid in the diagnosis of myocardial infarction and measuring the degree of myocardial necrosis. Since low levels of CK-MB can be detected in the blood of healthy persons, any CK-MB value above the 95th percentile may be indicative of some degree of myocardial necrosis. As the level of CK-MB is not cardiac specific, the results of a single test are not indicative of a myocardial infarction. Typically a myocardial infarction is diagnosed based on the pattern of CK-MB analyses taken at 3 hour intervals for a 6- to 9-hour period or at 6 to 8 hour intervals for a 24-hour period.[30,31]

BNP is one of a family of structurally similar peptide neurohormones whose function is to regulate blood pressure, electrolyte balances, and fluid volume. BNP is synthesized, stored, and released primarily by the ventricular myocardium in response to volume expansion and pressure overload.[32] BNP has an important role in the prognostic assessment of patients with heart failure.[33,34] BNP is a powerful prognostic indicator for patients with congestive heart failure at all stages of the disease and seems to be a better predictor of survival than many traditional prognostic indicators.[35] BNP has also been shown to predict morbidity and mortality in other cardiovascular conditions, such as acute coronary syndromes and acute myocardial infarction.[33,36] Patients with acute coronary syndrome with increased BNP levels have a higher rate of cardiac complications and higher mortality post myocardical infarction.[36] The i-STAT BNP assay is specific (has no cross-reaction with the other member of the neurohormone family), sensitive (has an analytical sensitivity of 14 pg mL^{-1}), and has a total imprecision of <11%, which allow this assay to be used for clinical settings. It has a clinical sensitivity and specificity of 80% and 85%, respectively, for the laboratory diagnosis of heart failure.[37]

In different clinical settings, such as cardiopulmonary bypass, interventional cardiology procedures, and hemodialysis, the need exists for adequate anticoagulation and its rapid assessment to prevent thrombosis of circuits used during extracorporeal circulation. The current practice is to rapidly monitor the degree of heparin-induced anticoagulation and its reversal by means of activated clotting time, which is performed with automated bedside devices. The i-STAT Kaolin ACT or Celite ACT tests reflect the activity of heparin administered during any cardiovascular procedure, and they have a widespread clinical use for cardiopulmonary bypass, interventional cardiology procedures, and hemodialysis.[19,38]

The i-STAT PT, a prothrombin time test, is used for monitoring patients receiving oral anticoagulation therapy. The test result is reported as INR (International Normalized Ratio) because this is the WHO recommended method of result reporting for monitoring of oral anticoagulant therapy. The INR is a dimensionless ratio between the measured prothrombin time of the patient and the mean normal i-STAT prothrombin time both in seconds.

11.5 Conclusions

POCT provides a revolutionary diagnostic technology in improving patient outcomes. Although developing rapidly and being more and more accepted by hospitals and patients, POCT still faces some challenging problems. This is mainly due to the fact that POCT as a laboratory trend is much more than just a reliable analyzer. Due to the present technology POCT devices are extremely user friendly. Therefore they can be brought to the end-user (patient) but the device cannot bring the expertise of laboratory professionals for the quality control and interpretation of the results. Lack of test standardization is still a big problem to be solved. Different POCT systems and different assay generations may yield different results for the same analyte, because of the use of different assays, reference materials, assay imprecision, and setting of different reference and decision limits. Harmony between the results obtained in central laboratory and those obtained by various POC assays remains an important issue. In addition, only a reliable quality-management system can insure valid laboratory test results, and the laboratory professionals have to be responsible for this quality control system. Further development of POCT instruments has to result in in-built quality control which can help to solve this problem. However, due to the fact that these instruments will be used by untrained end-users it will leave the pre- and post-analytical factors unsolved. The most important post-analytical problem is the interpretation of the results. "Telemedicine" is increasingly important from this point of view. Instruments used to monitor the patient can, in the right circumstances, send the data obtained by a POCT device directly to the relevant medical advisory service point, which can validate and interpret the results and intervene if necessary.

References

1. C. D. Chin, V. Linder and S. K. Sia, *Lab Chip*, 2007, 7, 41.
2. M. C. Winter, *Point of Care*, 2010, **9**, 12.
3. S. E. F. Melanson, *Point of Care*, 2011, **10**, 63.
4. P. B. Luppa, C. Müller, A. Schlichtiger and H. Schlebusch, *Trends Anal. Chem.*, 2011, **30**, 887.
5. C. M. Girardin, C. Huot, M. Gonthier and E. Delvin, *Clin. Biochem.*, 2009, **42**, 136.
6. P. D'Orazio, R. W. Burnett, N. Fogh-Andersen, E. Jacobs, K. Kuwa, W. R. Külpmann, L. Larsson, A. Lewenstam, A. H. J. Maas, G. Mager, J. W. Naskalski and A. O. Okorodudu, *Clin. Chem. Lab. Med.*, 2006, **44**, 1486.
7. E. Alarousu, J. Hast, M. R. J. McNichols and G. L. Coté, *J. Biomed. Optics*, 2000, **5**, 5.
8. J. M. Burrin and K. G. Alberti, *Diabet. Med.*, 1990, 7, 199.
9. World Health Organization, Geneva, 1999.
10. American Diabetes Association, *Diabetes Care*, 2005, **28**, 4.
11. K. Gijzen, D. Moolenaar, J. Weusten, J. Pluim and A. Demir, *Clin. Chem. Lab. Med.*, 2012, Epub ahead of print.
12. F. G. Cunningham, K. J. Leveno, S. L. Bloom, J. C. Hauth, L. C. Gilstrap and K. D. Wenstrom, *Diabetes. Williams Obstetrics*, McGraw-Hill, New York, 22nd edn, 2005, p. 1169.
13. American Diabetes Association, *Diabetes Care*, 2003, **26**, 5.
14. M. Y. Kirillin, A. V. Priezzhev, M. Kinnunen, E. Alarousu, Z. Zhao, J. T. Hast and R. A. Myllylä, *Proc. SPIE*, 2004, **5325**, 164.
15. I. R. Lauks, H. J. Wieck, M. P. Zelin and P. J. Blyskal, US Pat., 5,096,669, 1992.
16. N. Peled, *Pure Appl. Chem.*, 1996, **68**, 1837.
17. I. R. Lauks, *Acc. Chem. Res.*, 1998, **31**, 317.
18. P. G. Hattersley, *JAMA*, 1966, **96**, 150.
19. R. Panniccia, S. Fedi, F. Carbonetto, D. Noferi, P. Conti, B. Bandinelli, B. Giusti, L. Evangelisti, P. Pretelli, M. F. G. Palmarine, R. Abbate and D. Prisco, *Anesthesiology*, 2003, **99**, 54.
20. K. A. Erickson and P. Wilding, *Clin. Chem.*, 1993, **39**, 283.

21. E. Jacobs, E. Vadasdi, L. Sarkozi and N. Colman, *Clin. Chem.*, 1993, **39**, 1069.

22. D. A. Adams and M. Buus-Frank, *J. Pediatr. Nurs.*, 1995, **10**, 194.

23. A. L. Blomkalns, M. Sperling, S. E. Ronan, A. M. Hochhausler and C. J. Lindsell, *Point of Care*, 2009, **8**, 4.

24. N. I. Shapiro, C. Fischer, M. Donnino, L. Cataldo, A. Tang, S. Trzeciak, G. Horowitz and R. E. Wolfe, *J. Emerg. Med.*, 2010, **39**, 89.

25. C. Papadea, J. Foster, S. Grant, S. A. Ballard, J. C. Cate, W. M. Southgate and D. M. Purohit, *Ann. Clin. Lab. Sci.*, 2002, **32**, 231.

26. Z. Yang and D. M. Zhou, *Clin. Biochem.*, 2006, **39**, 771.

27. R. H. Christenson and M. E. Azzazy, *Clin. Biochem.*, 2009, **42**, 150.

28. U. Friess and M. Stark, *Anal. Bioanal. Chem.*, 2009, **393**, 1453.

29. M. I. Mohammed and P. Y. Desmulliez, *Lab. Chip*, 2011, **11**, 569.

30. J. S. Alpert and K. Thygesen, *J. Am. Coll. Cardiol.*, 2000, **21**, 1502.

31. F. S. Apple, R. Ler, A. Y. Chung, M. Berger and M. A. M. Murakami, *Clin. Chem.*, 2006, **52**, 322.

32. A. M. Richards, M. G. Nicholls, T. G. Yandle, H. Ikram, E. A. Espiner, J. G. Turner, R. Buttimore, J. Lainchbury, J. Elliott, C. Frampton, I. Crozier and D. Smyth, *Heart*, 1999, **81**, 114.

33. R. J. Rodeheffer, *J. Am. Coll. Cardiol.*, 2004, **44**, 740.

34. J. A. Doust, E. Petrzak, A. Dobson and P. Glasziou, *Br. Med. J.*, 2005, **330**, 625.

35. I. S. Anand, L. D. Fisher, Y.-T. Chiang, R. Latini, S. Masson, A. P. Maggioni, R. D. Glazer, G. Tognoni and J. N. Cohn for the Val-HeFT Investigators, *Circulation*, 2003, **107**, 1278.

36. A. M. Richards, M. G. Nicholls, E. A. Espiner, J. G. Lainchbury, R. W. Troughton, J. Elliott, C. Frampton, J. Turner, I. G. Crozier and T. G. Yandle, *Circulation*, 2003, **107**, 2786.

37. S. J Wieczorek, A. H. B. Wu, R. Christenson, P. Krishnaswamy, S. Gottlieb, T. Rosano, D. Hager, N. Gardetto, A. Chiu, K. R. Bailly and A. Maisel, *Am. Heart J.*, 2002, **144**, 834.

38. D. Prisco and R. Panniccia, *Thromb. J.*, 2003, **1**, 1.

CHAPTER 12

Medimate Minilab, a Microchip Capillary Electrophoresis Self-test Platform

STEVEN S. STAAL,*[a] MATHIJN C. UNGERER,[a]
KRIS L. L. MOVIG,[b] JODY A. BARTHOLOMEW,[c] HANS KRABBE[c]
AND JAN C. T. EIJKEL[d]

[a] Medimate BV, De Veldmaat 10, 7522 NM Enschede, the Netherlands;
[b] Department of Clinical Pharmacy, Medical Spectrum Twente,
Haaksbergerstraat 55, 7513 ER Enschede, the Netherlands; [c] Medlon BV,
Ariensplein 1, 7500 KA Enschede, the Netherlands; [d] Bios Lab-on-a-Chip
Group, University of Twente, Drienerlolaan 5, 7500 AE Enschede,
the Netherlands
*Email: steven.staal@medimate.com

12.1 Introduction

The Medimate Minilab is a new technology platform for the measurement of
ionic species in blood by lay users, which was introduced in the market
in 2013.

The Minilab technology is based on microchip capillary electrophoresis, a
well-known technology for laboratory analysis of ionic species. Historically,
companies like Beckman Coulter Life Sciences, Helena Biosciences, and
others provide bench-top capillary electrophoresis devices for clinical analysis.
Specialized lab personnel can identify and quantify species by performing
capillary electrophoresis with the aid of dedicated software programs.

RSC Nanoscience & Nanotechnology No. 36
Microfluidics for Medical Applications
Edited by Albert van den Berg and Loes Segerink
© The Royal Society of Chemistry 2015
Published by the Royal Society of Chemistry, www.rsc.org

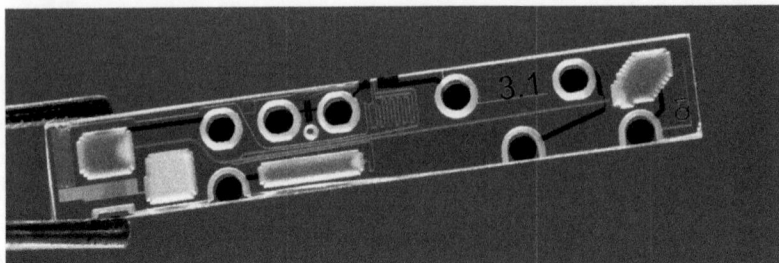

Figure 12.1   Photograph of the microfluidic chip.

Medimate has now successfully transferred this technology from the clinical laboratory to a microfluidic chip and made it available as a self-test platform. Their device, the Medimate Minilab, enables diagnostic measurements by a lay user at any location. Figure 12.1 shows a photograph of the microfluidic chip, which is the diagnostic core of this new platform.

Looking at the present self-test market it is possible to identify a relatively small number of technologies capable of performing quantitative measurements at home. The measured quantities are for instance temperature, weight, size, blood pressure, blood cholesterol level, and blood glucose level. The largest and most famous example that is also closest to the Medimate Minilab is the blood glucose meter. The glucose meter was first marketed around 1981 by the companies Bayer and Roche with the Glucometer and Accucheck respectively.[1] The blood glucose meter caused a clinical revolution comparable to the cell phone revolution in telecommunication in the 1990s. From 1981 onwards, patients were given the ability to measure glucose at home, to measure it more often and at each specific moment in time. This change caused a huge increase in quality of life of patients with diabetes mellitus and increased the life expectancy of the patients towards the normal life expectance of healthy people.[2] Now, with the transfer of capillary electrophoresis to a self-test, a new, more complex platform enters the market able to quantify species for which home measurements were never possible. Table 12.1 sums up the advantages of a self-test compared to conventional laboratory practice. Of course the use of the self-test is not only restricted to home use. The ease of operation and robustness of the test make it also suitable for use at the point of care, in first aid and even in laboratories where measurements are less frequent and where facility and training costs are to be reduced.

This chapter will first explain the technology in more detail, for example how a measurement is performed, how species are measured and which achievements have been realized during the transfer of the technology towards a self-test. After explaining the technology the lithium validation results will be presented to demonstrate the capabilities of the platform as an *in vitro* diagnostic device and more specifically as a self-test. First a brief explanation of lithium medication is given as background information to make it easier to understand certain choices made in the design of the

Table 12.1   Advantages of a self-test compared to conventional laboratory practice.

	Self-test	Conventional laboratory practice
Measurement time	Minutes	Hours
Location	Anywhere	Hospital
Hardware	Simple	Complex
User	Anyone	Lab technician
Indirect costs	Low	High
Sample	Droplet	Cuvette
Maintenance	No	Required
Telecommunication	Yes	No
Patient empowerment	Strong	No

validation studies. Subsequently the validation method, results, and conclusions are discussed in detail. Thereby two studies are presented, an in-house study by Medimate to define the performance and a study by an external party to verify the performance. Studies were performed according to the latest protocols from the Clinical and Laboratory Standards Institute (CLSI) including information about the performance by a lay user. In view of the fact that the technology platform is suitable to measure more species, this chapter will end with an overview of the platform capabilities. Capabilities will be demonstrated to measure sodium in urine, creatinine in blood, and even calcium in cows' blood. Finally the future potential will be discussed.

12.2   Microfluidic Capillary Electrophoresis as a Self-test Platform

In this section it is explained how the technology platform of the Medimate Minilab enables the measurement of different ionic species in aqueous solutions. After explaining how a measurement is performed, we will explain how the technology is applied for sample analysis, species concentration calculation, and communication of the result. Finally the inventions that were required to transfer capillary electrophoresis (CE) from a lab technology[3] to a self-test are explained. The main parts of this section have also been published.[4]

12.2.1   Conducting a Measurement

The Medimate Minilab enables the measurement of different ionic species in aqueous solution. For a measurement different components are required, which are shown in Figure 12.2. Each measurement is performed on a one-time use cartridge called the Lab-chip. The Medimate Multireader, equipped with a voltage adapter, initiates the actual measurement, calculates the concentration, and communicates the final result. In case of a fingerstick an extra fingerstick pen is used.

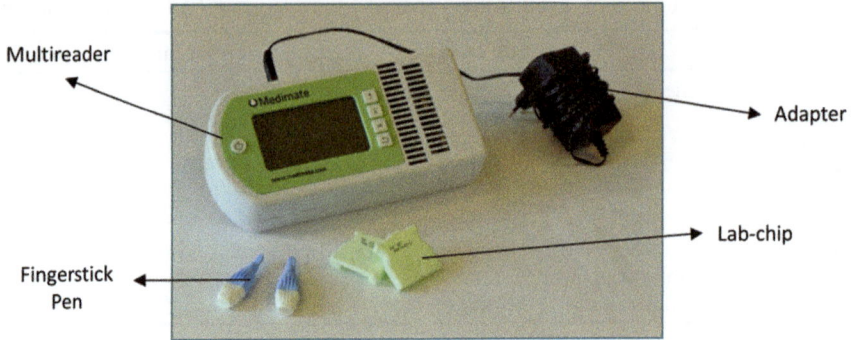

Figure 12.2 The Medimate Minilab.

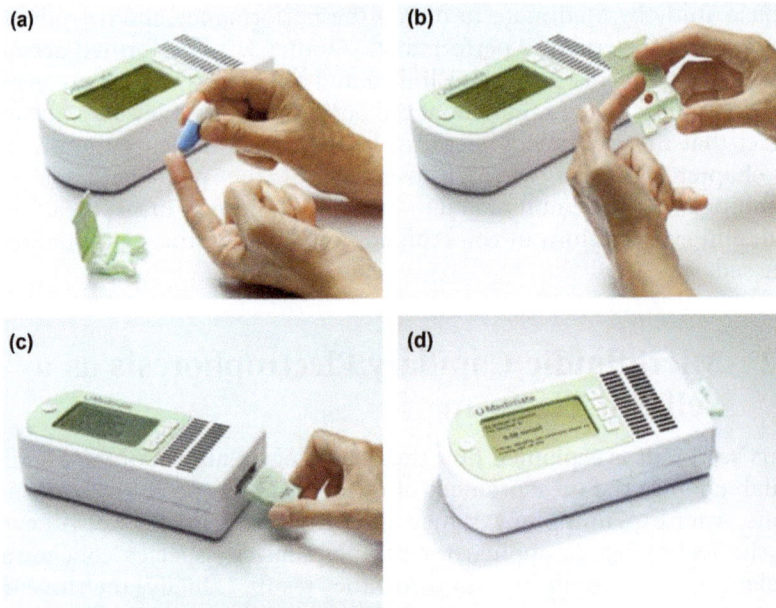

Figure 12.3 Performing a measurement. A fingerstick is performed with the fingerstick pen (a), after which the blood droplet from the finger is applied to the Lab-chip (b). Then the Lab-chip is inserted into the Multireader (c), and finally after a few minutes the concentration value is shown on the screen (d).

A measurement with the Medimate MiniLab using the fingerstick proceeds as shown in Figure 12.3. The measurement is performed in four basic steps: sampling, application, Lab-chip placement, and analysis.

12.2.2 Measurement Process

The measurement process is automatically performed by the Medimate Multireader and the Lab-chip. The Lab-chip consists of three parts: a

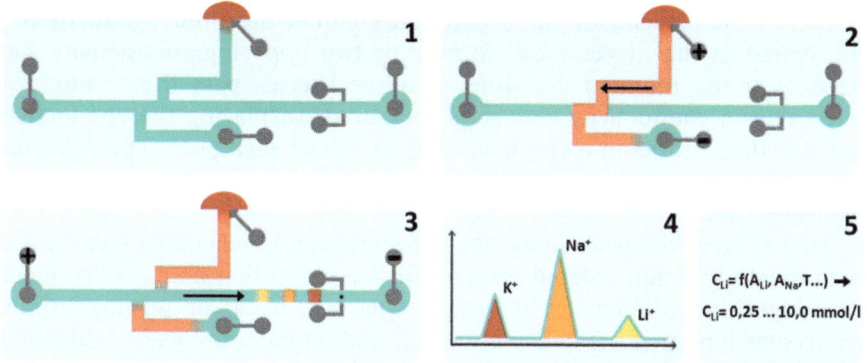

Figure 12.4   Schematic representation of the measurement process. Red: sample droplet. blue: prefilled channels and reservoirs; gray: electrodes and electrode pads.

microfluidic glass chip, housing, and a seal. The measurement process can be divided into eight consecutive steps; sampling, injection, separation, detection, calculation, error detection, displaying the result, and (optionally) telecommunication.

The core of the technology is the microfluidic glass chip. This glass chip is 30 mm long and 4 mm wide and consists of a network of channels, electrodes, and reservoirs. It comes prefilled with an analysis solution. A schematic representation of this network is given in Figure 12.4. The first five measurement steps are explained with the aid of this scheme.

Step 1: After removing the seal and performing a fingerstick, a sample droplet is placed at the sample entrance of the channel system.

Step 2: An electric field is applied along the injection channel. By applying a high positive voltage at the sample electrode, the positive ionic species from the sample will move towards the other end of the injection channel.

Step 3: An electric field is applied along the separation channel. When the high voltage electrode near the conductivity detector cell is negative, the positive ionic species in the tiny volume defined by the cross-section of the two channels will move to the conductivity detector cell at the end of the separation channel. Due to their velocity differences in the electrical field, the different ionic species are separated from each other. The velocity differences of the ionic species can be characterized by the ion velocity equation shown in eqn (12.1).

$$v_i = \mu_i E \tag{12.1}$$

Here v_i (m s^{-1}) is the velocity of ionic species i, μ_i (m^2 V^{-1}s^{-1}) its mobility in an electrical field, and E (V m^{-1}) the applied electric field. The mobility determines which species is detected first and which later. The mobility depends on the viscosity of the surrounding fluid, η (Pa s), the ionic radius, r_i (m), of the individual ionic species, and its charge, z_iq(C), as shown in eqn (12.2).

$$\mu_i \approx \frac{z_iq}{6\pi\eta r_i} \tag{12.2}$$

Step 4: At the end of the separation channel the fluid conductivity is measured in the detector cell formed by two opposing conductivity electrodes. At the moment the different ionic species pass the conductivity electrodes a conductivity change is measured. Displaying the conductivity against time results in a graph with peaks, where each peak represents one type of species and where the size of the peak is a measure for the concentration.

Step 5: From the peak signal the concentration is calculated based on an empirical algorithm created by multivariate regression analysis. To create this algorithm differences in sample type (*e.g.* blood or serum), lithium concentration, disposable batch, device, and temperature were deliberately included. As reference the IL943 flame photometer was used. The final algorithm used for the serum outcome is derived from a regression with solely serum measurements, the algorithm for the fingerstick outcome from a regression with serum, EDTA whole blood, EDTA fingerstick, and fingerstick measurements.

Step 6: An error detection protocol is initiated on the basis of implemented error detection criteria. Criteria are amongst others based on instable sample contact, instable Multireader Lab-chip contact and electrophoresis instability.

Step 7: The lithium value is shown on the display and stored inside the Multireader. The analysis time for the lithium measurement is about 9 minutes and for the sodium measurement about $2\frac{1}{2}$ minutes.

Step 8: *Via* USB it is possible to connect the Multireader to the internet and transfer measurement data towards a central laboratory or hospital system, including sample, order, and patient information. It is also possible to connect to the Medimate server for software updates and other support. It is expected that the Multireader will have Bluetooth support in the near future.

12.2.3 From Research Technology to Self-test Platform

Robustness and ease of operation are key requirements for any self-test. The Medimate Minilab was developed so that both requirements would be met. The main achievements in the transfer from research to self-test are shown in Table 12.2.

Table 12.2 Important features needed for the use of the Medimate Minilab as self-test.

✓ No sample mess	✓ Easy handling	✓ Low manufacturing costs
✓ No sample cleaning	✓ No calculations	✓ No storage conditions
✓ No reagent filling	✓ Handheld	✓ Fast analysis times
✓ No market calibration	✓ No venous sample	✓ Telecommunication
✓ Reliable	collection	functionalities
✓ Small sample volume	✓ Location	✓ Multiple species analysis
	independent	

Table 12.3 Main inventions enabling the required self-test features.

✓ Microfluidic chip	✓ Closed channel system	✓ Electric isolation
✓ Protected sample	✓ Evaporation chamber	✓ No calibration
✓ Pre-filled	✓ Sample detection	✓ Standard electronics
✓ Gas bubble		

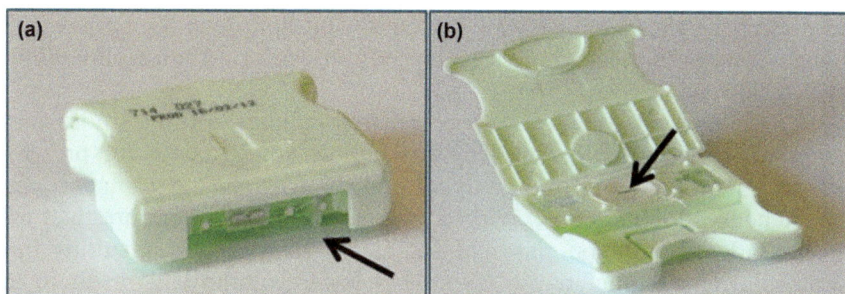

Figure 12.5 (a) Closed Lab-chip, the arrow points at the glass chip, (b) open Lab-chip shown with seal removed, the arrow points at the location where the sample will be applied.

To realize a platform fulfilling these requirements, inventions for the transfer of the laboratory technology to a self-test were required, see Table 12.3. The main inventions are patented by the company.[5,6,7]

In Figure 12.5 a photograph is shown of a Lab-chip with the glass chip visible in the front. Figure 12.6 shows a schematic of the network structure in the glass chip, with the different functional parts indicated. For the clarity of explanation the schematic is not an exact copy of the real structures, which can be seen in Figure 12.1. The inventions are explained with the aid of Figures 12.5 and 12.6.

Microfluidic chip: Using a microfluidic chip enables analysis times of a few minutes. Other advantages of the microfluidic chip are the low sample volume requirements of 10 microliters and the low cost perspective for mass production. Prefilled channels and reservoir as shown by I and II in Figure 12.6 enable the capillary electrophoresis separation process.

Protected sample: The sample is placed inside the disposable Lab-chip cartridge with a cover lid that can be closed. This ensures clean working conditions and minimal chance of contamination of the Multireader connector.

Pre-filled: The glass chip comes prefilled with measuring solution and is suitable for storage and ready for sample application. These features are realized by designing the glass chip cartridge in such a way that a polymer seal can be fitted closing the sample opening. The seal is designed to keep the fluid stored over time and can easily be removed prior to measurement. Specifically the location of the sample opening at the side of the glass chip is very convenient, see Figure 12.5, since it allows a vertical orientation of the chip when the sample is applied at the opening.

Closed channel system: Solution flow has to be minimized to prevent interference with the electrokinetic injection and separation during the measurement. For this reason a fully closed channel system was chosen with solely one (sample) opening blocking solution flow in all directions (see the channel and reservoir structure in Figure 12.6). Because temperature changes during the measurement can still cause solution flow, the temperature change during the measurement is limited by design and reservoir volumes are kept low.

Gas bubble: A gas bubble has been created inside the furthermore completely filled channel system. This gas bubble compensates for pressure differences originating from temperature changes during transport and storage between 0 and 60 °C. Without the gas bubble temperature rises will force solution out of the glass chip at higher temperature, resulting in a lower solution volume in the chip. This lower volume will result in an instable sample contact at the time of measurement. The gas bubble is contained in a separate reservoir (number 1 in Figure 12.6).

Evaporation chamber: To ensure sufficient sample handling time between the moment the seal is removed and the moment of sample application, the sample channel has to be kept completely filled. This is realized by a design with two parallel channel entrances. The difference in opening diameters ensures that capillary forces keep the smaller sample entrance filled while the larger evaporation opening will slowly empty. The openings are indicated at location 2 in Figure 12.6. In case of evaporation at the small sample opening, the solution in the injection channel will be replenished from the solution in the larger opening *via* a connecting channel.

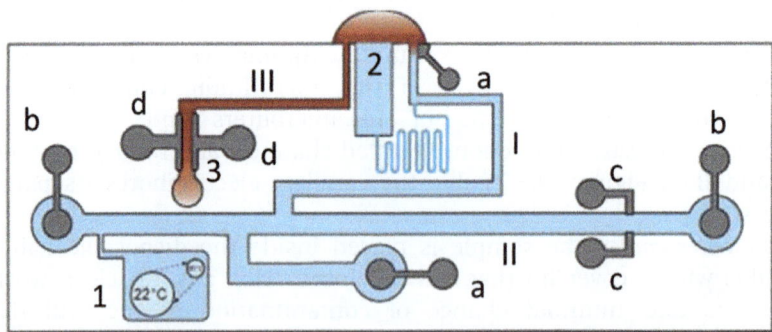

Figure 12.6 Schematic structure of the Lab-chip with indication of the different functional units. (a) High voltage electrodes for field application along the injection channel (I), (b) high voltage electrodes for field application along the separation channel (II), (c) conductivity electrodes, (d) sample electrodes to measure the sample presence after filling the sample channel (III). (1) Gas bubble reservoir, (2) evaporation reservoir, (3) sample detection electrodes.

Sample detection: The sample electrodes at location 3 in Figure 12.6 can be used to detect the presence of sample that entered the microchip *via* channel III. Additionally they can aid in the calculation of the lithium concentration by conductometric measurements.

Electric isolation: The combination of contact conductivity detection and high separation voltages in the fluidic network in the glass chip can cause electrolysis and bubble formation at the conductivity electrodes. To avoid bubble formation the electric current flowing between both circuitries has to be minimized. This was realized by electrically isolating both electric circuits from each other by clever design, as both circuits are positioned inside one small embodiment of the Multireader.

Standard electronics: The Multireader is factory-calibrated and market calibration is not necessary. Additionally, it is not necessary to enter correction factors in the Multireader for different disposable batches as in some cases is required for glucose measurements.

12.3 A Lithium Self-test for Patients with Manic Depressive Illness

As explained in the introduction, the lithium self-test is the first application of the Medimate Minilab in the market. To judge the performance of the platform as a diagnostic tool validation tests needed to be performed. The next sections will be about the validation results of lithium. To understand why lithium should be measured, in which concentration range the measurements are expected, and why patients will benefit from self-testing, this section will first provide some background information on lithium medication.

Manic depressive illness affects around 0.7% of the population worldwide.[8] In the Netherlands it was estimated that 88 400 persons had this illness in 2009.[9] Lithium continues to be the main medication given in manic depressive illness despite risks associated with high blood levels. Lithium is a highly effective medicine with a positive response of approximately 80%.[10] Therapeutic drug monitoring is a norm while using lithium. Monitoring of lithium is necessary in clinical practice as lithium levels are below or above the therapeutic range in a substantial number of patients. Regular lithium level monitoring helps in uncovering unsuspected poor compliance and improves clinical management. Lithium has one major drawback which is its toxicity above a certain blood concentration level. When intoxication is not treated it will lead to serious toxic effects such as neurologic damage and even coma and death. Current therapy prescribes daily intake of lithium combined with a regular check at the clinical laboratory. The number of laboratory visits varies per patient and ranges between 2 and 24 visits annually, with an average during the setting phase of 9 to 13 visits.[11] In 2012 30 934 lithium users were registered at the GIP databank, which is the drug information system of the health care insurance board of the Netherlands.[12] The annual number of measurements for patients using lithium is estimated between 200 000 and 300 000.

The lithium blood concentration level in the patient can change over time even when the medication intake is kept constant. The lithium level depends on several different factors such as changes in a person's activity, eating and drinking behavior, and kidney function. It also depends on interaction with other medication that is taken. For this reason even at normal lithium medication intake, the lithium concentration can still reach toxic levels.

The current approach with measurements at the hospital has serious drawbacks. It is for instance not possible to verify the lithium level at the time intoxication is expected, and the patients always need to travel to the laboratory themselves. This traveling takes extra time and can in some cases be costly. Patients are furthermore restricted in traveling to holiday destinations in foreign countries. Finally the psychiatrists, who are most frequently the treating physicians, will not have an actual lithium level available at the time of the consult, making it impossible to accurately judge the current lithium toxicity risks.

On the basis of the considerations given above, it is expected that the availability of a lithium self-test will decrease the number of intoxications and thus the number of hospitalizations. Furthermore it is expected that therapy loyalty and commitment will improve, increasing the effectiveness of the medication and quality of life of the patient. Telecommunication can furthermore aid in the treatment by creating an easy communication channel with the psychiatrist in combination with possible distant medication control by clinical chemists and pharmacists. In conclusion it is expected that the use of lithium self-tests will have a positive impact on society and specifically on patients with manic depressive illness.

The normal therapeutic lithium range in the blood is between 0.40 and 1.20 mmol L^{-1}, see Figure 12.7.[13,14,15] Lithium levels higher than 1.20 mmol L^{-1} should be prevented and if found reduced as soon as possible. Levels above 1.50 mmol L^{-1} are considered toxic and hospitalization should follow. Lithium has a mean half-life time in the human body of approximately 12–48 hours. This figure varies with for example age, kidney function, and time of total lithium exposure. In case of lithium intake once a day, reference levels should be measured 12 hours after medication intake. In case of lithium intake three to four times a day, the measurement should be within one hour before medication is taken.[13]

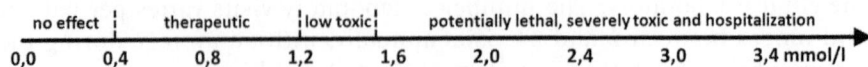

no effect	therapeutic	low toxic	potentially lethal, severely toxic and hospitalization

| 0,0 | 0,4 | 0,8 | 1,2 | 1,6 | 2,0 | 2,4 | 3,0 | 3,4 mmol/l |

Figure 12.7 Effects of different concentrations of lithium in the human body.

12.4 Validation Method

In order to meet customer demands and regulatory requirements it is amongst others required to test the product performance *via* pre-defined validation studies. For point of care *in vitro* diagnostic self-tests this means fulfilling the EU and USA regulatory requirements and fulfilling the expectations of the clinical chemists, physicians, and pharmacists.

The CLSI is the world leading institute for publishing evaluation protocols in strong collaboration with industry, government, and health care professionals. These standards are widely accepted and describe the validation methods in a broad applicable range suitable to be used for many different diagnostic tests. To validate the Medimate Minilab for lithium, the validation plans of the CLSI have to be worked out in detail. The ISO standard for glucose monitoring is the only *in vitro* diagnostic product for which this has already extensively been done and has even been accepted as an ISO standard. This is the ISO 15179:2013 with the title: "In vitro diagnostic test systems–Requirements for blood-glucose monitoring systems for self-testing in managing diabetes mellitus". This standard describes in detail what is important for glucose testing, which tests are to be performed and how the tests are to be conducted. The main approach and principles of this standard are therefore adopted as a guideline for the lithium validation studies performed on the Medimate Minilab when detailing the CLSI protocols.

One specific aspect when it comes to self-testing is the accuracy and reliability of the test when handled by a lay user such as a patient. For glucose monitoring it is well known that the performance by a lay user outside the laboratory is worse compared to the performance of a laboratory technician.[16] For this reason additional tests were carried out to identify the performance under lay user conditions and to identify differences in measurement outcome between professional and lay user settings.

The validation studies can be divided into two separate studies, both intended to obtain the same performance characteristics. The first study is performed by Medimate personnel in-house and the second study is performed by the medical hospital "Medisch Spectrum Twente (MST)" together with Medlon BV, a clinical laboratory group, facilitating clinical laboratory testing for hospitals (including MST) and primary care.

For both studies the Medical Ethical Committee Twente reviewed and approved the study protocols (reference NL34961.044.10). Detailed study reports are available at Medimate.[17,18]

12.4.1 Applied Guidelines

In Table 12.4 the applied CLSI evaluation protocols are shown, including the explanation of the parameters investigated in each protocol. EP-Evaluator is a software program from Data Innovations LLC, which is specifically developed to evaluate and measure clinical laboratory performance. Where possible, EP-Evaluator 9 version 9.6.0.466 is used to evaluate acceptance criteria.

Table 12.4 Selection of analytical protocols applicable to the validation.

ID	Title	Validation study
Analytical guidelines followed		
EP5-A2	Evaluation of precision performance of quantitative measurement method	Repeatability, reproducibility, extreme conditions, and equivalence studies
EP6-A	Evaluation of the linearity of quantitative measurement procedure	Linearity
EP7-A2	Interference testing in clinical chemistry	Interference
EP9-A	Method comparison and bias estimation using patient samples	Method comparison, lay user study
EP17-A	Protocol for determination of limits of detection and limits of quantification	Detection limit, limit of blank, limit of quantification
EP25-A	Stability testing of in vitro diagnostic reagents	Stability, isochronous design
Guidelines taken into consideration		
NEN-EN-ISO 15197 2011 (DRAFT)	In vitro diagnostic test systems – Requirements for blood glucose monitoring systems for self-testing in managing diabetes mellitus	Fingerstick protocol, design of validation studies

Table 12.5 Main lithium acceptability criteria as taken from the Rhoads Table for the Total Allowable Error.

Organization	Lithium acceptance criteria
CLIA, WLSH, AAB, CAP	$\pm$ 20% or 0.30 mmol L^{-1}
NYS	$\pm$ 15% or 0.30 mmol L^{-1}
RCPA	0.30 mmol L^{-1}

12.4.2 Acceptance Criteria

General acceptance criteria for lithium monitoring are defined by different organizations (see Table 12.5).[19] The acceptance criteria for the validation studies are defined as Total Allowable Error (TEA).

The criteria most referred to are the Clinical Laboratory Improvement Amendments (CLIA) criteria.[20] The CLIA is part of the US Federal Register and it results from a partnership between the US Food and Drug Administration (FDA), the US Center for Medicare & Medical Services and the US Center for Disease Control and Prevention.[21]

The FDA states the following about the CLIA: "*Congress passed the Clinical Laboratory Improvement Amendments in 1988 establishing quality standards for all laboratory testing to ensure the accuracy, reliability and timeliness of patient test results regardless of where the test was performed.*"[22]

Medimate has estimated the performance of the Minilab for lithium determination in serum in terms of TEA as 10% and in fingerstick as 15%. This results in the following acceptance criteria for TEA:

- serum   TEA of 0.10 mmol L^{-1} or 10% lithium
- fingerstick   TEA of 0.15 mmol L^{-1} or 15% lithium

The TEA has a 95% confidence interval.

There is not one single test that can be used to verify the TEA criteria. For this reason an approach is chosen to divide the TEA into two separate sections, called error budgets. Both sections are investigated in specific tests. This approach is also advised in the CLSI Evaluation Protocols.[23]

The TEA is divided into two separate budget sections, one section for the random error and one section for the systematic error. In Figure 12.8 the

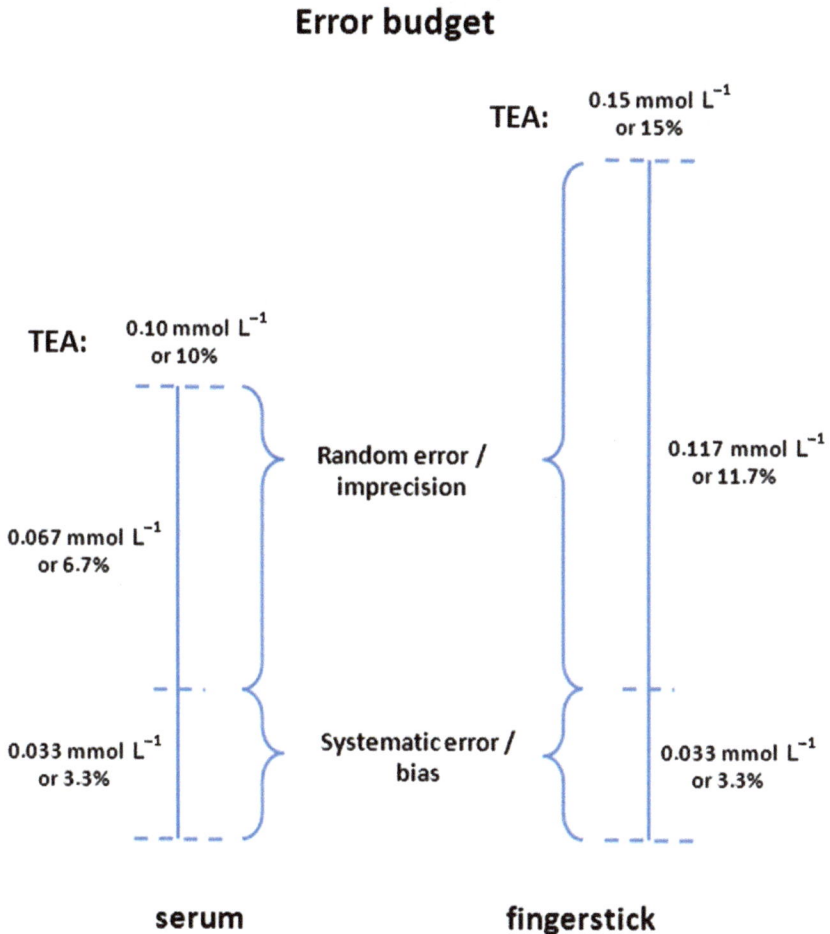

Figure 12.8   Division of the error budget for the Medimate Minilab for serum and fingerstick.

Table 12.6 Error budgets for the main lithium tests.

Study type	Type of error	Lithium acceptance criteria at 95% CI	
		Serum	Fingerstick
Precision	Random	0.067 mmol L^{-1} or 6.7%	0.117 mmol L^{-1} or 11.7%
Non-linearity	Systematic	0.033 mmol L^{-1} or 3.3% 0.4–4	0.033 mmol L^{-1} or 3.3% 0.4–4
Method comparison	Random and systematic	0.1 mmol L^{-1} or 10%	0.15 mmol L^{-1} or 15%

budget is shown for the Medimate Minilab for serum and fingerstick. The width of each section is based on internal knowledge about the expected differences.

Based on the TEA budgeting shown in Figure 12.8 the acceptance criteria for each test can now be defined, see Table 12.6.

All criteria are based on 95% confidence intervals (CI). Since the 95% confidence interval is used, the standard deviation of the lithium outcome as a verification parameter for precision tests can be used. For precision studies where the acceptance criterion is based on standard deviation, it is called the Acceptable Random Error (ARE) and defined as half the precision budget.

12.4.3 Sample Availability, Preparation, and other Considerations

During the studies different sample types are used, which are defined as follows:

- Serum — Venous serum prepared with clot activator and gel for serum separation, silicone coated interior
- Fingerstick — Capillary whole blood taken from the fingers by a fingerstick pen, no anticoagulant
- EDTA fingerstick — Capillary whole blood taken from the fingers by a fingerstick pen with K-EDTA
- Whole blood — Venous whole blood, no anticoagulant
- EDTA whole blood — Venous whole blood with K-EDTA

All therapeutic levels investigated during the studies are obtained from normal subjects taking lithium medication. Blank levels are obtained from non-lithium taking subjects. Toxic levels did not occur in the patient population and are obtained by spiking samples with lithium chloride. All lithium levels are verified by serum or fingerstick plasma measurements using the IL943 analyzer at the clinical lab of the hospital as reference method. K-EDTA is used in whole blood samples as an anti-coagulant to prevent coagulation. Whole blood lithium samples are not stable over time, which is ascribed to differences between the lithium concentration in intracellular and extracellular fluid. When spiking whole blood samples for validation studies this has to be taken into account. To prevent a decrease of the lithium concentration and sample instability, EDTA whole blood and EDTA fingerstick measurements should be performed within one day but preferably within two hours from sampling. When performing multiple fingerstick measurements, lithium blood level deviations should be taken into account. Deviations can be more than 10% within one hour and are primarily caused by lithium clearance by the kidneys.

12.5   Validation Results

In this section results of individual validation studies are described. Detailed results will be described for the following tests: reproducibility, linearity, method comparison, home test. Summarizing results will be described for the following tests: repeatability, limit of quantification, stability, interference, matrix comparison, lay user usage, extreme temperatures, extreme humidities, and extreme waiting times.

12.5.1   Reproducibility

Reproducibility studies are performed with a design based on CLSI, EP5-A2: "Evaluation of Precision Performance of Quantitative Measurement Methods". The goal of these studies is to evaluate the performance in independent settings, which is simulated by using different Multireaders, Labchip batches, and operators over several days. The measurements are performed at different lithium levels relevant in the therapeutic and toxic range.

The acceptance criterion for precision is used, see Table 12.6.

Normal procedure prescribes measurements with the same samples over multiple days. However, in the case of fingerstick the samples cannot be stored reliably. For this reason it was chosen to perform the fingerstick study with extra Multireaders at more than one location.

Tests performed by Medimate in-house are for serum (during 10 days 4 samples are analyzed on 3 Multireaders in duplicate resulting in a total of 240 measurements) and for EDTA fingerstick (4 samples are 3 times analyzed on 10 Multireaders in duplicate resulting in a total of 240 measurements). EDTA fingerstick measurements were spread over two locations and measured in one day. EDTA fingerstick samples were collected by collection of multiple blood droplets from a finger puncture in a K-EDTA Micro Blood Collection Tube. All measurements were performed by 5 different operators on 8 different disposable batches.

Tests performed by the clinical laboratory are for serum (20 days 2 samples on 4 Multireaders in duplicate resulting in a total of 320 measurements) and for fingerstick (on day 1 and day 20, 2 patients are tested with 4 fingerstick punctures, from each of which 5 droplets are measured resulting in 80 measurements). Measurements were performed at two different locations and with 5 Multireaders and 5 disposable batches.

Table 12.7   Lithium reproducibility test results.

		#	Lithium (mmol L^{-1})	ARE	4SD outliers	Std. Dev.	%CV	PASS
Medimate	Serum	60	0.34	0.033	0	0.021	6.0%	YES
		58	1.27	0.042	1	0.032	2.5%	YES
		54	1.67	0.055	0	0.033	2.0%	YES
		57	2.64	0.087	0	0.044	1.7%	YES
	Capillary	57	0.34	0.058	0	0.016	4.8%	YES
		60	1.25	0.073	0	0.026	2.1%	YES
		57	2.68	0.155	0	0.080	3.0%	YES
		58	4.08	0.237	0	0.106	2.6%	YES
Clinical laboratory	Serum	160	0.33	0.033	0	0.025	7.5%	YES
		160	1.06	0.035	5	0.034	3.2%	YES
	Capillary	20	0.30	0.058	0	0.026	8.7%	YES
		20	0.32	0.058	0	0.023	7.3%	YES
		20	0.86	0.058	0	0.053	6.2%	YES
		20	1.07	0.062	0	0.042	3.9%	YES

Table 12.7 gives an overview of the test results obtained for both studies. The measurements pass the quality criteria when the standard deviation (Std. Dev.) below 1.0 mmol L^{-1} or coefficient of variation (%CV) for levels higher than 1.0 mmol L^{-1} is equal to or smaller than the allowable random error (ARE).

The conclusion of both reproducibility tests is that they meet the acceptance criteria.

12.5.2   Linearity

To evaluate linearity the CLSI EP-6 is performed. The protocol is performed with multiple inter-dilutions of spiked patient samples with zero-level samples. For this experiment EDTA fingerstick samples and serum samples were measured. The sample for the highest concentration was spiked with lithium chloride and from this spiked sample a concentration range was prepared by mixing it with blank lithium sample in different ratios. The linearity is verified in the full range *versus* an allowable non-linearity of 0.03 mmol L^{-1} or 3%.

Tests performed by Medimate personnel in-house consisted of 12 samples for serum and 11 samples for EDTA fingerstick in a range from 0.30 to 12.5 mmol L^{-1}. Each lithium sample is analyzed in duplicate on 3 Multi-readers resulting in 72 measurements and 66 measurements, respectively. Measurements were performed on one day with 4 disposable batches.

Tests performed by the clinical laboratory consisted of 7 samples in a range from 0.40 to 10 mmol L^{-1}. Each sample is measured 5 times resulting in 35 measurements. Measurements are performed on 3 different Multi-readers and 3 different disposable batches.

Full range results are shown in Figure 12.9.

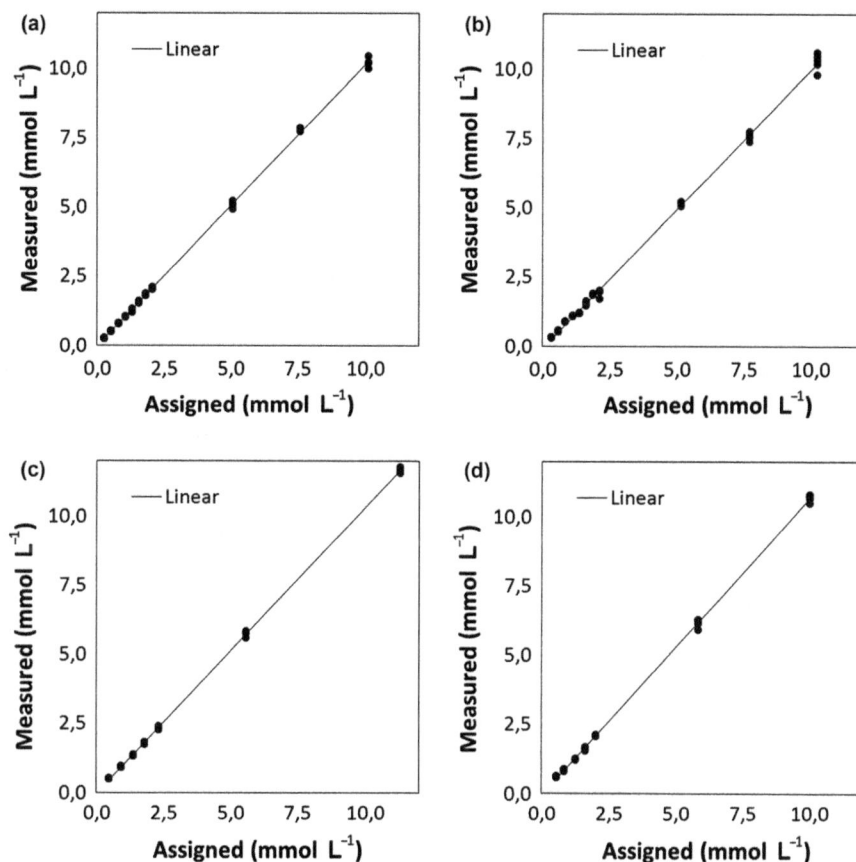

Figure 12.9 Scatterplot linearity results full range, *y*-axis Medimate Minilab lithium measurement, *x*-axis lithium reference concentration; (a) Medimate serum results, (b) Medimate EDTA fingerstick results, (c) clinical laboratory serum results, (d) clinical laboratory EDTA fingerstick results.

Table 12.8 Linear regression fit and linearity verification, performed by EP-Evaluator.

		High range	Linear within 0.03 mmol L^{-1} or 3%
Medimate	Serum	$Y = 1.019x - 0.022$	YES
	Fingerstick	$Y = 1.004x - 0.047$	YES
Clinical laboratory	Serum	$Y = 1.024x + 0.011$	YES
	Fingerstick	$Y = 1.067x - 0.044$	YES

In Table 12.8 the linear regression lines are stated including the evaluation of pass criteria. The pass criterion for linearity is evaluated *via* EP-Evaluator. Verification of linearity is performed by polynomial fit analysis indicating the linear fit as best fit.

As conclusion it can be stated that all data are found to be linearly related within the acceptable non-linearity of 0.03 mmol L^{-1} or 3%.

12.5.3   Method Comparison

Method comparison studies are performed with a design based on CLSI, EP-9. Different samples are analyzed and compared to the reference value. Serum and fingerstick samples are both measured with the Medimate Minilab and compared with serum samples measured by the reference method. The methods are evaluated in the range between 0.4 and 1.2 mmol L^{-1} Li^+. Test results for fingerstick are much more precise when the samples are pooled and then pipetted from one volume on different Lab-chips; however, to simulate the actual fingerstick practice it is chosen to take individual droplets. To decrease the random error from the reference method the reference value is the averaged value.

For the Medimate study there are too few data points to be evaluated *via* EP-Evaluator. As an alternative the methods are considered equivalent when both the means at 0.40 and 1.20 mmol L^{-1} Li^+ are within 0.05 mmol L^{-1} or 5% from the reference.

For the clinical laboratory study, acceptance is evaluated by EP-evaluator with for serum a TEA of 0.1 mmol L^{-1} or 10% Li^+ and for fingerstick a TEA of 0.15 mmol L^{-1} or 15% Li^+.

During the test performed by Medimate in-house, 20 patients were measured twice taking the measurements with an interval of minimal 2 hours. Each sample is measured 5 times on the Medimate Minilab by a Medimate professional, while the IL-943 reference method was performed by a professional from the clinical laboratory. All measurements are spread over 7 days. For the serum comparison 24 samples are included and for the fingerstick comparison 40 samples.

During the tests performed by the clinical laboratory 41 patients are included. To identify possible differences between users, the fingerstick measurements are performed by three different parties. The patient applies three drops of blood on three different chips and then afterwards a physician officer and a laboratory technician perform the same procedure. The serum samples are measured by one operator.

In Figure 12.10 measurements results are shown of the Medimate study. In Table 12.9 the linear fit result is shown as well as the evaluation result of the bias at 0.4 and 1.2 mmol L^{-1}. The fingerstick result shows two samples with a large difference between individual measurements. Further investigation of these measurements led to the conclusion that this could not have been caused by hemolysis and was most likely caused by patients who had ringed fingers. Additional testing with ringed fingers showed that ringed fingers interfere with the measurements. Up to now an explanation for this phenomenon has not been given but on the basis of these experiments, the pre-analytical protocol was updated with the restriction that ringed fingers are excluded.

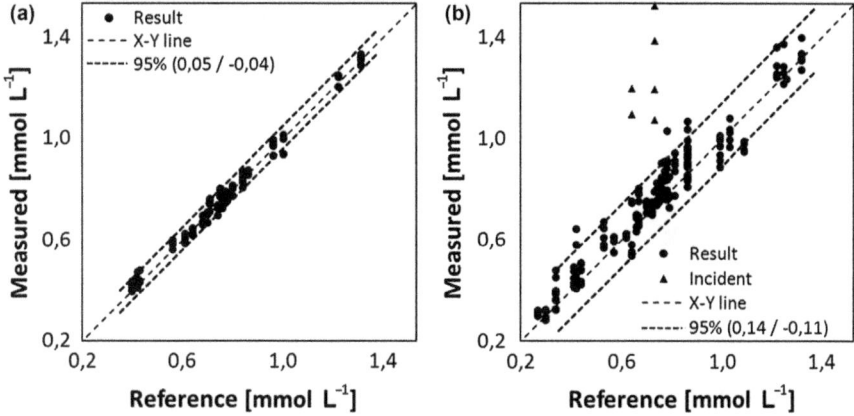

Figure 12.10 Method comparison results by Medimate, *x*-axis: lithium result IL943, *y*-axis: lithium result Medimate Minlab; (a) serum *versus* serum, (b) fingerstick *versus* serum.

Table 12.9 Method comparison results for lithium by Medimate.

Medimate	Outlier removal for fit and R	Fit y	R	Bias @0.4	@1.2	Evaluation
Serum	0	0.982x + 0.0154	0.995	0.01	− 0.01	PASS
Fingerstick	6	0.986x + 0.0455	0.977	0.04	0.03	PASS

In Figure 12.11 measurement results are shown of the clinical laboratory test. Analysis is performed by EP-Evaluator, and the main results are summarized in Table 12.10. All data meet the pass criteria except for the R value of the fingerstick results. These have a correlation coefficient smaller than 0.95 indicating too small a data range, which is due to the problem of the non-availability of an adequate sample population with low and high lithium concentrations.

Importantly, one-way ANOVA analysis indicates with a calculated *p*-value of 0.995 that there exists no difference between the lab technician, physician officer, and lay user when performing a fingerstick.

12.5.4 Home Test

The Medimate Minilab is besides the clinical setting mainly intended to be used by patients at home. For this reason a test is carried out in the home environment. In this case the patient stays as much as possible in his own surroundings with minimal interference from the study itself. Each of the five patients performs over a period of five days three consecutive measurements within one hour and minimally 10 hours after medication intake. The patient receives training prior to the first day according to the normal training method. The patients were allowed to get familiar with the Minilab.

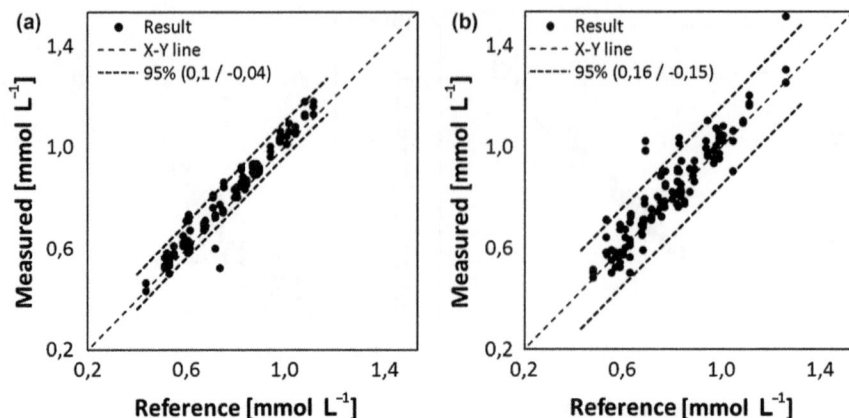

Figure 12.11 Method comparison results by the clinical laboratory. *x*-axis: lithium result IL943, *y*-axis: lithium result Medimate Minilab, (a) serum *versus* serum, (b) fingerstick *versus* serum by patient.

Table 12.10 Method comparison results for lithium by the clinical laboratory.

Clinical laboratory	Result range x method	Fit y	Std. err. est	R	TEA	Evaluation
Serum	0.54 – 1.16	1.02x – 0.014	0.029	0.987	0.10 mmol L^{-1} or 10%	PASS
Fingerstick Lab technician	0.53 – 1.26	1.24x – 0.161	0.082	0.927	0.15 mmol L^{-1} or 15%	PASS
Fingerstick Physician officer	0.53 – 1.26	1.12x – 0.065	0.086	0.899	0.15 mmol L^{-1} or 15%	PASS
Fingerstick Patient	0.53 – 1.26	1.11x – 0.067	0.064	0.945	0.15 mmol L^{-1} or 15%	PASS

Five extra Lab-chips were supplied for this purpose. One patient performed five test measurements at different time intervals prior to starting the measurement series, the other patients directly started with the measurements according to protocol. After the patient has finished all measurements, the Minilab was returned to Medimate and the stored measurement results were read out from the history of the Multireader.

A comparable precision is expected as with measurements in a clinical setting. Taking into account precision statistics of only three measurements and a small lithium change in the blood level the following acceptance criteria is chosen: 95% of the measurements should be within 0.117 mmol L^{-1} or 11.7% from its own average. This results in an allowable random error of 0.058 mmol L^{-1} or 5.8%.

Figure 12.12 shows the results obtained in the home test. Patient 5 day 5 result did not meet the minimal 10 hour criteria and is therefore excluded from further analysis.

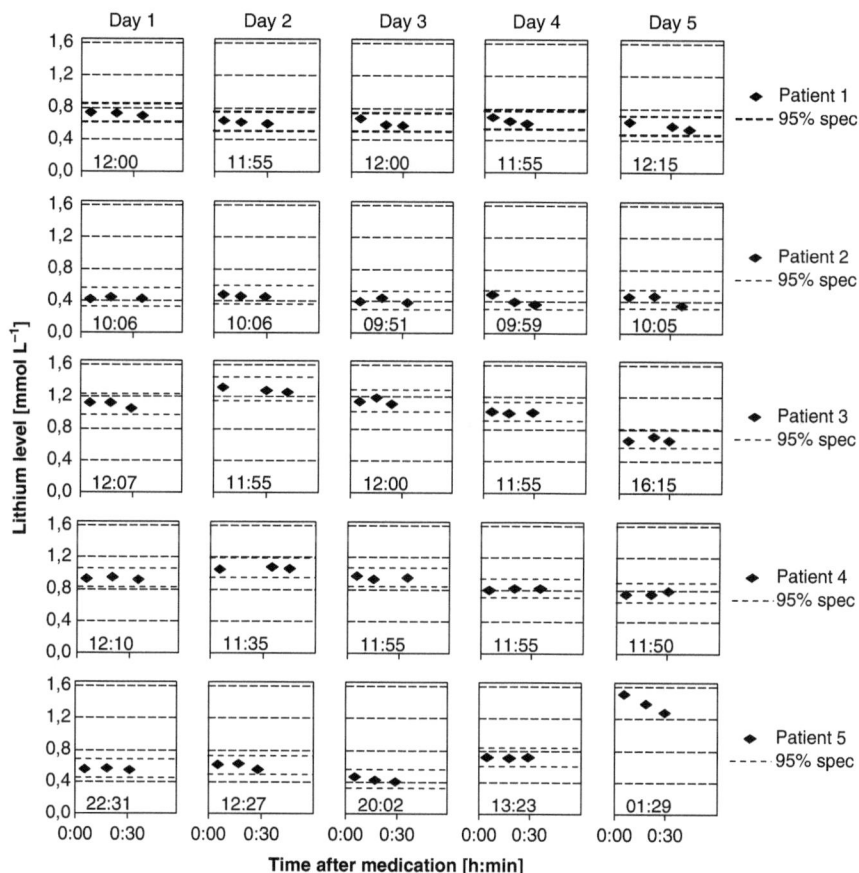

Figure 12.12 Measurement results from 5 patients in home environment. Each patient measured 3 times at 5 different days. In each graph in the bottom left corner the time between medication intake and the start of the graph is indicated. The width of each graph is 1 hour. Error bars are not indicated for patient 5 day 5. This series did not fulfill the minimal 10 hour time delay after medication intake.

Table 12.11 shows the analysis results. For each patient the minimum and maximum difference per patient from its own mean is calculated as well as the pooled standard deviation. The standard deviation is verified against the pass criteria. Each patient showed a significantly better result than the pass criteria.

In a separate calculation an average concentration decrease between two consecutive measurements on the same day is calculated to be 0.02 mmol L^{-1}. A probable explanation is that this is due to the decrease of the lithium concentration in the blood between two consecutive measurements because of the progressive lithium clearance from the blood.

These preliminary results show that a patient is well capable to measure lithium at home with the Medimate Minilab.

Table 12.11   Patient self-test results.

	Mean mmol L^{-1}	Measure-ments #	Min. dif. mmol L^{-1}	Max. dif. mmol L^{-1}	Pooled st. dev. mmol L^{-1}	ARE mmol L^{-1}	Evaluation Pass/No pass
Patient 1	0.65	15	-0.04	0.06	0.035	0.058	Pass
Patient 2	0.44	15	-0.07	0.07	0.037	0.058	Pass
Patient 3	1.05	15	-0.05	0.04	0.030	0.071	Pass
Patient 4	0.91	15	-0.02	0.03	0.017	0.058	Pass
Patient 5	0.75	12	-0.04	0.03	0.021	0.058	Pass
Total	n.a.	72	-0.07	0.07	0.028	n.a.	Pass

12.5.5   Other Study Results

Repeatability tests were performed by both Medimate in-house and by the clinical laboratory. The experiment did not show unexpected results when compared to the reproducibility results. No significant differences are found between operators, Multireaders, or Lab-chip batches.

The *limit of quantitation* (LOQ) is calculated to be 0.25 mmol L^{-1} Li^+ for both capillary blood and serum samples. The Medimate Minilabs shows a "<0.25 mmol L^{-1}" message on the display if the concentration calculation results in a value below 0.25 mmol L^{-1}. It was verified that without the message the LOQ would be 0.16 mmol L^{-1} for fingerstick.

Two real-time *stability* studies were performed with serum based on the CLSI EP-25 protocol. Changes in average and standard deviation of multiple measurements were monitored. All Lab-chips were stored at room temperature. The study at Medimate lasted 15 months, the study performed by the clinical laboratory lasted 7 months. During the test no evidence of instability was found.

Interference is studied according to CLSI EP7 for the following species/ effects: sodium, potassium, ammonium, calcium, magnesium, zinc, urea, iron, hemoglobin, albumin, creatinine, bilirubin, the Lipid Index and hemolysis. The first experiment is performed by Medimate as a screening to identify interference. The second experiment is performed by the clinical laboratory to verify the interference. Both experiments did not show interference for the species under investigation except for sodium and hemolysis.

The results of the Medimate study and the clinical laboratory study for sodium and hemolysis are combined and are shown as calculated corrected lithium in eqn (12.3) and (12.4), respectively.

$$[Li^+] = [Li^+_{meas}] + 0{,}0071 * [Li^+_{meas}] * (140 - [Na^+]) \qquad (12.3)$$

$$[Li^+] = [Li^+_{meas}] + 0{,}0031 * [Li^+_{meas}] * H \qquad (12.4)$$

Here $[Li^+]$ (mmol L^{-1}) is the corrected lithium concentration, $[Li^+_{meas}]$ (mmol L^{-1}) is the measured lithium concentration, $[Na^+]$ (mmol L^{-1}) the

Table 12.12   Effects on the lithium concentration of deviations in the sodium concentration and hemolysis percentages.

Sodium mmol L^{-1}	Lithium deviation %	Hemolysis %	Lithium deviation %
120	14.2	5	1.6
130	7.1	10	3.1
140	0	20	6.2
150	− 7.1	50	15.5
160	− 14.2		

sodium concentration, and $H(\%)$ the hemolysis percentage. Further investigation indicated that the hemolysis interference is most likely caused by the sodium dependency.

Table 12.12 summarizes the effect of the interference of sodium concentration deviations and hemolysis on the lithium outcome.

Matrix comparison between serum and fingerstick was performed on the clinical laboratory results from the method comparison as shown in Table 12.10. For the analysis also whole blood results were included. One-way ANOVA analysis indicates with a calculated p-value of 0.112 that no difference exists between measurements in serum, whole blood, and fingerstick. In general it can, however, be stated that fingerstick measurements are more prone to higher lithium concentration values because fingerstick samples are more vulnerable for hemolysis, which will lead to a positive lithium offset.

Lay user usage has been verified in accordance with NEN-EN 13532:2002: general requirements for *in-vitro* diagnostic medical devices for self- testing, and indicated no problems for lay users to use the Medimate Minilab.

Extreme temperature studies at 15 and 30 °C showed acceptable performance. 139 measurements were performed on different lithium levels and showed a standard deviation equal to or better than 0.040 mmol L^{-1} Li^{+}. Average deviations were evaluated with an acceptance criterion of 0.05 mmol L^{-1} or 5%. No average differences were observed.

High humidity tests above 80% showed a decreased performance in terms of standard deviation. Two times 40 measurements were performed and resulted in a standard deviation of 0.048 and 0.065 mmol L^{-1} Li^{+}, respectively. Average deviations were evaluated with an acceptance criterion of 0.05 mmol L^{-1} or 5%. No average differences were observed. It can be concluded that it is possible to measure at high humidity but that for best results this is not advised. *Low humidity* tests did not show any deviation from the normal conditions.

Serum waiting times of approximately 2 minutes did not show a decrease in performance. 160 measurements were performed and showed a standard deviation equal to or better than 0.033 mmol L^{-1} Li^{+}. Average deviations were evaluated with an acceptance criterion of 0.05 mmol L^{-1} or 5%. No average differences were observed.

Table 12.13 Acceptability summary for lithium measured on the Medimate Minilab.

Organization	Lithium acceptability criteria	Serum	Fingerstick
CLIA, WLSH, AAB, CAP	$\pm$ 20% or 0.3 mmol L^{-1}	PASS	PASS
NYS	$\pm$ 15% or 0.3 mmol L^{-1}	PASS	PASS
RCPA	0.3 mmol L^{-1}	PASS	PASS
Medimate	10% or 0.1 mmol L^{-1} serum 15% or 0.15 mmol L^{-1} fingerstick	PASS	PASS

Fingerstick waiting times at extreme temperatures were evaluated. A $2\frac{1}{2}$ minute waiting time prior to sample application and $2\frac{1}{2}$ minutes after sample application was tested. Two times 80 fingerstick measurements at 16 and 30 °C were performed. Results showed a standard deviation equal to or better than 0.034 mmol L^{-1} Li^{+}. Average deviations were evaluated with an acceptance criterion of 0.05 mmol L^{-1} or 5%. No average differences were observed.

12.5.6 Final Evaluation

In Table 12.13 the conclusions are stated with respect to the acceptance criteria and the Medimate Minilab performance.

12.6 Platform Potential

12.6.1 Current Platform Capabilities

A literature review in 2006 indicated that many ($\sim$150) species can be measured by microchip capillary electrophoresis with conductivity detection. To explore this wide range of possible analytes, in this section Medimate Minilab results are shown for species other than lithium. At the end, the limitations and future prospects of the platform will be discussed.

12.6.1.1 Sodium in Urine

A substantial number of kidney and heart patients are on a salt restricted diet. To aid the therapy by feedback on the salt intake, a sodium measurement in urine is developed.

The sodium measurement was validated in a comparable manner as shown for lithium by the clinical laboratory from Leiden University Medical Center (LUMC), demonstrating the performance within the TEA for sodium in urine of 28.8%.[19] Following this study, changes have been implemented, improving the performance. A new linearity and reproducibility study was performed by Medimate in-house based on these improvements. The results are shown in Figure 12.13 and Table 12.14.

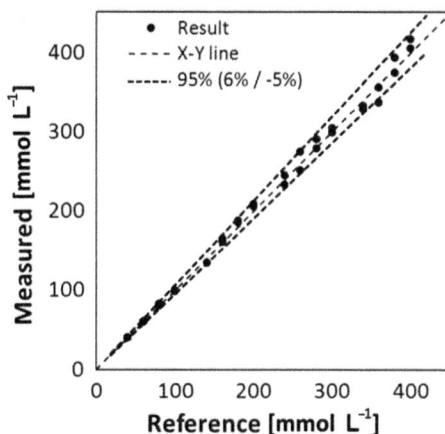

Figure 12.13 Medimate in-house sodium results, *x*-axis sodium reference and *y*-axis measured sodium concentration.

Table 12.14 Reproducibility results for sodium measurements at three levels.

Sodium mmol L^{-1}	Std. dev. mmol L^{-1}	CV %
60	2.7	4.5
160	8.4	5.3
300	8.0	2.7

12.6.1.2 Creatinine in Whole Blood

Creatinine is a marker for kidney function. A feasibility study to measure creatinine in serum using the Medimate Minilab was performed and published in 2013.[24] In a subsequent study, whole blood samples with an estimated creatinine level of 100 μmol L^{-1} creatinine were spiked with creatinine up to 900 μmol L^{-1}, and multiple measurements were performed on each level. The measurement results after calibration are shown in Figure 12.14 and the linearity analysis is shown in Table 12.15.

On the basis of these results the limit of quantification is calculated as 190 μmol L^{-1} creatinine. This level will enable the detection of a decreased kidney function. To monitor the creatinine level in a healthy population, a limit of quantification is desired of approximately 40 μmol L^{-1} creatinine. Improvements tested in the company have shown that a factor of 7 decrease in limit of quantification is possible, which will make it also possible to monitor creatinine in a healthy population.

12.6.1.3 Ionized Calcium in Cow Whole Blood

A feasibility study for ionized calcium was performed by Blue4Green BV. Blue4Green uses the same technology as Medimate for the veterinarian

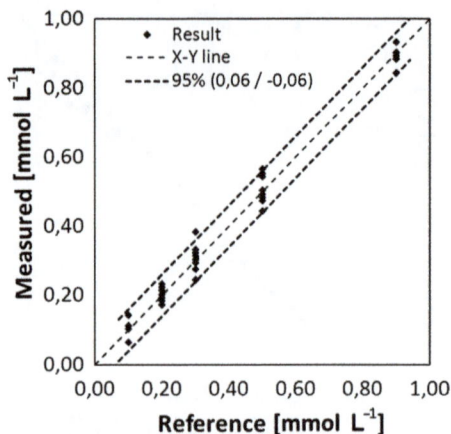

Figure 12.14   Creatinine measurements in whole blood, *x*-axis reference concentration, *y*-axis measured concentration.

Table 12.15   Linearity and lower limit results for creatinine.

Range (μM)	Fit	R^2	LOD (μM)	LOQ (μM)
50–900	0.986x + 0.01	0.985	50	190

Figure 12.15   Blue4Green Labbook using the Lab-chip to measure calcium.

market and the same Lab-chip/cartridge with a proprietary measurement device called the Labbook. The Labbook is battery powered and enables ionized calcium measurements in a temperature range of 5 to 30 °C, see Figure 12.15.

12 Lab-chip batches and 3 Labbooks were used to perform 110 measurements on 8 samples over multiple days at temperatures between 17 and 21 °C. Based on these results a calibration model was constructed.

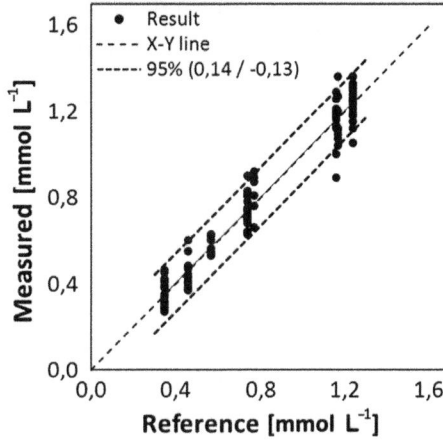

Figure 12.16   Blue4Green study results for ionized calcium, *x*-axis plots the reference method (I-Stat), *y*-axis the concentration as calculated by the Labbook.

Table 12.16   Ionized calcium in cow blood study results: mean, standard deviation, and coefficient of variation per sample.

Reference mmol L^{-1}	Mean mmol L^{-1}	St. dev. mmol L^{-1}	CV %
0.35	0.35	0.06	17
0.46	0.46	0.06	14
0.57	0.57	0.04	7
0.74	0.74	0.07	9
0.77	0.82	0.09	11
1.16	1.16	0.10	8
1.17	1.17	0.10	9
1.24	1.24	0.08	6

The individual measurement results after calibration are shown in Figure 12.16. As a reference device whole blood was measured on the I-Stat from Abbott Point of Care.

The results were analyzed for standard deviation and coefficient of variation, see Table 12.16. Based on these results it is concluded that it is possible to measure ionized calcium. Combined with the Medimate experience of modeling and performance it is expected that 0.20 mmol L^{-1} or 20% iCa^{2+} performance is feasible for calcium with the current status of technology. New developments are initiated to further improve this performance.

12.6.1.4   Species that Indicatively Have Been Measured on the Minilab

Table 12.17 shows other species that have been measured on the Minilab but that have not yet been tested on quantification.

Table 12.17   Different species measured on the Medimate Minilab.

Positive ions		Negative ions	
Potassium	Strontium	Chloride	Nitrate
Calcium	Bromide	Phosphate	Sulfate
Magnesium	Ammonium	Carbonate	Acetate

12.6.2   Future Possibilities and Limitations

The Medimate Minilab platform is able to measure many different ionic species in aqueous solutions of widely varying composition. It is expected that the product portfolio will be extended by the following new developments

- Quantifying more species in human bodily fluids
- Quantifying more species in other samples, like ground or tap water, beverages, livestock and process fluids
- Quantifying multiple species during one measurement
- Identifying samples

To detect species, two main conditions have to be met. The first condition is that the species to be analyzed should be ionic. In some cases it is thereby possible to influence the species charge by adjustment of sample or solution pH or other adjustments. A second condition is that the concentration of the species should be above the detection limit. As a consequence of using capillary electrophoresis the detection limit scales with the sample conductivity. This makes it possible to measure species in for instance deionized water but also in very salty seawater. Based on our experience with lithium in whole blood we expect that it is possible to estimate the detection limit as a percentage of the conductivity. This is, however, at this point beyond the scope of this chapter.

New feasibility studies have indicated that with improved conductivity detection the measurement signal for surface area detection can be increased. Further calculation showed an improvement of the signal to noise ratio by a factor of 7. With this estimation the detection limit is expected to be lowered by the same factor. It is expected that this enables creatinine measurements in blood at low level. Other results have shown measurement times below 1 minute as being feasible. Improving these characteristics as well as reliability improvements is a continuous effort within the company.

12.7   Conclusions

A new microchip capillary electrophoresis self-test platform to quantify ionic species in aqueous solution is successfully introduced.

It is shown that the Medimate Minilab is capable of measuring lithium in serum and fingerstick whole blood and sodium in urine, both accurately and in the clinically relevant range. The test can be performed by a lay user as a self-test at home or at the physician's office. Besides the flexibility in handling, location and sample type, the technology also gives results of comparable quality to state-of-the-art laboratory tests, specifically when it concerns serum measurements. This makes the platform also suitable for laboratories, especially when measurements are rare, when time is critical, or when costly equipment and staff are not available.

Specifically, in two extensive lithium validation studies it is shown that there are no user- or location-related differences in the determination result and that the Minilab is well capable of measuring lithium concentration in fingerstick whole blood.

This new technology platform enables patients with manic depressive illness to measure more flexibly, at any location and at any time. It is expected that this will increase the quality of life, improve the therapy outcome, and improve the well-being of the patient in society in general. Last but not least a reduction in societal cost is expected.

Acknowledgements

The European fund for regional development from the European Union and the province of Overijssel and Gelderland are gratefully acknowledged for their financial support.

Dimence, the East Netherlands mental healthcare organization, is acknowledged for its support with organizing patients.

References

1. Wikipedia, *Blood glucose meter*, http://en.wikipedia.org/wiki/Blood_glucose_meter#History, visited 2014-01-04.
2. Diabeticsconfessions, Life Expectancy for Diabetics – It's Longer Than You Think, http://diabeticonfessions.tumblr.com/post/32253739020/life-expectancy-for-diabetics-its-longer-than-you, visited 2014-01-04.
3. E. X. Vrouwe, R. Luttge and A. van den Berg, *Electrophoresis*, 2004, **25**(10–11), 1660–1667.
4. A. Floris, S. Staal, S. Lenk, E. Staijen, D. Kohlheyer, J. Eijkel and A. van den Berg, *Lab on a Chip*, 2010, **10**, 1799–1806, DOI: 10.1039/c003899g.
5. S. Staal, J. Floris and A. van den Berg, *Ion sensor for fluid and method for manufacture*, Application Number WO2006EP11148, File date 21 November 2006.
6. S. Staal, J. Floris, M. Blom and J. Oonk, *Test chip with plug for measuring the concentration of an analyte in a liquid, housing for test chip and socket for plug*, Application number WO2007EP04468, File date 18 May 2007.
7. S. Staal, J. Floris and S. Lenk, *An apparatus for the measurement of a concentration of a charged species in a sample*, Application number WO2009EP51874, File date 17 February 2009.
8. K. R. Merikangas, R. Jin, J. P. He, R. C. Kessler, S. Lee, N. A. Sampson, M. C. Viana, L. H. Andrade, C. Hu, E. G. Karam, M. Ladea,

M. E. Medina-Mora, Y. Ono, J. Posada-Villa, R. Sagar, J. E. Wells and Z. Zarkov, *Arch. Gen. Psychiatr.*, 2011, **68**(3), 241–292, DOI: 10.1001/archgenpsychiatry.2011.12.

9. R. de Graaf, M. Ten Have and S. Van Dorsselaer, *NEMESIS-2: De psychische gezondheid van de Nederlandse bevolking. Opzet en eerste resultaten*, Utrecht, 2010, Trimbos-instituut, www.trimbos.nl, visited 2014-01-04.

10. I. Wilting, E. R. Heerdink, A. C. G. Egberts and W. A. Nolen, *Pharm. Weekbl.*, 2004, **10**(139), 328–322.

11. D. J. Miklowitz, *Hulpgids Bipolaire stoornis, alles wat jij en je omgeving moet weten over manisch-depressiviteit*, Nieuwezijds, Amsterdam, 2004. ISBN: 90-5712-181-6.

12. College for Health Insurance Companies, Netherlands, *GIP databank, The Drug Information System of the Health Care Insurance Board*, http://www.gipdatabank.nl/databank.asp, visited 2013-01-04.

13. Nederlandse Vereniging van Ziekenhuisapothekers Commissie Analyse & Toxicologie, *Guideline therapeutic drug monitoring TDM-monografie LITHIUM*, version 28-4-2011.

14. P. Moleman, Praktische Psychofarmacologie, Bohn Stafleu van Loghum, 3rd edn, 1998, ISBN: 90 313 2081 1.

15. F. K. Goodwin and K. R. Jamison, *Manic Depressive Illness, Bipolar Disorder and Recurrent Depression*, Oxford University Press, 2nd edn, 2007, ISBN-13: 978-0-19-513579-4.

16. A. Rebel, M. A. Rice and B. G. Fahy, *Journal of Diabetes Science and Technology*, 2012, **6**(2), 396–411.

17. S. S. Staal and M. C. Ungerer, *Medimate MiniLab for lithium analysis in venous serum, venous whole blood and fingerstick whole blood, Technology description and detailed validation results*, Medimate, version 1.02, Dec 2013.

18. J. Krabbe, J. A. Bartholomew and K. L. L. Movig, *Validation report Medimate Minilab*, Medlon BV & Medisch Spectrum Twente, expected Feb 2014.

19. Data innovations LLC, *Total Allowable Error Table*, http://www.dgrhoads.com/db2004/ae2004.php, visited 2013-12-06.

20. Centers for Medicare & Medicaid Services, *Clinical Laboratory Improvement Amendments*, http://www.cms.gov/Regulations-and-Guidance/Legislation/CLIA/index.html?redirect = /clia/, visited 2013-12-06.

21. Centers for Disease Control and Prevention, *Clinical Laboratory Improvement Amendments*, partnership: http://wwwn.cdc.gov/CLIA/default.aspx, visited 2013-12-06.

22. US Food and Drug Administration, *Clinical Laboratory Improvement Amendments*, http://www.fda.gov/medicaldevices/deviceregulationandguidance/ivdregulatoryassistance/ucm124105.htm, visited 2013-12-06.

23. Clinical and laboratory standards institute, *CLSI evaluation protocols*, http://www.clsi.org/, visited 2014-01-04.

24. M. Avila, A. Floris, S. Staal, A. Rıos, J. Eijkel and A. van den Berg, *Electrophoresis*, 2013, **34**(20–21), 2956–2961.

Subject Index